Biotechnology in Our Lives

What Modern Genetics Can Tell You about Assisted Reproduction, Human Behavior, Personalized Medicine, and Much More

Sheldon Krimsky and Jeremy Gruber, editors

Skyhorse Publishing

This book is dedicated to members of the Council for Responsible Genetics Board of Directors and Advisory Board who are no longer with us but whose work and vision continue to inspire new generations of social and environmental justice activists: David Brower, Barry Commoner, Marc Lappé, Anthony Mazzocchi, Albert Meyerhoff, Bernard Rapoport, George Wald, and William Winpisinger.

Skyhorse Publishing books may be purchased in bulk at special discounts for sales promotion, corporate gifts, fund-raising, or educational purposes. Special editions can also be created to specifications. For details, contact the Special Sales Department, Skyhorse Publishing, 307 West 36th Street, 11th Floor, New York, NY 10018 or info@skyhorsepublishing.com.

Skyhorse® and Skyhorse Publishing® are registered trademarks of Skyhorse Publishing, Inc.®, a Delaware corporation.

Visit our website at www.skyhorsepublishing.com.

10 9 8 7 6 5 4 3 2 1

Library of Congress Cataloging-in-Publication Data

Krimsky, Sheldon.
Biotechnology in our lives: what modern genetics can tell you about assisted reproduction, predicting criminal behavior, and much more / Sheldon Krimsky, Jeremy Gruber.
pages cm
ISBN 978-1-62087-573-5 (hardcover: alk. paper) 1. Medical innovations–Social aspects. 2. Genetic engineering–Social aspects. 3. Medical technology–Social aspects. I. Gruber, Jeremy. II. Title.
RA418.5.M4K75 2013
610.28'4–dc23

2013005417

Printed in the United States of America

CONTENTS

ACKNOWLEDGMENTS

We are grateful for the public interest commitment and superb work of *GeneWatch* editors and editorial boards past and present: Sam Anderson, Philip Bereano, Kostia Bergman, Phil Brown, Sujatha Byravan, Nancy Connell, Donna Cremans, Christopher Edwards, Leslie Fraser, Phyllis Freeman, Judith Glaubman, Terri Goldberg, Barbara Goldoftas, Colin Gracey, Susan Gracey, Daniel Grossman, Jeremy Gruber, Ruth Hubbard, Kit Johnson, Brandon Keim, Sophia Kolehmainen, Sheldon Krimsky, Herbert Lass, Evan Lerner, Gary Marchant, Wendy McGoodwin, Shelley Minden, Eva Ng, Rayna Rapp, Barbara Rosenberg, Peter Shorett, Seth Shulman, Christine Skwiot, Martin Teitel, Suzanne C. Theberge, Jonathan B. Tucker, Shawna Vogel, Nachama Wilker, Kimberly Wilson, and Susan Wright.

We would like to thank Holly Rubino, editor-at-large for Skyhorse Publishing, who provided editorial guidance that helped the editors clarify the book's goals and audience.

We also wish to acknowledge the substantial support of both the Cornerstone Campaign and the Safety Systems Foundation.

INTRODUCTION

Thirty years ago a group of scientists, public and occupational health advocates, disability rights activists, environmentalists, and union leaders formed the Council for Responsible Genetics, an independent nonprofit organization that has thus far taken no funding from government or corporations. Today, the CRG remains the only biotech public interest organization that is explicitly dedicated to examining the best science, interpreting the results, assessing the implications, communicating them to a general audience, and facilitating meaningful, measureable change in the field of biotechnology.

Among the first projects undertaken by the CRG was the publication of *GeneWatch*, a magazine that provided timely information and commentary on social and ethical implications of advances in genetics and their applications to biotechnology. The CRG was the first public interest organization to bring the social issues raised by genetic technology to a broad public readership before the appearance of the Internet.

As biotechnology grew into a multi-billion-dollar financial sector, genetic technology began to enter almost every area of human life: from the genetically modified foods that we eat to the biodiversity of our ecosystem, from human health and reproductive technologies to the operation of the criminal justice system. From its inception, *GeneWatch* has filled the critical role of analyzing the rush of information and opinion resulting from the rapid growth in genetic research and technology. *GeneWatch* expanded from laboratory and occupational safety to biological weapons, human genetic engineering, cloning, reproductive technologies, genetically modified crops, and genetic privacy and discrimination. Contributors to *GeneWatch* had a gift for communicating complex science to popular audiences while situating the science in a social context.

Apart from *GeneWatch*, the CRG has also contributed to public discussion by framing issues in biotechnology from a social justice perspective. It has issued policy statements and published books such as *Preventing a Biological Arms Race, Changing the Nature of Nature*, *Rights and Liberties in the Biotech Age*, *Genetic Explanations: Sense and Nonsense* and *Race and the Genetic Revolution*. The CRG has also worked with other organizations for the passage of national legislation that has resulted in turning the international treaty banning biological weapons into domestic law and in passing the Genetic Information Nondiscrimination Act. The CRG has held workshops, organized conferences, and published many articles covering agricultural biotechnology. It has raised public understanding about the patenting of seeds, herbicide resistant crops, genetically modified food, bovine growth hormone, and biodiversity. The CRG staff and board members regularly assist policy makers and have testified before Congress, federal agencies, regulatory bodies, and state legislatures.

CRG issued the *Genetic Bill of Rights*, which consisted of ten principles for a humane biotechnology. Among the principles was a call for a ban on human genetic engineering, a consumer right to purchase food unaltered by foreign genes, and a right to be protected from genetic discrimination in all forms of civil society. As the preamble to the *Genetic Bill of Rights* states, "People everywhere have the right to participate in evaluating the social and biological implications of the genetic revolution and in democratically guiding its applications."

For this book, the editors of *GeneWatch* have selected the best essays published over its thirty-year run. Among the contributors are leading scientists and social activists who demystify the science while giving readers insight into how genetic technologies affect their lives. Readers will find essays that cover a variety of subjects, like assisted reproduction, the potential for error and discrimination in DNA testing, how biotechnology has been portrayed in science fiction, personalized genetic medicine, how genetics can affect human behavior, and much more. The distinguished contributors include scientists, consumer advocates, women's rights activists, and law professors. The book takes us from where modern genetic technology began, where it is today, and where it might be headed. For this volume we have selected articles from six categories of biotechnology: Genetic Privacy and Discrimination, Assisted Reproductive Technologies, Forensic DNA, Behavioral Genetics, Genetics and Popular Culture, and Genetics in Medicine. The articles not only provide a historical view of the public discourse in biotechnology over three decades but also highlight enduring moral questions faced by modern society. The essays emphasize the authors' commitment to the social control of technology, to the critique of genetic reductionism and the view that our genes are our destiny, and to democratic participation in technological choices.

The eminent Columbia University sociologist of science, Robert Merton, spoke about organized skepticism as a core value in science. Scientists must question results until they are convinced they have rooted out error and bias. A second level of skepticism must be adopted by the citizenry: namely skepticism that every new technology should be embraced as an advancement of the human condition. That result must be demonstrated to an informed citizenry and should not be accepted by default.

We are indebted to the many authors who have contributed essays to *GeneWatch* over the years. They received no financial compensation for their writing but rather offered their work as a contribution to public education. We also wish to acknowledge the founding members of the Council for Responsible Genetics who, through a collective vision, recognized the importance of a scientifically literate society.

Sheldon Krimsky and Jeremy Gruber
November 2012

PART I

Genetic Privacy and Discrimination: Can Your Genes Keep You from Getting Insurance?

Artist: Peter Baril

The concept of "genetic discrimination" only recently entered our vocabulary. But the problem is well documented. In hundreds of cases, individuals and family members have been barred from employment or lost their health and life insurance based on an apparent or perceived genetic abnormality. Many of those who have suffered discrimination are clinically healthy and exhibit none of the symptoms of a genetic disorder. Often, genetic tests deliver uncertain probabilities rather than clear-cut predictions of disease. Even in the most definitive genetic conditions, which are few in number, there remains a wide variability in the timing of onset and severity of clinical symptoms. Yet discrimination continues. The Burlington Northern Santa Fe Railroad Company, for example, was revealed to be conducting genetic tests on its employees without their informed consent, as a means of counteracting workers compensation claims for job-related stress injuries.

On May 21, 2008, President George W. Bush signed the Genetic Information Nondiscrimination Act (GINA) into law. Senator Ted Kennedy hailed GINA as the "first civil rights bill of the new century." GINA provides strong new protections against access to genetic information and genetic discrimination in both the health insurance and employment settings.

But GINA does not address all possible forms of genetic discrimination. For example, it does not address life insurance, disability insurance, or long-term care insurance. GINA does not protect symptomatic individuals as well. In other words, any observable symptoms of a disease that has clear genetic origins can be used to discriminate against an individual under the law. GINA is a strong and essential first step in the fight against genetic discrimination and misuse of medical information more generally, but it is far from the battle.

Ten years after the mapping of the human genome was completed, the genetic revolution has led to a tsunami of DNA data created by genetics research and the commercialization of such research. As more and more of this personal information becomes public knowledge, it can be bought and sold by any company interested in predictive information about an individual's future health status. A recent survey by Cogent Research found that 71 percent of Americans continue to be concerned about access to and use of their personal genetic information.

There simply is no comprehensive genetic privacy law in the United States or any other country. And so the access to and misuse of personal genetic information continues.

From health insurance companies to employers, from student testing to direct-to-consumer genetics and biobank administration, these chapters explore those concerns.

1

New Genetic Privacy Concerns

BY PATRICIA A. ROCHE AND GEORGE J. ANNAS

Patricia A. Roche *and* **George J. Annas** *are professors in the Department of Health Law, Bioethics and Human Rights at Boston University School of Public Health.* *This article, first published in* GeneWatch, *volume 20, number 1, January-February 2007, is adapted from the authors' "DNA Testing, Banking and Genetic Privacy," which appeared in* The New England Journal of Medicine 2006; *355:545-6.* *George Annas is a former member of the Board of the Council for Responsible Genetics.*

Everyone wants a piece of us, specifically a DNA sample. Private marketers assure us that our DNA will unlock useful secrets that can make our lives better by helping to determine what food we should eat, what drugs we should take, and even whether or with whom we should have children. We are encouraged to have our DNA decoded to discover our ancestry, and even mass marketed magazines, like National Geographic, encourage readers to submit their DNA samples for analysis to determine "where we come from." Federal and state officials discuss a national, or even international, DNA databank that can be used to hunt for criminals and sometimes also to help exonerate criminal suspects. The promises of DNA testing seem endless to both individuals and society. But DNA testing has a dark side as well. As the source of genetic information, the DNA molecule is also a separate entity that can be collected and stored for multiple, currently unknown (or at least unconsented to) uses. Since multiple copies can readily be made, once an individual's sample is obtained, no further contact with the person is required to procure additional material for testing. Should this be of concern to us, or do existing federal and state laws adequately protect our genetic privacy?

In May 2006, the Secretary's Advisory Committee on Genetics, Health, and Society (SACGHS) issued a draft report on a proposed national biobank at the National Institutes of Health (NIH).[1] Similar to biobanks in other countries, such as the U.K., Estonia, and Canada, this repository would contain tissue samples and health information from 500,000 to 1,000,000 individuals to be utilized for studying the gene-environment interactions underlying common, complex diseases. During the preparation of its final report, SACGHS asked for public comments on several issues, including the privacy implications of biobanking and ownership of samples and data. How public input will affect the federal government's decisions regarding a national biobank remains to be seen. Federal officials seem to want to exercise caution before collecting DNA from a significant part of the population and appear committed to protecting the interests of Americans who contribute samples and information for use in research. It would be a mistake, however, for the public to think that a future national biobank presents the principal risk to their genetic privacy or that the federal government has taken, or will take, sufficient action to protect their interests in regard to DNA banking. The fact is that DNA samples are currently routinely collected and used in multiple arenas with little legal restraint or regulatory oversight at the federal level. In this article we focus on developments in DNA banking in relation to genetic research and the expansion of commercial testing services to illustrate why current laws are inadequate to protect individual interests affected by DNA banking.

Banking Practices in Genetic Research

Throughout the 1990s, those most actively involved with collecting DNA samples from individuals (other than law enforcement officials and the U.S. military) were researchers intent on discovering genetic markers, primarily associated with relatively rare diseases. Typically, researchers collected DNA samples and information from families at risk for the disease of interest and analyzed the samples in an effort to understand the heritability of the disease. In the event that their endeavors proved successful and a disease marker was identified, additional research using the same samples would inevitably follow. Therefore, it became customary for researchers to seek consent when samples were collected to store DNA beyond the completion of initial research for use in subsequent studies of the same disease. As knowledge about the human genome advanced and attention became focused on exploring the role of genetics in more common diseases, research studies were structured in keeping with that broader goal. Consequently, researchers sought samples from a larger segment of the general population, anticipating that they would be useful in a variety of studies. The exploratory nature of this research made it difficult to describe any specific details about the focus of secondary studies and exactly how samples would be used in the future. Consequently, consent documents relating to the initial sample collection customarily stated that, by consenting to participate in research, subjects acknowledged that samples they provided would become the property of the researcher (or the institution where the research took place), who could use the samples for their own benefit. In this way, individuals who relinquished their samples were transformed from research subjects into donors and the consent-for-research forms morphed into documents effectuating gifts. To appreciate the significance of this phenomenon it is important to consider the nature of DNA samples, principles traditionally applied to research and the court's ruling in the case of *The Washington University v. Catalona.*[2]

The DNA sample can be viewed as a "future diary," or even a coded probabilistic medical record.[3] Possession of an identifiable DNA sample gives the possessor access to a wealth of information about the individual as well as his or her genetic relatives. This includes information that has been made derivable due to advances in genetics that occur after the sample has been relinquished. It is therefore unlikely that a person can be cognizant of, or fully appreciate, the informational value of his or her DNA sample at the point in time that it is relinquished to others. It also means that, as long as personally identifiable samples are stored, there is the possibility that those with access to the samples will discover things about individuals that the individuals themselves do not know, and, most importantly, would not want others to know about them. In other words, unless those with access to samples refrain from using samples for any unauthorized purposes and destroy samples as soon as authorized uses have been completed or upon request of the sample's source, invasions of genetic privacy are inevitable.

The most basic principle of research with human subjects is that participation must be voluntary and the subject free to withdraw at any time. Researchers are therefore legally and ethically obligated to abide by requests from subjects to withdraw their tissue samples from research. However, this may no longer be the case, at least according to one lower federal court judge. In 2003, patients who had previously provided tissue samples for prostate cancer research conducted by Dr. William Catalona at Washington University wrote to the University requesting that their samples be removed from the University's tissue bank and released to Dr. Catalona, who had moved to Northwestern University.[4] In response, Washington University brought legal action seeking a determination that it owned the samples and was not obligated to comply with the subjects' instructions. In ruling in favor of the University, the judge relied on the act of handing samples over to researchers as evidence that a "gift" had been made under Missouri law.

It was determined that ownership of the samples had been transferred to the University, and the judge was dismissive of provisions in the relevant consent forms. In keeping with established research rules, the forms included statements about the voluntary nature of participation and procedures for termination of participation. In regard to those provisions, the judge stated that "the right to discontinue participation in a research project means nothing more than that the [research participant] has chosen not to provide any more biological materials pursuant to one or more research protocols; *i.e.*, not to make any more inter vivos gifts of donated biological materials to [Washington University]."[5] Since this was a ruling in federal district court, the case has little legal precedent and it may very well be that other courts faced with similar claims may not adopt the flawed reasoning of this court. Nevertheless, it presents a cautionary lesson to all individuals who have provided samples for genetic research under similar circumstances: if you relinquished a DNA sample to be banked for research, you may very well have made a gift under the law of your state and thereby lost control over your sample.

There is nothing in the Catalona case to indicate that the outcome would have been any different if the patients had given their samples to a repository operated by a for-profit corporation rather than a private academic research institution. So the thousands of individuals who have contributed samples to entities like the Ardais Corporation or DNA Sciences, which obtain samples either from patients at medical centers or through appeals over the Internet for the express purpose of selling them to researchers, should also be concerned about what may happen to their DNA samples (and thus their genetic information) as a result of this case.

Those who have contributed samples for research might take some comfort from the fact that ownership of a DNA sample by a researcher does not necessarily relieve the researcher of his or her obligations (at least under the federal regulations governing research) to maintain confidentiality and protect the privacy of subjects. Thus, federal law may yet provide some protection against unauthorized disclosures of private genetic information by researchers. Nevertheless, those disclosure rules would not put limits on what information the owner could derive from the sample. In that sense, invasions of genetic privacy are very likely to result from donation of samples. This is important because one of the basic reasons for protecting genetic privacy is to guard against secret genetic testing that would result in someone else knowing more about an individual's genetic status than that individual knows her or himself, or would choose to have others know.

Commercial Testing and DNA Banking

Direct-to-consumer marketing of genetic tests, and the collection of samples in relation to such testing, presents additional opportunities for invasions of genetic privacy. In this context federal laws that might restrict misuse of samples by researchers are of course irrelevant and no federal law addresses genetic privacy in other contexts. Consequently, individuals who send samples directly to laboratories for testing must look to state laws for rules governing the collection and uses of DNA samples. By 2000, a majority of states had enacted laws regulating genetic testing and fair uses of genetic information. These statutes are almost exclusively antidiscrimination laws that govern the behavior of insurers, employers or both. The provisions in these laws primarily address what happens to information after samples have been collected and analyzed, but a few states, such as New Jersey, include broader privacy protections by prohibiting unconsented-to collection and testing of samples generally and by defining requirements for consent to testing. Only about half a dozen states require explicit consent for sample storage, or require the destruction of samples after the purpose of their collection has been achieved.

Thus, laws that address DNA collection and banking activities do not generally apply to companies that sell genetic testing services directly to consumers. Such testing is a growth industry and includes tests for genetic diseases and susceptibilities as well as tests for non-medical purposes such as genealogical research and for marketing nutritional supplements and skin-care products to the public.

Without adequate protections for genetic privacy, individuals will not be free to discover and use genetic information for their own benefit and purposes. Laws that regulate uses of genetic information after the fact are necessary but not sufficient to achieve that goal. DNA collection, banking and analyses are expanding rapidly within the world of biomedical research and beyond. But in the absence of a comprehensive federal law addressing genetic privacy, those who want to explore their genetic information may hesitate to do so, and those who do relinquish their DNA, assuming that they have control over its uses, may discover that they have given it all away. Until comprehensive laws governing DNA banking are enacted, the best that individuals can do to protect themselves is to only utilize testing services that guarantee to destroy the DNA sample on completion of the specified test.

2

Position Statement on Genetic Discrimination

BY THE HUMAN GENETICS COMMITTEE OF THE COUNCIL FOR RESPONSIBLE GENETICS

This article was published in GeneWatch, *volume 6, number 6, in 1988. The Human Genetics Committee at that time was made up of the following members, whose affiliations are listed from the original date of publication: Ruth Hubbard, Professor of Biology, Harvard University; Philip Bereano, Professor of Engineering and Public Policy, University of Washington; Paul Billings, Director of the Clinic for Inherited Diseases, New England Deaconess Hospital; Colin Gracey, Head of the Religious Life Office, Northeastern University; Mary Sue Henifin, Deputy Attorney General, State of New Jersey; Sheldon Krimsky, Associate Professor of Urban and Environmental Policy, Tufts University; Richard Lewontin, Alexander Agassiz Professor of Zoology, Harvard University; Karen Messing, Professor of Biology, Université du Québec à Montréal; Stuart Newman, Professor of Cell Biology and Anatomy, New York Medical College; Judy Norsigian, Co-Director, Boston Women's Health Book Collective; Marsha Saxton, Director, Project on Women and Disability; Nachama L. Wilker, Executive Director, Council for Responsible Genetics.*

During the past decade there has been a dramatic expansion in the number and range of genetic tests designed to predict future health. Whereas ten years ago tests were only available for a few inherited conditions, now tests exist to diagnose cystic fibrosis, Huntington's disease, and several other gene-based diseases. Physicians are even projecting that they may be able to diagnose genetic predispositions for complex conditions such as cancer, cardiovascular disease and mental disorders.

As tests become simpler to administer and their use expands, a growing number of individuals will be labeled on the basis of predictive genetic information. This kind of information, whether or not it is eventually proved correct, will encourage some sectors of our society to classify individuals on the basis of their genetic status and to discriminate among them based on perceptions of long-term health risks and predictions about future abilities and disabilities. The use of predictive genetic diagnoses creates a new category of individuals—the "healthy ill"—who are not ill now but have reason to expect they may develop a specific disease sometime in the future.

While the new diagnostics will provide identification of genetic factors that may be responsible for evoking certain diseases or disabilities, it is not at all obvious how rapidly and to what extent this information will lead to treatments or cures for the diseases in question. Diagnoses unaccompanied by cures are of questionable value. This is especially true when the diagnosis can be made long before the person in question begins to notice any symptoms of disability or disease, as is often the case. Many genetic tests predict—often with limited accuracy—that a disease may become manifest at an undetermined time in the future. And although the severity of many genetic diseases varies widely among those individuals who develop the disease, the diagnoses usually cannot predict how disabling a specific person's disease will be. To this extent, the situation is similiar to the experience of people diagnosed to be infected with the human immunodeficiency virus (HIV), who know that they will probably develop one or more AIDS-associated diseases, but not when or which ones.

This kind of "predictive medicine" raises novel problems for affected individuals and they, together with their physicians and counselors, will have to learn how to approach them. Meanwhile the exaggerated emphasis on genetic diagnoses is not without its dangers because it draws attention away from the social measures that are needed in order to ameliorate most diseases, including equitable access to health care. Once socially stigmatized behaviors, such as alcoholism or other forms of addiction or mental illness, become included under the umbrella of "genetic diseases," economic and social resources are likely to be diverted into finding biomedical "cures" while social measures will be short-changed.

Individuals labeled as a result of predictive genetic tests face the threat of genetic discrimination. They and their families are already experiencing discrimination in life and health insurance and employment because genetic information is being generated much more quickly than our legal and social service systems can respond. As our abilities to label individuals on the basis of genetic information increase, particularly through the efforts of the Human Genome Initiative,[1] there will be an even more urgent need to address these problems.

Employment Discrimination

The tragedies of race and sex discrimination illustrate the dangers of basing employment decisions on inborn characteristics. Like these, discrimination on the basis of genetics ignores the present abilities and health status of workers and substitutes questionable stereotypes about future performance.

Basing employment decisions on genetic status opens the door to unfounded generalizations about employee performance and increases acceptance of the notion that employers need to exercise such discrimination in order to lower labor costs. Indeed, without countervailing equitable forces, employers face economic pressures to identify workers who are likely to remain healthy. Less absenteeism, reduced life and health insurance costs, and longer returns on investments in employee training all reduce the costs of labor. To the extent that employers believe that genetic information can help identify workers who have a "healthy constitution," they have strong economic incentives to screen applicants and workers.

Employer discrimination on the basis of antibodies to HIV and of previous cancer history, despite current ability to work, demonstrates that employers take health status into account when making employment decisions to the detriment of individuals labeled as being at increased risk of ill health in the future.[2] Even more revealing is the history of discrimination on the basis of perceived genetic hypersusceptibility to occupational diseases.[3] For example, African Americans who are healthy but have what is called sickle cell trait have been denied certain jobs despite the absence of scientific proof that any genetic characteristics are predictive of industrial diseases.[4]

Such policies victimize all workers. In the case of sickle cell trait, African Americans have been "protected" out of jobs involving exposures to certain industrial chemicals, while remaining workers continue to be at risk from these chemicals. Discrimination against individuals with particular genetic characteristics harms all workers by diverting attention from the need to improve and, if possible, eliminate workplace and environmental conditions that contribute to ill health for everyone. Moreover, such genetic discrimination masks the fundamental need for adequate leave policies and insurance coverage as well as for reasonable workplace accommodations for all workers who experience temporary or permanent disabilities, for whatever reasons.

Basing employment decisions on genetic status may run afoul of the patchwork of state and federal laws that protect the employment rights of individuals with disabilities. To date, federal laws only cover workplaces receiving federal funds.[5] No state or federal court has ever determined whether such laws apply to the employment rights of individuals discriminated against because of their genetic status. Although a bill is pending in Congress that would provide comprehensive protection to workers who are disabled, there is disagreement among legislative experts over whether this bill would prohibit genetic discrimination.[6]

Screening individuals for genetic risk of late-onset diseases raises particularly difficult problems because such individuals may not be considered disabled at the time they are discriminated against and therefore may not be afforded protection under present or proposed federal and state laws protecting the rights of disabled individuals.[7] Ironically, someone who is stigmatized for being at risk for future genetic illness may, due to his or her asymptomatic status, fall outside the protection of laws prohibiting discrimination on the basis of disability. A clearly worded federal law is needed to prohibit discrimination on the basis of such information and to protect the privacy of genetic information.

The need for laws to protect the privacy of genetic information can be illustrated by the secrecy with which employers may use medical information. There are few limits, for example, on employer discretion in deciding what pre-employment medical tests to perform on job applicants. Thus, once a sample of blood is taken from an applicant during the pre-employment physical, it can be tested for many conditions, including pregnancy, sickle cell trait, HIV antibodies, cholesterol, or drugs. Since employers do not have to give a reason for refusing to hire an applicant, many individuals never realize that they have been denied employment because of their medical status. Although it might be possible to challenge an employer's hiring policies that discriminate on the basis of medical status, it is very difficult to document such discriminatory practices.

Insurance Discrimination

Insurers also face strong economic incentives to identify individuals perceived to be at increased risk for ill health in the future. Historically, such inherent characteristics as race and sex were used to deny African Americans and women insurance coverage.[8] Some insurance companies did not end the practice of using explicit racial classifications in setting rates and benefits until the early 1960s. And, in the early 1970s, healthy African Americans who were identified as having "sickle cell trait" once again experienced insurance discrimination, when some insurance companies charged them higher rates, despite the lack of evidence that such individuals were at greater risk than usual of ill health or shortened life span.

Life and health insurance companies are regulated by the states, and a patchwork of laws governs how rates are set and what types of discrimination are permissible. For example, Maryland and New Jersey, which limit unjustified discrimination, may permit discrimination on the basis of genetic status if increased actuarial risk of disease or decreased life span can be demonstrated.[9] Insurance companies argue that they have the right to make appropriate business and financial decisions based on their objective statistical determination of group risk. However, it is not equitable to stigmatize individuals on the basis of group risk, nor is it sound public health policy to deny life and health insurance generically to individuals with risk factors.

Without legislation mandating that all insurers cover populations at risk without discrimination, those who do provide comprehensive coverage are at a financial disadvantage. Insurance companies have successfully staved off legislative interference with their decisions to deny coverage based on actuarial risk and there is every reason to believe that they would lobby aggressively against laws that would prohibit genetic discrimination. The actions of the insurance industry regarding HIV antibody status are revealing. For example, states that have tried to regulate against discrimination on the basis of antibodies to HIV have met vigorous legal challenges by insurance companies, and several such state regulations have been invalidated by the courts.

In their survey of discrimination as a consequence of genetic screening, Paul R. Billings, Mel A. Kohn, Margaret de Cuevas and Jonathan Beckwith of Harvard Medical School illustrate how "data banking" of genetic information can lead to future abuses not only against at-risk individuals, but also against their relatives.[10] Already companies that manage medical information for insurers track individuals identified as having specific genetic conditions so that such people may be denied insurance whether or not they reveal the relevant genetic information on their applications. In addition, government agencies have the capacity to retain records of "DNA fingerprints" on individuals who have been charged with committing violent crimes.[11]

Data banking increases the risk that genetic information will be used in ways that violate individual privacy and encourage irresponsible genetic epidemiology. To examine the full impact of genetic data banking we need to answer three questions: 1) What information is stored, 2) who has access to the information, and 3) how can such information be used?

An individual's right to refuse genetic screening is eroded when employers and insurers require such information as a precondition for employment or for life or health insurance. Even more chilling are instances where insurers have attempted to manipulate individual decisions about childbearing. Insurers have pressured potential parents to be screened or to have their fetuses screened, and then have tried to manipulate their procreative decisions by threatening to withdraw benefits to those who choose to give birth to children at risk of genetic disabilities.

Proposed Actions

The dangers of genetic discrimination may be lessened if advocacy groups and the relevant public and private agencies take the following actions:

- **Develop fact sheets that describe what is known** about genetic screening and why genetic status does not necessarily identify an individual's health or abilities. The fact sheets should be written by health and disability rights advocates and geneticists. They should encourage discussion of the dangers of stigmatizing individuals on the basis of future risks of ill health or disability.
- **Offer short courses on the uses and abuses of genetic screening** to the general public and to journalists, health care professionals, teachers, labor unions, and scientists by public interest groups, educational institutions, cable television, and other media.
- **Draft model laws that can be proposed at local, regional, and,** where appropriate, state and federal levels. These laws would prohibit discrimination in education, employment, insurance, housing, public accommodations, and other areas, based on present or predicted medical status or hereditary traits.
- **Design proposals to end disability discrimination** in all its forms, including proposals that will afford access and participation in all aspects of public life by individuals who are disabled. Coalitions should be encouraged between groups concerned with civil liberties, disability rights, women's rights, procreative rights, occupational health and safety, workers' rights, and the right to health care.
- **Propose absolute and legally binding guarantees of confidentiality** to protect information obtained from genetic screening. The information

should not be released to anyone without the informed consent of the screened person or her/his legal guardian.

- **Advocate nonbiased counseling about the option to refuse tests** and about the benefits and risks of doing so to every individual offered genetic testing. Appropriate consent and refusal forms must explicitly state that refusal to undergo genetic testing will not lead to termination of medical care or insurance, denial of services, or to other discriminatory practices.

When Science Fiction Became Fact

BY SAMUEL W. ANDERSON

Sam Anderson *is the editor of* GeneWatch. *This article originally appeared in* GeneWatch, *volume 22, number 2, April-May 2009.*

In the summer of 2000, Burlington Northern Santa Fe Railroad Corp. was dealing with two train derailments in central Nebraska. Workers went out to make track repairs, equipped with recently adopted hydraulic wrenches. The new tools were faster than their manual predecessors but harder on workers' wrists: over a hundred Burlington Northern employees filed reports that year that their work had caused carpal tunnel syndrome. Gary Avary was one of the workers who repaired the Nebraska tracks, and in September he was diagnosed with carpal tunnel. The railroad authorized surgery, and by the end of October, Gary had recovered from the operation and returned to work with full use of his hands—but his ordeal was just beginning.

In December the company sent Gary a letter instructing him to travel to Lincoln for a mandatory medical exam. Gary and his wife, Janice, found this strange, since Gary's carpal tunnel had already been successfully treated. Stranger yet, a co-worker returning from the same exam reported that the doctor had taken seven vials of blood. This raised a red flag for Janice, a registered nurse, and she called the company's medical liaison to see why so much bloodwork was necessary.

"She gave me a list," Janice says, "just standard lab tests, and I said, 'I'm in the medical field, and those do not require that many tubes of blood.' I got her flustered because I knew what I was talking about—and she just let it slip accidentally that they were going to do a genetic test."

When Janice protested that those tests could not be conducted without Gary's knowledge and consent, she was told that refusal would be considered insubordination and could result in dismissal.

The Avarys went on to fight Burlington Northern's practices and to testify on Capitol Hill to advocate for genetic nondiscrimination legislation. In 2001, Burlington Northern settled with the Equal Employment Opportunity Commission (EEOC), prohibiting the railroad company from carrying out further genetic testing. A settlement with the railroad workers' union required the company to destroy the blood samples it had acquired from at least 20 employees along with the records of the tests carried out on that blood. Burlington Northern also pledged to support a federal genetic nondiscrimination bill. Those results were hardly crippling for the company, but for their part, the Avarys were most interested in pushing for the legislation which would later become the Genetic Information Nondiscrimination Act. Finally, seven years later, GINA was signed into law.

"Everything's changed," Gary says. "I do feel that we made a difference—not just Janice and me, a lot of us."

Janice agrees, even if Burlington Northern got off easy in the lawsuits. "It was kind of bad the way it ended, but for me, we got the genetic testing stopped, the railroad was publicly humiliated for what they did, and as you can see, we're still talking about it eight years later."

"The gene for carpal tunnel"

Why would a company go to lengths to undertake secret genetic testing of its employees? Burlington Northern insisted that the DNA test results were never intended to factor into hiring and firing decisions. According to company statements, the tests focused on determining whether or not employees carried the genetic marker for carpal tunnel syndrome. If so, the company might be able to shed responsibility for employees' carpal tunnel complaints—and avoid paying workers compensation.

The first problem? Dr. Phillip Chance, the scientist who discovered the gene the railroad was testing for, pointed out that the genetic marker in question has little use as a predictor of carpal tunnel injuries. In fact, the gene is not linked to carpal tunnel specifically, but rather to a rare neuromuscular condition for which carpal tunnel is one of several symptoms. The railroad's medical department was either oblivious to this or perhaps simply disagreed. Either way, the brains behind the secret genetic tests seemed to believe they were doing something exciting—and something ethical.

"When Gary went for the initial interview with the carpal tunnel doctor, the doctor started talking about genetic implications for carpal tunnel disease," Janice says. "Now this is what that man does for a living, carpal tunnel surgery—and he was going on and on and on with Gary about how he was trying to get involved in different protocols to prove genetic backgrounds for carpal tunnel disease."

Once the case was under way, Gary met personally with the chief medical examiner behind the railroad's testing protocol. "I looked him right in the eye and said, I understand what you were doing, Dr. Michaels, and you can run it in front of medical people and they wouldn't see any problem with what you were doing, with the legalities—that you have the right to ask a mandatory medical exam any time you see fit. But anyone else would say, 'You're doing what? And you're not telling these people what you're doing?' And he couldn't really give me a good answer there. He just thought, in his own mind, I think, that he was doing everything right. They'd been to these conferences on genetics, and they were bent on looking for that marker [for carpal tunnel]."

Considering the cost of putting together a system for secretly testing employees—and considering that none of the at least 20 employees tested were found to have the marker in question—it would hardly seem an economical program. Indeed, Janice says the company had bigger plans for the DNA testing scheme it referred to as a "pilot program."

"They were in the process of designing a protocol to eliminate corporations from having to be responsible for any injury on the job. If they had been able to prove that back injuries, heart injuries, carpal tunnel, and other things along that line had a genetic background to them, they no longer would have to take responsibility for those payments. Then, had they been able to prove this, they would have been able to patent the protocol and sell it to other large companies."

Keeping the secret

"The entire system knew that there was this network of deceit out there," Janice says. "All of the people that had gone to this testing because it was mandatory had no idea that a genetic test was going to be included in the bloodwork. They were actually doing genetic testing without people's consent or knowledge."

Behind the scenes, railroad officials monitored a detailed DNA testing program.

"The day I didn't go to the mandatory exam, the chief medical examiner called me and said, 'Sir, I see you didn't make your appointment. Why is that?'" Gary says. "And I thought to myself, how would you know so quickly, and why would you need to know so quickly? Well, it turns out they had couriers there to take the blood immediately to the lab. The doctors were just there to draw the blood, and couriers were there to pick up my blood after they drew it."

Knowledge of the program extended throughout Burlington Northern's offices. Janice says that the medical liaison she first spoke with "knew exactly what was going on. She knew there was genetic testing going on, she knew exactly which doctor employees were to be referred to for the testing, and the doctor they had set up to do the testing in Lincoln knew exactly what was going on.

"You get sent to a doctor in Lincoln who's doing the testing, who knows full well you're going to be genetically tested; the medical liaison knew there was a protocol; but none of the employees knew. No one had ever notified the employees." If the Avarys sound cynical about the company's attitude, it's for good reason. During the lawsuit, the EEOC discovered documents sent between employees in Burlington Northern's offices revealing their knowledge of the testing program, even referring to it as "the guinea pig trail."

Burlington Northern's employee handbook—which was over 500 pages long— included a rule allowing the company to require mandatory medical testing, but no obligation for the company to reveal the results of those exams to employees. "That's what they were banking on," Janice says. "There was no legal recourse—they were going by that one rule in the employee handbook."

By refusing to submit to DNA tests, Gary broke the rule. For the railroad, this amounted to insubordination and grounds for dismissal. The company informed Gary that he was being investigated for disciplinary reasons and was scheduled for a hearing. The railroad later denied that he had been scheduled for a disciplinary hearing, redubbing it "an investigation," "a meeting," and "a discussion." From Gary's point of view, though, the message was clear.

"My supervisor had been on vacation, and when he came back he said, 'Gary, what have you done?' He wasn't up to speed on what had happened. He said, 'They want me to fire you!'"

"They were going to fire him, and actually did start it," Janice says. "They actually canceled our medical insurance." At least one other Burlington Northern employee who had raised concerns about the testing reported having his insurance dropped as well.

The company might have continued their secret testing if not for one seeming oversight. "I'd already had plenty of exams—my doctor released me, and I was 100 percent," Gary says. "So how badly did they want my blood? I was already released from the doctor, I was already back to work, and they still send me this letter—I kind of wonder sometimes if it wasn't a foul-up."

"They were trying to fire him for refusing to take a medical test for a problem that had already been fixed," Janice says. "It was just bizarre, really. They just picked the wrong person, I guess, to try to force and coerce into doing something—I mean, how can you force someone to have medical tests on something you've already had fixed?"

Ultimately, though, it was Janice's medical know-how that made her question the bloodwork, setting the whole case in motion. "Look how naïve I'd have been if my wife wasn't a registered nurse!" Gary says.

National attention

The Avarys' case attracted a flurry of news coverage. "When you expose big companies to the media . . . the railroad can't take that. They really want to keep things out of the paper," Gary says. In which case, the timing couldn't have been worse for Burlington Northern: the EEOC suit was filed only days from researchers' announcement of a draft map of the human genome, so the potential use and misuse of genetic information was a hot topic. The first reporters to interview the Avarys told them up front: "It's going to get crazy."

With all the unwanted attention it received, did the railroad shift its attitude? "No," Janice says flatly. "They were basically embarrassed because they got caught. And that was my feeling all the way around. They were very embarrassed because they got caught."

After the suit was filed, Burlington Northern spokesman Richard Russack insisted that the company had "good intentions" and said, "I don't think there's anything wrong with a company trying to do something for the benefit of its employees."

He did admit, though, that "it could have been handled better."

"They were so candid, and the people they got the genetic information from had no idea it was even going on," Janice says. But she also says that attitude was not surprising. "If you look at the history of the railroad, the employees have always been disposable. So that was not surprising, that they'd have the arrogance to talk about it that way. Not at all."

It's the same reason, Gary says, that the railroad adopted the hydraulic tools that led to the same carpal tunnel surgeries it was trying to avoid paying for. "The railroad knew the kind of injuries they were going to get into when they started going to hydraulic tools instead of manual. They were looking at production versus what it would do to employees."

As part of Burlington Northern's settlement with the EEOC, the company would not have to admit guilt for secretly testing its employees. However, the settlement also required the company to send an apology letter to all of its employees, Gary says. "Now how do you do that without admitting guilt?"

Advocacy

The Avarys became involved in the push for genetic nondiscrimination legislation, flying to New York and Washington, D.C., meeting with legislators, testifying before Congress, and raising awareness so that their situation would not be repeated. They found out that it already had.

"We got a lot of calls from the East Coast, from the West Coast—people who did not want to tell their name, but wanted to tell us what happened—because they had actually applied for jobs and that was part of the pre-employment examination," Janice says.

"A lot of people wouldn't come forward," Gary says, "because they didn't want to get fired."

Even with GINA's passage, the Avarys note that fear of repercussions may deter individuals from reporting cases of illegal genetic testing and discrimination, especially in difficult economic times.

"You have a right to ask why," Janice says. "People need to know that one phone call to the EEOC would put the flags up, and that company would be immediately investigated. If someone were to call there today and say, 'I was applying for a job and part of the requirement to get hired was genetic testing,' you can bet in a New York minute that the EEOC would be all over them."

"What most people, even today, do not do is ask questions. Why are you drawing the blood? What blood tests are you doing, and what are they for? Ask them explicitly: 'I need to know exactly what this blood test is for.' And ask for the entire list. You never know what might show up in there."

Life goes on

The Avarys couldn't have imagined where their cause would take them, Janice says.

"We're in the middle of Nebraska, so it was this whirlwind thing for us, very exciting, flying back and forth to Washington, testifying before Congress, to think that we might actually assist in getting that law put into place."

The plane trips stopped early in the Bush presidency, but the Avarys stayed in the loop. They celebrated GINA's passage but maintain a cautious view of its impact; after all, Gary says, "Laws are broken every day."

When the excitement died down, Gary and Janice spent some time on the road, seeing the country in their RV. Gary doesn't work for the railroad anymore. Burlington Northern didn't try to fire him again, but they changed his position to one that involved a good deal of traveling—knowing, Gary suspects, that he would quit rather than spend that much time away from his family.

After all they've been through, though, the Avarys don't waste time being bitter. Gary is working at a railroad museum, and he and Janice are still in the middle of Nebraska, about two hours west of Lincoln, "enjoying the simple things: unobstructed sunrises, grandchildren, and the daily miracles that God provides."

4

How the Genetic Information Nondiscrimination Act Came to Pass

BY JEREMY GRUBER

Jeremy Gruber, J.D., *is President and Executive Director of the Council for Responsible Genetics. This article originally appeared in* GeneWatch, *volume 22, number 2, April-May 2009.*

Just last year, the Genetic Information Nondiscrimination Act (GINA), hailed by Senator Kennedy as "the first civil rights bill of the new century," was signed into law. I had the privilege of working on GINA from the very beginning. Before GINA, it had been almost twenty years since Congress had passed civil rights legislation.[1] Indeed, GINA provides a promising departure from past civil rights laws that were passed to correct historic injustices that had festered within American society for generations. Perhaps the most obvious example is the Civil Rights Act of 1964, which protects against discrimination based on race and gender, among other categories, and which was one of the crowning achievements of the American Civil Rights Movement. GINA represents the first time that Congress has passed such legislation before the discrimination it is meant to address has become permanently ingrained in the country's social fabric, and in that respect I think we should all be proud.

GINA: A History

The story of how GINA came to be enacted is a classic lesson in American politics and perseverance. By 1995, the need for federal genetic discrimination legislation was becoming increasingly clear. Early reports from the Congressional Office of Technology Assessment identified the potential for the misuse of genetic information for purposes of discrimination.[2] This was followed by a number of documented cases of genetic discrimination and several case studies that supported the credibility of many of the claims, such as those completed by CRG Board member Paul Billings.[3] At the same time, the Human Genome Project, one of the largest scientific research projects ever undertaken, was racing toward its goal of producing a reference sequence of the human genome. A working group associated with the Project that included CRG Board member George Annas[4] announced model genetic discrimination legislation that same year. A number of states were already enacting legislation in this area; more than 15 states had enacted protections by 1995 and many were soon to follow.[5] Many of these state laws are deficient, however, due either to limited coverage or to insufficient enforcement mechanisms. Many states never passed comprehensive genetic discrimination legislation.

In 1996, Congress passed the *Health Insurance Portability and Accountability Act* (HIPAA) which, among other things, prohibited the use of genetic information in certain group health-insurance eligibility decisions. New legislation was necessary to plug the holes in health insurance protections that HIPAA left behind and extend protections to other potential areas of discrimination that HIPAA wasn't designed to cover.

A number of individuals, myself included, began to raise a chorus of concern on Capitol Hill over the lack of legal protections against genetic discrimination and various media outlets began to echo them. Genetic nondiscrimination legislation was first introduced in the House of Representatives in late 1995 by Rep. Louise Slaughter, D-N.Y., during the 104th Congress. Congresswoman Slaughter would go on to become the primary champion of genetic discrimination legislation on Capitol Hill through GINA's passage. In 1996, Senator Olympia Snowe, R-Maine, and Senator Ted Kennedy, D-Mass., introduced similar legislation in the Senate. Both bills specifically addressed discrimination in health insurance. Employment discrimination was not covered in this early legislation, and for some in Congress, such as Senator Jim Jeffords, R-Vt., it was unclear whether employment protections were necessary. Opponents of the legislation claimed that genetic discrimination in employment was already covered by the Americans with Disabilities Act (ADA), one section of which allows for a discrimination claim when an individual is regarded as disabled and discriminated against as a result. The EEOC further complicated the situation by supporting this interpretation of the ADA early on.[6] As the ADA made its rounds of court decisions, though, it became clear that the courts were if anything taking a very limited view of the expansiveness of the ADA, and in most cases it would be highly unlikely to be available to support a genetic discrimination claim.[7]

There certainly was no dispute that the ADA offered limited protections against the collection of genetic information. I, and others, continued to build the case that employment discrimination provisions were necessary. With our persuasion, Senator Jeffords held a Congressional hearing on genetic discrimination in the workplace[8] and the argument eventually succeeded. Employment provisions were added shortly after to the legislation that would later become GINA.

GINA suffered in obscurity for a number of years as the result of a Republican-led Congress that was hostile to adding additional restrictions on the insurance industry and employer communities. Powerful lobbies, such as the Chamber of Commerce, also ensured this to be the case. Despite this adversity, momentum continued to build. In 1997, the American Civil Liberties Union and organizations such as the Alpha-1 Association, Genetic Alliance, Hadassah, National Partnership for Women & Families and National Society of Genetic Counselors recognized the need to unite and create an equally powerful organization to fight for genetic discrimination protections. Representing the ACLU, I joined representatives from these organizations in founding the Coalition for Genetic Fairness (CGF) that same year. The CGF would consist of more than 500 organizations by the time GINA was enacted.

From that day forward, the CGF became the leader of a coordinated campaign to enact genetic nondiscrimination legislation. We targeted the White House and members of the Senate and House of Representatives one by one and began to build support within Congress for genetic nondiscrimination legislation. These efforts began to bear fruit in 2000, when then President Clinton signed an Executive Order extending genetic discrimination in employment protections to federal employees.[9] Following this breakthrough the Senate passed genetic discrimination legislation in 2003 and 2005. The House of Representatives, though, is a body more strictly controlled by majority leadership. These same years, despite growing support in the House (in 2003 and 2005 GINA had the support of 242 and 244 co-sponsors respectively), genetic discrimination legislation failed to reach the floor of the House for a vote. Support among House Republicans for this legislation was beginning to grow, though, largely as the result of the CGF working with Rep. Judy Biggert, R-Ill., to become a primary co-sponsor of the legislation. It was also during this time period that the CGF released a highly influential report on the issue of genetic discrimination.[10]

With more bipartisan teams in place in both the House and Senate, GINA was reintroduced in 2007. A change in party leadership in the House gave many of us hope that there would be a major breakthrough on GINA. Its fate in the House remained precarious, though, as three separate committees[11] maintained jurisdiction and therefore many opportunities for roadblocks existed. To overcome them, many of us worked tirelessly during this period to educate members of the appropriate House committees. Finally, the House passed GINA for the first time on April 25, 2007. Having passed the legislation twice before, most expected GINA to pass the Senate soon thereafter. But one last hurdle remained. Senator Tom Coburn, R-Okla., put a hold[12] on the legislation in the Senate, preventing it from coming to the floor for a vote. Coburn's Senate voting record is extremely conservative and he has a well-deserved reputation for stalling measures in the Senate, to the frustration of members of both major parties. Coburn's given objection to GINA was an alleged lack of clarity on protections for embryos and fetuses, although most experts did not believe this to be a legitimate concern, and his hold lasted long after this issue was specifically addressed in the legislation. Support from the White House allowed Senator Coburn to become intransigent despite the administration having earlier issued a Statement of Administration Policy in support of GINA.[13]

The groundswell of support for GINA nonetheless continued to mount and the CGF put significant pressure on Senator Coburn to relent. In early 2008, Senator Coburn released his hold and on April 24, 2008 the U.S. Senate passed GINA. Finally on May 21, 2008 President Bush signed GINA into law.

What does GINA do?

GINA provides protections against genetic discrimination in both the health insurance and employment settings and puts additional limitations on the access to and disclosure of genetic information.

A discussion of GINA necessarily requires that we first take a look at the type of genetic information that GINA is meant to protect. Generally, a person's genetic information is defined as information obtained from the individual's genetic tests, the individual's family member's genetic test, or the individual's family health history. A genetic test as defined by the statute is a process that analyzes human DNA, RNA, chromosomes, proteins, or metabolites, and that detects genotypes, mutations, or chromosomal changes. If the test does not detect genotypes, mutations, or chromosomal changes, or if the analysis directly relates to a health condition that could be reasonably detected without such a test, it is not considered a genetic test under GINA. The definition of genetic information includes "the occurrence of a disease or disorder in family members of the individual" because family medical history could be used to identify genetic information. GINA applies to asymptomatic individuals only. Manifested health conditions are not protected, even if they have genetic origins. The underlying genetic information, though, remains protected by GINA.

Title I of GINA deals with genetic nondiscrimination in the issuance of health insurance. Health insurance companies (including those that sell group policies to employers, non-group policies to individuals and families, and Medicare supplemental policies to Medicare enrollees) are not allowed to determine eligibility (including continued eligibility) for coverage based on genetic information, charge higher or lower premiums based on genetic information, or consider genetic information as a pre-existing condition.

In addition, health insurers are prohibited from asking about genetic information as part of the application process. Once a person is covered, insurers may not ask about or use that person's genetic information for "underwriting purposes," for example, to determine whether to raise premiums when an individual renews his or her coverage. In this Title GINA builds on the *Health Insurance Portability and Accountability Act,* extending protections to the individual market to confirm that individuals in all types of health insurance plans have the same protections. But it goes even further than HIPAA in another important respect with regards to privacy of genetic information. One of the primary tenets of HIPAA is that there is some privacy to your health information, but to the extent that your health insurer needs that information for underwriting, they could have it, as underwriting is a permitted use of health information under HIPAA. This was not a genetics-specific issue. While the HIPAA rule remains in place for most health information, GINA prohibits health insurers from collecting or using your genetic health information for underwriting.

Title II of GINA deals with genetic discrimination in the employment context. GINA makes it an unlawful employment practice for an employer (applying equally to an employment agency, labor organization, or joint labor-management committee controlling job training) to "fail or refuse to hire . . . discharge . . . or otherwise to discriminate against any employee with respect to the compensation, terms, conditions, or privileges of employment" because of the genetic information of the employee. In this respect, GINA adds genetic information as a protected category to the existing body of federal civil rights law.

GINA makes it an unlawful practice for an employer to "request, require or purchase genetic information with respect to an employee or family member of an employee." Subject to the exceptions I will discuss below, this is a total ban that will affect an employer's ability to access such information even under circumstances where it was previously authorized under other statutes. GINA prohibits employers from requesting genetic information even in the rare case where it is arguably job related (though examples to date of such have been extremely difficult to demonstrate). In terms of the Americans with Disabilities Act, this poses a change in how employers collect medical information, at least in terms of genetic information. Under the ADA employers are authorized to collect medical information relevant to meeting a reasonable accommodation request or otherwise authorized as part of a post conditional offer of employment medical evaluation. Additionally, worker's compensation laws would similarly be implicated.

GINA lays out specific exceptions to the prohibition on acquisition of genetic information:

1. Where an employer "inadvertently requests or requires family medical history of the employee or family member of the employee." While the chosen terms are somewhat unartful, it is clear from the intent of this exception and the legislative history of it that it is meant to address so-called "water-cooler conversations" where an employee might voluntarily offer such information up to the employer.
2. A second exception applies to situations where an employer offers health or genetic services including a wellness program. Many employers have adopted wellness programs which begin with extensive health questionnaires. In such situations the statute requires the voluntary consent of the employee and limits access of the information obtained.
3. The third exception applies to the genetic monitoring of the biological effects of toxic substances. Some employers have programs that conduct genetic tests of workers in specific hazardous environments to determine if they need to be reassigned before they become symptomatic of adverse exposure. GINA now requires this testing to be voluntary.
4. Additional exceptions include Federal or state Family and Medical Leave Act compliance, commercially and publicly available records (but not medical databases and court records) and law enforcement purposes.

GINA requires that genetic information be kept as part of the employee confidential medical record and prohibits the disclosure of genetic information unless:

1. There is a written request of an employee for such information (or a family member if the family member is receiving a genetic service).
2. To a health researcher in compliance with applicable law.
3. In response to a court order, but only if the genetic information is specifically authorized and only with the knowledge of the employee.
4. FMLA compliance and information of a manifested disease or disorder that poses an imminent hazard of death or life-threatening illness.

GINA maintains strong enforcement mechanisms in both Titles consistent with existing law in their respective areas.

The sophistication of many subjects does not lend itself well to comprehensive legislation. Therefore in these cases the Congress will designate federal agencies for additional rulemaking. At present, agencies charged with administering GINA are undertaking the rulemaking process.[14] While such rules do not have the same standing as law, they are highly influential. My first act as President of the Council for Responsible Genetics was to testify before the Commissioners of the Equal Employment Opportunity Commission for strong regulations, and I continue to work on comments to address lingering concerns. As this process unfolds we are sure to learn even more about GINA. Much of the regulations will center on further clarifying key terms. My hope is that they will do so in a strong and unambiguous manner. They will also discuss in more detail the affirmative steps that covered entities (be they insurers or employers) will need to take to remain in compliance with GINA and how both Titles will interact. Finally, they will further define the exceptions in Title II on the prohibition of collection of genetic information. It is our hope and expectation that they will construe these exceptions narrowly to be consistent with the intent of Congress.

Conclusion

Otto von Bismarck said well over a hundred years ago that "politics is the art of the possible." GINA provides strong new protections for all Americans against discrimination and access to genetic information, but to pass any law involves compromise. GINA, for example, does not address genetic discrimination in life insurance, disability insurance, or long-term care insurance. Nor does GINA address other, at least theoretical, areas where there might be genetic discrimination in the future. Certainly many of us who worked on GINA regret dearly that GINA does not protect symptomatic individuals. The inclusion of this class of individuals would have brought up a larger debate on the limitations of the Americans with Disabilities Act because in many cases the extent of their condition would not be significant enough to qualify for its protections. It would also have implicated an ethical debate on "genetic exceptionalism" by offering protections to symptomatic individuals with genetic conditions but not to individuals for whom no such linkage could be demonstrated. There was no support in Congress for these discussions and they would have derailed GINA.

No single law solves every issue that it might implicate. The Civil Rights Act of 1964 was no less a watershed moment in the history of our country because it didn't cover discrimination in housing or based on age and disability. These protections came later. Nor was the Civil Rights Act of 1964 any less significant because it remains ineffective at addressing areas where race and gender discrimination remain institutionalized in this country from education to the criminal justice system. GINA is a strong and essential first step in the fight against genetic discrimination and misuse of medical information more generally, but it is not our last battle. The precedent of GINA, as well as the improved level of education on Capitol Hill as the result of the process of enacting it, will allow us to build upon the foundation that GINA now provides. We must continue to seek out and address discrimination in every corner and ensure that strong protections are in place to address it.

5

Politics and Perseverance

INTERVIEW WITH LOUISE SLAUGHTER

Rep. Louise Slaughter *has served western New York State in the U.S. House of Representatives since 1987. She introduced some of the first federal genetic nondiscrimination legislation in 1995. In 2006, Rep. Slaughter became Chair of the influential House Rules Committee and is currently the ranking minority member. This interview took place in 2009.*

GeneWatch: If any legislator can be called a champion of GINA, it would have to be you. You've pushed for genetic nondiscrimination legislation for seven sessions. Why did you stick with it for so long?

Rep. Slaughter: Because of the importance of it. We had an opportunity, when they first announced that they were going to decode the genome, to have science and political policy go hand in hand. Obviously we recognized how marvelous the science could be, but we also recognized that there were pitfalls there for people who were already being discriminated against because of genes, and people were terrified to know what their genetic makeup was—people who had Alzheimer's in their family, a lot of cancer in their family, women who thought they might have a gene for breast cancer or were afraid to find out—and their doctors told them not to have a test until the bill passed.

It was a very strange odyssey. It passed the Senate unanimously twice but didn't have a single hearing in the House of Representatives for all those years, until Nancy Pelosi became speaker and George Miller became chair [of the House Education and Labor Committee]. Both of them struggled all these years, they were cosponsors of the bill, and I can't praise George Miller enough for what he did to make it possible to get this passed.

GeneWatch: So is that the main reason it came about?

Slaughter: Politics. There's no question about it. The National Association of Manufacturers, Small Business Association, Chamber of Commerce, and some drug companies were very much against it. Drug companies, I think, because what they like to do for research is rifle through medical records and see who's got the most of what. And I don't object to that, but I saw no reason for them to know who this person was, where they lived or where they worked.

GeneWatch: Did you ever hear a justification for why they would need to know those details?

Slaughter: No, they never spoke to me. All those years we had no real contact.

GeneWatch: What sorts of justifications did you hear from other opponents?

Slaughter: They didn't give us any. Well, they thought it would be a handicap for employers, and that it would result in tremendous numbers of lawsuits. But it took so long that a number of states had already passed legislation—which isn't good enough, because it shouldn't be luck of the draw, based on where you live, whether you're going to be protected or not.

GeneWatch: What do you think are the next pressing genetics issues?

Slaughter: There's so much more happening out there. There's the epigenome, which we all need to understand better. But let me tell you the most important thing that we're going to find from this debate, in my book. We ought to be able to do individualized health care—your genes will tell you what medicine will help you, individually. Dr. (Elias) Zerhouni, who was head of the NIH, told me that 80 percent of patients on Lipitor are not benefited. But that is the drug of choice, so 20 percent benefit and it costs an awful lot of money. Obviously, by doing specialized medicine we can not only keep people in better shape medically, but we're also going to save a lot of money.

GeneWatch: One concern about GINA is that part of what it does is reassure people to get genetic tests.

Slaughter: Well, everybody doesn't have to do that—we're by no means encouraging everybody to run out and get a genetic test.

GeneWatch: In terms of public awareness of genetic discrimination issues in general . . . a lot of people, when I talk to them about this. . . .

Slaughter: They don't even think about it.

GeneWatch: They don't think about it at all. Do you think that's changed?

Slaughter: No, I'm not sure it has. It's there, it's a tool for science, it's critically important. But personally, nobody comes up to me to say, "Thank you for making it so that I can get a genetic test."

I take great pride in the fact that we have done this, and I know that it's critically important. I think Sen. Kennedy said it's the first civil rights bill of the century, and Francis Collins said it's the equivalent of splitting the atom. So I'm perfectly aware of the importance of it, but I don't have much reason to believe that the public knows or cares what we've done here.

GeneWatch: It seems that it's pretty difficult for the average person to notice when they're being discriminated against, or to avoid it—to know to think twice about that blood test.

Slaughter: I think the day will come when if you're going to be treated for high blood pressure or arthritis, that we would be able to, because of your genetic makeup, have the specific medicine to help you. But again, when the genome first came out, we were very much concerned that they were going to sell the information or it would fall into private hands. That would not be a good thing. And if you're alluding to the testing that's popping up all over the place, I think it needs to be regulated. It is really an opportunity, I think, for a lot of scams.

GeneWatch: Is there much political interest in regulating consumer genetic tests?

Slaughter: Well, I think we're going to do it. Whether there's political interest or not, I think we can get it done. When it comes to any kind of regulation, they get skittish around here . . . but it's coming.

I remember trying to get this bill passed early, talking about breast cancer, and Collins was saying, "We don't want to micromanage medicine." I think they felt that breast cancer was a women's disease. And they felt that it was really a sad thing. But I didn't get much indication from a lot of them that it was anything they were really concerned about.

So I think it was really all the outside groups. Your group, Hadassah, all the others—we had enough outside support to represent at least half the population of the country. And the Senate, as I said, passed it unanimously twice.

GeneWatch: Considering that it passed unanimously in the Senate, and when it did pass in the House there was what, one dissenting vote—it seems like something everyone could get behind.

Slaughter: We always knew it would pass. The chairs of the committee just wouldn't let it out because of the groups who were opposed to it. So it never had a hearing here until George Miller was the chair.

GeneWatch: What about the lobbying on the other side?

Slaughter: They were mostly groups, individualized groups. Hadassah worked so hard and they were so generous. We would have never known so much about the gene, the BRCA1 gene, if they hadn't been so generous to let themselves get tested for it. I know a woman here who had seven major operations because she had the gene and breast cancer, and I think she's going to be one of our best advocates for young women to get tested.

GeneWatch: But ultimately, getting GINA passed came down to the politics.

Slaughter: Let me tell you right out the outset, it was the leadership change.

And I should say about President Bush, on his behalf—he said in his State of the Union that he wanted to sign this bill. And he did sign the bill. It was the House leadership that needed to change, and if we had not changed it, we would be in bad shape.

GINA's Beauty Is Only Skin Deep

BY MARK A. ROTHSTEIN

Mark A. Rothstein, J.D., *is Director of the Institute for Bioethics, Health Policy and Law at the University of Louisville School of Medicine, where he holds the Herbert F. Boehl Chair of Law and Medicine. This article originally appeared in* GeneWatch, *volume 22, number 2, April-May 2009.*

It is hard to be critical of the Genetic Information Nondiscrimination Act of 2008 (GINA). After all, it's the first federal law enacted to prohibit genetic discrimination, and passing it took 13 years of work by people whose goals I share. In analyzing the law, however, it is apparent that GINA fails to resolve or even address many of the basic concerns that drove the legislative effort. It is also clear why, despite 13 years of wrangling on Capitol Hill, the final version of GINA was passed unanimously in the Senate and received only one negative vote in the House of Representatives—and that from inveterate naysayer, Representative Ron Paul.

GINA was *not* enacted in response to a wave of genetic discrimination, defined as the adverse treatment of an individual based on genotype. There have been very few documented cases of such discrimination. To some degree, GINA was enacted to prevent genetic discrimination in the future when health records will routinely contain genetic information and genetic testing will be so inexpensive that it's cost-effective to perform it on a widespread basis. The real reason for enacting GINA was to assure people that they could undergo genetic testing without fear of genetic discrimination. As any clinical geneticist or genetic counselor will tell you, these fears are real.

According to section 2(5) of GINA, federal legislation "is necessary to fully protect the public from discrimination and allay their concerns about the potential for discrimination, thereby allowing individuals to take advantage of genetic testing, technologies, research, and new therapies." The phrase "fully protect the public" is a curious choice of wording. GINA *did* extend protection against genetic discrimination to the few states that had not previously enacted a law prohibiting genetic discrimination in health insurance or the one-third of the states without a law banning genetic discrimination in employment. Yet, the method of protection, similar to state approaches, is not fully protective in any way. Therefore, it would be more accurate to say that GINA "fully covers" the public; it certainly does not provide "full protection."

There are three major flaws with GINA. First, it applies only to two aspects of the problem: discrimination in health insurance and employment. To allay public concerns about genetic discrimination, it's necessary to prohibit the adverse treatment of individuals in numerous settings. GINA does nothing to prohibit discrimination in life insurance, disability insurance, long-term care insurance, mortgages, commercial transactions, or any of the other possible uses of genetic information. It remains to be seen whether GINA's limited applicability, coupled with its inadequate protections in health insurance and employment, will be enough to reassure the public that undergoing genetic testing will not endanger their economic security.

Second, GINA's prohibition on genetic discrimination in health insurance is largely a mirage. The Health Insurance Portability and Accountability Act (HIPAA) contains a little-known provision prohibiting employer-sponsored group health plans from denying individuals coverage, charging them higher rates, or varying their coverage based on "genetic information." Significantly, HIPAA prohibits discrimination by group health plans on the basis of *any* health information. Because HIPAA prohibits genetic discrimination for the largest source of private health coverage (group plans), GINA's main value is to cover people with individual health insurance policies in the few states that did not previously enact a state genetic nondiscrimination law.

Unfortunately, the protections afforded individuals under either state laws prohibiting genetic discrimination in health insurance or GINA are not particularly robust or valuable. (Because state laws and GINA are similar in substance, for simplicity I'll merely refer to GINA.) The problem is that GINA only applies to asymptomatic individuals. There are few incentives for health insurers to discriminate against asymptomatic individuals and few laws to prohibit them from discriminating against symptomatic individuals.

An example will bring this problem more clearly into focus. Under GINA, it is unlawful for an individual health insurance company to refuse to offer coverage, charge higher rates, or exclude certain conditions on the basis of genetic information, including the results of a genetic test. For example, it would be unlawful to deny coverage to a woman with a positive test for one of the breast cancer mutations. Now, suppose some months or years later, the woman develops breast cancer. GINA simply does not apply. The insurance company's permissible response would depend on state insurance law. In virtually every state, the health insurance company could lawfully react to the changed health status of the individual by refusing to renew the policy (at its typically annual renewal date), increase the rates to reflect the increased risk (and the rates might double or triple), or renew the policy but exclude coverage for breast cancer.

GINA does have some limited value in this scenario. Because of GINA, an at-risk woman is no worse off in terms of insurability due to having a genetic test, and there might be psychological or medical benefits from being tested, depending on the results. Yet, the overall picture in terms of health policy remains bleak. So long as individual health insurance is medically underwritten at the initial application and for renewals, individuals who are ill or more likely to become ill are extremely vulnerable. Many advocates and policy makers have concentrated on the issue of genetic discrimination in health insurance, but the issue is much broader and cannot be resolved by such a narrow focus. To state the obvious: Under any system of universal access to health care, the issue of genetic discrimination in health insurance disappears.

Third, the employment provisions of GINA are ineffective, but for different reasons. As with health insurance, the employment provisions only apply to individuals who are asymptomatic. The Americans with Disabilities Act (ADA) covers individuals who have substantially limiting impairments and therefore the ADA would prohibit discrimination against symptomatic individuals, regardless of the cause of their condition.

GINA makes it unlawful for an employer to request, require, or purchase genetic information regarding an applicant or employee. This is an important issue, because individuals are concerned with employers merely having access to their genetic information. The problem is that the provision is infeasible and therefore is not being followed.

Under the ADA, after a conditional offer of employment, it is lawful for an employer to require individuals to undergo a preplacement medical examination and to sign an authorization releasing all of their medical records to the employer. In effect, GINA now qualifies this by saying that employers can require the release of all medical information except genetic information.

GINA defines genetic information as the genetic tests of the individual, genetic tests of the individual's family members, and family health histories. Because this information is commonly interspersed in medical records there is no practical way for the custodians of the health records (*e.g.*, physicians, hospitals) to send only non-genetic information. In practice, when presented with a limited or unlimited request, the custodians usually send the entire records.

The development and adoption of electronic health records (EHRs) and networks hold the possibility of using health information technology to limit the scope of health information disclosed for any particular purpose. Unfortunately, there have been no efforts undertaken to design health records with the capacity to segment or sequester sensitive health information (including but not limited to genetic information) to facilitate more targeted access or disclosures. Without such efforts, health privacy will decline precipitously with the shift to EHRs because records increasingly will be comprehensive (*i.e.*, containing information generated by substantially all health care providers) and longitudinal (*i.e.*, containing information over an extended period of time). Thus, when employers and other third parties require access to an individual's health records the amount of information they receive will be much more extensive than they receive today.

GINA represents an incremental approach to problems that do not lend themselves to incremental approaches. Numerous entities have economic interests in learning about an individual's current or likely future health. GINA consists of halfway measures limited to health insurance and employment that do not provide adequate assurances to the public that genetic information will not be used to their detriment in other ways. GINA prohibits genetic discrimination in individual health insurance against people when they are asymptomatic but fails to provide them with what they need most—health coverage when they are ill. GINA prohibits employers from requesting or requiring the release of genetic information in comprehensive health records at a time when it is infeasible to separate genetic information from other health records.

If GINA serves to declare the unacceptability of genetic-based discrimination and begins a process of careful consideration of a wide range of health-related issues, then it will be valuable. But it is far from clear that GINA will have such an effect. It is not clear that GINA, by singling out genetic information for special treatment, will not increase the stigma associated with genetics and encourage other condition-specific, rather than comprehensive, legislation.

It is not clear whether GINA will be the first step to meaningful legislation or cause legislative fatigue based on the erroneous assumption that the issues already have been resolved. It is not clear whether consumers will understand GINA's limitations or mistakenly rely on its presumed protections. In the short term, the worst thing that could happen is for advocates of genetic rights and fairness in health care to be satisfied with GINA or exult in its enactment.

7

Lessons Learned: How Berkeley Came to Abruptly Change Its Genetic Testing Program

BY JEREMY GRUBER

Jeremy Gruber, J.D., *is the President of the Council for Responsible Genetics. This article originally appeared in* GeneWatch, *volume 23, number 4, August-September 2010.*

Last month, the University of California, Berkeley changed course after months of intransigence and made significant changes to its controversial "Bring Your Genes to Cal" freshman genetic testing program. The most prominent modification was the elimination of any individually identifiable analysis of student DNA. While this change of course appeared abrupt, it was the result of several months of significant and mostly behind-the-scenes work. The lessons learned from successfully pressuring the University to revise its program can provide a blueprint for future successes in steering the development of biotechnology toward the advancement of public health, environmental protection, equal justice, and respect for individual rights.

In May of this year, the University of California, Berkeley announced that it would be sending incoming freshmen a cotton swab with which to send in a DNA sample to be tested for three gene variants that help regulate the ability to metabolize alcohol, lactose and folates as part of a program for the class of 2014 that would focus on genetics and personalized medicine. The announcement, including rough details of the program, brought swift condemnation from a number of quarters for the lack of due consideration for issues ranging from the privacy protections for the DNA samples and the data generated from them to issues of improper informed consent and conflicts of interest. Noted professors at Berkeley, from Kimberly Tallbear, Troy Duster, Charis Thompson and

David Winickoff to Nancy-Scheper Hughes, Paul Rabinow and Laura Nader, as well as academics outside Berkeley, such as Hank Greely at Stanford, George Annas at Boston University, and Debra Geenfield at UCLA, all organized and raised their voices within and without the university (and continued to speak out during the course of the summer). Organizations such as the Council for Responsible Genetics (CRG) and the Center for Genetics and Society issued serious point-by-point critiques of the program as well as brought public attention to the issue through significant media coverage during the first few weeks following the program's announcement. Many other members of the public and media spoke out as well.

Unfortunately, as is often the case, the initial outcry was not enough. While the University met with many of its academic critics, it failed to seriously consider and ultimately rejected their concerns. The media attention ran its course. Having weathered the criticism, indeed seemingly enthralled by it, Berkeley became even more dismissive and convinced of the virtue of its program.

And that is where the story would likely have ended, had many critics of the Berkeley program not urged the Council for Responsible Genetics to dig deeper. Raising your voice and taking a public position on an issue is the easy part of advocacy work. Seldom do organizations like ours have the resources to roll up their sleeves and conduct the incredibly time-consuming and unglamorous work that is generally required to truly move an issue forward. Fortunately for us, we had an incredibly bright and energetic intern assisting us for the summer and there was no greater motivating factor than Berkeley's arrogance!

With our charge of halting this highly problematic experiment, CRG began a multi-prong advocacy plan. We began building a loose coalition of organizations to work on the Berkeley program. We had learned our initial lesson, that a group of organizations and individuals focused exclusively on biotechnology issues was insufficient to effect change on its own. This time we recruited a much larger and varied group of organizations, ranging from civil rights and privacy groups such as the American Civil Liberties Union (ACLU), Electronic Frontier Foundation (EFF), Privacy Rights Clearinghouse and World Privacy Forum to consumer protection groups such as Consumer Watchdog as well as more focused organizations such as the Alliance for Humane Biotechnology.

We filed a number of public records requests with Berkeley under California's Sunshine Act, similar to the Federal Freedom of Information Act, in an attempt to learn more details about the program, including information regarding the program's undisclosed funding source. Berkeley often delayed, and in some cases failed to comply with our requests, necessitating us to file a formal complaint with the Fair Political Practices Commission (which had enforcement authority over some of our requests) that resulted in a rare rebuke to the University. We also sent out a staggered series of press releases to the media with new information as we discovered it, keeping the issue on their radar.

We conducted a significant amount of legal research in our attempt to stop the program from moving forward, evaluating every facet of Berkeley's plan to look for opportunities for a legal or regulatory challenge. This research resulted in an extensive memorandum outlining Berkeley's potential regulatory and legal breaches and the opportunities for legal action, ranking them in terms of likelihood of success. The strongest opportunity for challenge that we identified was in the California Business and Professions Code, which regulates clinical laboratory licensure. Genetic tests, the results of which are reported back to the individual, are considered clinical laboratory tests by the California Department of Public Health (DPH) pursuant to the Code. Therefore, any lab conducting such tests must first obtain a license from the DPH. Through our public records requests, we determined that Berkeley had plans to conduct genetic testing of incoming students at the Genetic Epidemiology and Genomics Lab at the School of Public Health. We discovered that neither the lab (nor its Director or technicians), had any clinical laboratory licenses from the DPH. We further determined that in fact the only lab certified for this work was the campus health clinic, which was insufficiently equipped to perform this type of testing. We further determined that no licensed medical professionals were involved in the program, as is also required by the Code.

In the course of researching our memorandum, we had obtained copies of letters sent by the DPH to several in-state direct-to-consumer genetic testing companies a couple of years earlier that warned them of the requirement of proper clinical lab licensure. As the letters were signed, it was relatively easy to identify the appropriate individual to contact at the Department of Public Health with our findings. While the DPH confirmed that our analysis was a correct interpretation of the regulatory requirements, the Department seemed wary of making public comments before having an opportunity to fully investigate. Given that the University would violate only California law upon the actual testing—not yet performed at that time—it was too early for us to argue Berkeley was operating illegally.

Then we caught a lucky break. Assemblyman Chris Norby, a moderate Republican in the California legislature, introduced a bill (AB 70) that, if enacted, would essentially shield the state from any lawsuits as a result of Berkeley's (or any other public university's) genetic testing program. The costs of any lawsuit would be borne by the individual university's general funds. While highly unlikely to pass, the bill proved an invaluable tool to renew media and public attention on Berkeley's program. We formed a close relationship with the sponsor's incredibly dedicated staff and began working closely with them on both substance and strategy.

With this new platform, we drafted multiple organizational sign-on letters with the help of our coalition partners both in support of AB 70 as well as urging the legislature to force a more general accounting of the Berkeley program. With these letters in hand, the ACLU became an essential ally in using its deep connections in the legislature to open doors for multiple meetings between themselves, CRG and EFF with staff in both the Senate and Assembly. Both the ACLU and EFF also assisted in broadening our analysis. The meetings created the necessary pressure to hold a hearing on the Berkeley program. The Alliance for Humane Biotechnology used its connection with the chairman of the committee holding the hearing to weigh in as well. CRG worked closely with staff to identify those individuals who could provide testimony beyond that narrowly offered by the representatives from Berkeley including Professors Greely and Scheper-Hughes. Lee Tien from EFF and myself representing CRG also testified at the hearing. Perhaps most importantly, the California Department of Public Health was called to testify. The added pressure on the agency was the final piece we needed.

Just prior to the hearing, the Governor's office contacted the DPH and ordered them to issue a statement at the hearing, rather than testify. Nevertheless, their statement reflected our initial analysis and the hearing elicited a promise from Berkeley, still unconvinced that the DPH was serious, to abide by the agency's interpretation. A meeting between DPH and Berkeley later that day convinced them that the agency was indeed serious. Berkeley hastily called a press conference the next day to announce significant changes in the "Bring Your Genes to Cal" program, bringing a media firestorm even larger than the one initially covering the announcement of the program. While some elements of the program have proceeded, it's safe to say that most institutions considering a genetic testing program in the future will remember Berkeley and invest far more effort into crafting an ethical and legally compliant protocol.

We learned many things from this small success. We learned that a successful advocacy effort requires many hands and diverse talents and that to actually be effective on our issues we need to continue to build relationships and alliances with organizations that do not exclusively work on biotechnology issues, ranging from the civil rights and privacy communities to the organized labor, disability, patient, consumer protection and environmental communities, to academia and beyond. We need to build strong relationships with legislators and other policy makers as well as the media. We learned that legislation, even when unsuccessful, can provide a valuable tool to bring attention to and impact our issues. And most importantly we learned that there is no substitute for rolling up your sleeves and doing the type of hard work that often goes unnoticed and is usually unsuccessful. With hard work (and a lot of luck!) we can successfully steer biotechnology development toward advancing the public interest.

Direct-to-Consumer Genetic Testing and Privacy Concerns

BY JEREMY GRUBER

Jeremy Gruber, JD., *is President of the Council for Responsible Genetics. This essay is a modified excerpt from testimony offered at the FDA public meeting on "Oversight of Laboratory Developed Tests and 'Direct-to-Consumer Genetic Testing'" in Silver Springs, Maryland, on July 20, 2010.*

DNA provides a rich digital source of medical information; as a result it has great scientific value. But it is also ripe for data sharing and has significant commercial value as well.

Purchasing genetic testing services in an online commercial marketplace raises significant privacy concerns, as consumers may turn over their DNA and other personally identifiable information to companies without a clear understanding of the privacy risks and without clear guidance as to their legal and regulatory rights in this area.

There are currently no clear guidelines on the ownership of genetic material and the information derived from it, nor are there clear guidelines with respect to the protection of customer privacy by the direct-to-consumer genetic testing industry. Indeed, consent forms and privacy policies vary widely within the industry and without standards can be unclear and often subject to change.

There are three specific areas where significant privacy concerns arise:

[A] Controls on DNA Submitted by Customers

Current practices related to ensuring that customers are submitting only their own DNA are insufficient. At present, commercial personal genomics companies do require customers to confirm they have the legal authority to submit DNA samples, yet such statements are not clearly and conspicuously posted but rather often hidden within larger privacy and consent documents that are often visible to the consumer only after the registration process has begun. Moreover, they do not explicitly warn customers of the possible issues raised by submitting another individual's DNA for analysis.

Considering how simple surreptitious collection of individual DNA can be, it is not hard to imagine how political, social and personal motivations could compel the improper submission of DNA samples. This is a particular concern since most of these companies allow for an individual to purchase multiple testing kits per order. Yet, few controls are offered beyond such statements to ensure that customers are actually complying with this requirement. No offer of proof is requested beyond the statement. This could easily be included as part of the sample submission process.

Security of genetic information

Customers are often not limited to providing a DNA sample as part of their participation in the personal genomics marketplace. They are also offered a variety of surveys, blogs and other tools where they can provide personally identifiable information. Whenever identifiable DNA samples are collected and stored, there is a high risk that violations of genetic privacy will follow. The methodology by which this information is secured is essential, yet without standards and oversight we still know very little beyond the assurances of the industry as to what specific controls are used.

Moreover, the privacy policies of DTC companies are not subject to the health privacy regulations issued pursuant to the Health Insurance Portability and Accountability Act (HIPAA) and there are few state and federal privacy laws that apply. It is essential that personal information should be protected by security safeguards appropriate to the sensitivity of the information.

Safeguards should include physical, technical and administrative measures to protect information and biological samples from unauthorized access, use, disclosure, alteration or destruction.

Almost all DTC company privacy policies make statements about security safeguards, though the degree of detail varies substantially. Mistakes and other breaches of security are not uncommon. Just this summer [2010], the DTC company 23 and Me [which provides direct-to-consumer genetic testing for health and ancestry] accidentally sent data of up to 96 individuals to the wrong customers.[1]

There is also no transparency as to the degree to which personally identifiable health information is de-identified. As the ability to share, store, and aggregate genomic data progresses, the capability of keeping this data anonymous becomes increasingly important. Because an individual's genetic information is so personal and specific, it is vital to protect it from any unwarranted access or use. There have been several instances where de-identified data has been re-identified and personal information linked back to its owner. One such study[2] achieved re-identification of DNA data and established identifiable linkages in 33 to 100 percent of surveyed cases, which focused on eight gene-based diseases. The researchers used anonymized DNA database entries and related the information to publicly available health information despite the fact that the database did not include any explicit identifiers, such as name, address, social security number, or any other personal information. Because not all de-identification techniques adequately anonymize data, it is important that the process employed by the industry is robust, scalable, transparent and shown to provably prevent the identification of customer information.

Third party disclosure of customer data

One significant unresolved issue relating to the DTC industry is exactly who owns the customer's data. Most DTC companies do not explicitly address this issue in their privacy policies. If the DNA sample and other information submitted by the customer are the property of the company, the company is free to sell or otherwise transfer that information to a third party.

Many DTC companies have adopted this approach as part of their business model without sufficiently explaining to customers the extent to which this may occur, what type of data is being transferred and the potential negative consequences. For example, 23andMe has partnerships with the Swiss firm Mondobiotech and the Parkinson's Institute and Navigenics is conducting studies with the Mayo Clinic and Scripps Institute.

Moreover, how such information is to be treated upon sale of a company or if a company enters bankruptcy proceedings, particularly when the entities potentially acquiring such information have significantly less strict privacy standards, is less than clear and is certainly not expressed to customers.

Most DTC companies do not ask for specific consent for these purposes. Some companies are moving in the right direction. 23andMe has recently begun asking for specific consent for participation in published research. However, they note that even by refusing to participate, "... we may still use your Genetic and/or Self-Reported Information for R&D purposes . . . which may include disclosure . . . to third-party non-profit and/or commercial research partners who will not publish that information in a peer-reviewed scientific journal."[3]

The degree to which these types of partnerships and others have proliferated within the industry is still largely unclear. What is clear is that it is essential that affirmative written consent must be required before DTC companies can use any customer generated genetic information in this way.

There is currently very little guidance on how consumers can protect their privacy. For example, the US Federal Trade Commission gives the following advice to consumers who are considering DTC genetic tests:

Protect your privacy. At-home test companies may post patient test results online. If the website is not secure, your information may be seen by others. Before you do business with any company online, check the privacy policy to see how they may use your personal information, and whether they share customer information with marketers.[4]

Such advisories are hardly satisfactory to ensure consumer privacy is protected.

It is essential that Congress, the Food and Drug Administration, the Federal Trade Commission, and the Centers for Disease Control all work together to help set privacy standards for the direct-to-consumer genetic testing industry and ensure that all issues regarding industry practice are adequately supervised to ensure compliance.

From the Cradle to the Lab

BY SAMUEL W. ANDERSON

Samuel W. Anderson *is editor of* GeneWatch. *This article originally appeared in* GeneWatch, *volume 24, number 1, February-March 2011.*

If you were born in the United States in the past forty years or so, getting your heel pricked by a nurse was likely one of your earliest experiences. The nurse collected several drops of your blood, probably on a paper card, and sent it to be screened for diseases. Nearly every baby born in the U.S. (and many other countries) gets the same pinprick. Newborn screening tests are mandated in most of the country—and the benefits are undeniable. The Centers for Disease Control estimate that each year screening programs catch around 3,000 severe disorders in infants, with many of these diagnoses allowing the disorder to be controlled through early treatment.

If you were born more recently, depending on the state (around 1991 in Massachusetts), there is also a significant possibility that some of your DNA, in the form of dried blood on that same paper card, is sitting in a storage facility to this day. Perhaps some of it has been used in research, or to validate the state laboratory's screening accuracy.

Yet even under the best of intentions, the notion of a government entity collecting and storing children's blood samples is sensitive enough to stir controversy, particularly if parents don't find out about their state's newborn blood spot storage program until after their own child's sample has been added to it. In the past few years, parents in both Texas and Minnesota took their state health departments to court on the grounds that the agencies violated their children's privacy by failing to acquire consent from the parents before storing samples and making them available to researchers. The Minnesota lawsuit (Bearder v. Minnesota) was dismissed, but in the Texas case (Beleno v. Tex. Dept. of State Health Servs.) the health department agreed to new rules for consent and transparency.

The screening itself is less contentious, though not without some controversy. The idea of the procedure being required by law may not sit well with some, but the practice has been in place for decades. Even when given the opportunity, very few parents turn down the screening. There's good reason for that, says Robert Green, Associate Director of the Harvard-Partners Center for Personalized Genetic Medicine and a geneticist at Brigham and Women's Hospital.

"Newborn screening has undoubtedly saved thousands, if not tens of thousands of lives by identifying treatable metabolic disorders early," Green says. "PKU [Phenylketonuria]—the disease that started newborn screening—is the most dramatic example. If you don't start treatment right away with special diet, the child suffers irreversible neurological damage; if you do treat from the beginning with a special diet, the child grows up without these problems." The complications begin to arise after the child's panel of tests has been completed. Only a portion of the dried blood on the newborn blood spot card is used up in the standard screening. Generally, the rest of the sample will be stored for a while and used only to confirm a diagnosis or for quality assurance testing. If a lab is transitioning to a new machine, for example, it can compare the new machine's reading of the stored samples to their known results.

In some states—and in many parents' minds—the labs are presumed (or required) to destroy the samples once they are done with them. This is not always the case, however. Newborn blood spot samples are coveted by scientists studying the causes of childhood disorders and a range of other epidemiological and population genetics questions. In some cases, such as studies on diseases that are fatal for infants, these blood spots may be essentially the only source of samples available to researchers looking for a cure. So, rather than throwing away a perfectly good sample, states are often permitted to catalogue the blood spots, assign a code in place of the child's identity, and send the cards to a storage facility. The samples can pile up quickly; Texas' biobank grew to over 5 million newborn blood spot samples by 2010.

These biobanks are created with infants' health in mind, but many also allow researchers to request samples, and some can even sell samples to private companies. For example, a medical researcher or a pharmaceutical company might ask for samples marked as female, Asian, and positive for PKU. The biobank would send blood cards to the researcher or company, including general medical and demographic information, but no personal identifiers, save for the code assigned to the sample. Labs use de-identification to prevent recipients of samples from accessing personal information about the original donor while retaining that possibility for the state, which keeps the list linking infants' names with their samples' code numbers. In other types of research biobanks, samples might be further de-identified, with the lab attempting to completely divorce the information. However, proponents point out, preserving that link makes it possible to use the sample in a missing person case or, perhaps, to retroactively solve the mystery of a child's death through a "metabolic autopsy."

The privacy concerns may not be strictly theoretical. Some high-profile DNA databanks have run into problems recently. A laptop was stolen containing personal information of participants in the world's largest stem cell bank, the Cord Blood Registry; New Zealand's national DNA database investigated a staff member's inappropriate disclosure of information from the database; and a thief made off with a National Institutes of Health laptop holding the personal and medical information of 2,500 research participants. NIH also decided in 2009 to stop making subjects' genomic data publicly available online, after researchers at Arizona's Translational Genomics Research Institute demonstrated how to identify individual donors within large collections of DNA profiles. A few years earlier, NIH Alzheimer's researcher Trey Sunderland was revealed to have secretly supplied the spinal fluid samples and clinical data of over 500 research participants to Pfizer in return for hundreds of thousands of dollars.

Many parents—perhaps most—would gladly agree to have their child's de-identified sample included in research that could save lives. Not all states are required to inform parents that their child's sample may be kept and used for research, and parents are often asked to sign the consent form amid the flurry of activity shortly after the child is born. As a result, Green says, "People don't really realize that their kids' blood spots are being stored and are potentially searchable."

Just how long are these samples stored? In the absence of a federal standard, each state handles newborn blood spot storage differently. Permitted retention time ranges from less than a month to "indefinitely." In 2009, New Hampshire shortened the retention time from "indefinitely" to six months, while Maine removed its five-year limit in favor of allowing indefinite storage of newborn blood spots. Since 2008, five states have changed their rules to decrease retention time and another seven have increased it.

States differ just as widely in actual storage practices. Iowa stores samples for one year at -80 degrees Celsius, then four years at room temperature; Utah keeps the samples at room temperature for a week before cooling them to -20 degrees Celsius. Mississippi keeps the blood spots in Ziploc bags in a freezer. Louisiana's samples spend 30 days refrigerated in "gas permeable bags." Six states do not report having any written policy for specimen storage and disposal.

At least twenty states have made significant changes to their newborn blood spot storage policies within the past three years, but even if you live in one of these states, chances are you haven't heard anything about it. For that matter, chances are you haven't heard about your state's blood spot storage program at all.

This may often be by design on the part of state labs. Researchers and administrators working with these samples know very well how alarming newborn blood spot biobanking can sound to the layperson, particularly when inserted into the 24-hour media cycle, as by a February 2010 CNN story on the Minnesota case, titled, "The Government Has Your Baby's DNA."

One could see, then, why clinicians, researchers and state labs would prefer these projects to keep a low profile. "Those involved with newborn screening are concerned that additional publicity may cause parents to refuse to participate in newborn screening, which would then put children at risk. So there may have been a tendency to moderate publicity on this matter," Green says. "But at this point, it's probably too late for that."

The Texas Department of State Health Services tried that approach, and it came back to haunt them. Internal memos indicate that when the agency was preparing to start making their store of newborn blood spots available to researchers in 2003, officials acknowledged that parents "never consented for blood spots to be used for research," but decided to sidestep the issue. When the agency contracted with Texas A&M to warehouse the rapidly growing collection, DSHS specifically asked the university not to publicize the partnership. "This makes me nervous," one official wrote. "A press release would most likely only generate negative publicity." Word got out eventually, though, and the investigative pieces published in the *Texas Tribune* certainly did generate negative publicity. With the help of the Texas Civil Rights Project, a group of parents took the state to court. DSHS settled quickly, agreeing to destroy over 5 million newborn blood spot samples it had stored and made available to outside parties without parental consent.

That outcome didn't sit well with many researchers, Green among them. "That, to me, was a tragedy. There is tremendous value for the laboratory and society from having a population-based record of all these samples. At the very least, laboratories need to be able to quality control the tests they're doing."

Green has plenty of reservations about the way some research biobanks handle their business; but when it comes to newborn bloodspot screening and storage, he sees many of the privacy concerns as missing the point, focusing on theoretical problems without giving enough weight to the concrete benefits. If you're caught up on the privacy concerns, "Come to metabolic clinic and see these badly damaged children and what the families have to deal with," he says. "The newborn screening programs in this country are one of the great successes of modern public health."

Can we assume newborn blood spot banks are insulated from privacy issues? Many parents—perhaps most parents—know little about where their child's sample will end up; what happens when they learn of it by way of a scandal? Ask the health department in Texas. In the words of Sharon Terry, President of Genetic Alliance, if newborn blood spot collection and research is "done well and done right, there will be an enormous benefit overall to the system." In Texas, there's a good chance they'll tell you that doing it well, and doing it right, starts with informed parental consent.

10

Time to Raise Some Hell

BY HENRY T. GREELY

Henry T. Greely, J.D., *is Director of the Center for Law and the Biosciences at Stanford University.*

Researchers, research institutions, health care organizations, and others are busily creating vast repositories of tissue samples and health information about hundreds of thousands, or millions, of people. Some of these repositories are mainly samples, some are mainly data (silicon-based), and some are both. I will refer to them all broadly as "biobanks."

These biobanks are being created for good reasons. Researchers are desperate to get more—more samples, more analyses, more health records, more data, from more people, with more diseases, in more different settings. And they are right to be eager. Massive quantities of data may allow us to tease out the causes of various diseases and give us leads toward prevention, treatment, or cures. More immediately, research use of detailed electronic health records may allow us to make today's medicine safer, cheaper, and more effective. (Research from an HMO's electronic health records, for example, helped reveal the heart risks of Vioxx.)

But these biobanks also hold risks. The people whose tissue, DNA, or health records are stored in biobanks could be harmed. And cases of such harm could prompt a backlash, ultimately slowing medical progress. In their desperate quest for data and samples, researchers, aided and abetted by narrow interpretations of the laws, regulations, and ethical precepts governing human subjects research, are riding roughshod over the reasonable expectations—and the appropriate *rights*—of the people whose data and materials they are using.

Those expectations relate to consent to research, control over what research is done, privacy protections, and sharing of significant results with those whose samples and health information make the whole enterprise possible.

Most people think they need to give their consent before they can be research subjects. This is not unreasonable. A U.S. military court ordered several Nazi researchers executed for violating the first tenet of what came to be known as the Nuremberg Code—no research without the research participant's consent. That broad rule was never entirely true, but it is being stretched increasingly thin, in troubling ways. For one thing, the U.S. Government has decided that if the research samples or data cannot easily be connected to an identified person by the particular researcher doing the work—even though someone knows whose data is involved and many people *could* be identified by people who really wanted to do so—it isn't "really" human subjects research. Parents in Texas and Minnesota discovered, to their surprise, that this included blood spots from their newborn babies, taken for medical screening, with no consent, but then used for research.

Of course, many people actually do consent for research, often on a particular topic that is of interest to them. Members of a family with a strong history of Alzheimer's disease or breast cancer will often volunteer for research on that disease. They may later discover, to their shock, that their data is being used for research on topics they did not consent to, did not know about, and do not like. In the case of the Havasupai, a Native American people in Arizona, this included research they found deeply offensive. This secondary research often proceeds not just without the participants' consent but without any oversight by the Institutional Review Board; if the data and samples are viewed as unidentified, it isn't human subjects research—even if it looks at the DNA and health records of real, live humans.

A recent study looked at people who had volunteered for a study of the genetics of Alzheimer's disease, who were then asked to let their data and material be used for broader research. Eighty-eight percent of them agreed to this broader use—although that's a pretty good number, the twelve percent who refused are not trivial. On the other hand, a survey of those who agreed to further research uses showed that ninety percent of them believed it was important that they had been asked.[1]

Some people may not care, as long as their privacy is protected. But is it? In some situations, research data can be legally obtained by the police, attorneys, or anyone else with a court order. In other cases, even protected data can be illegally, or accidentally, obtained for non-research purposes—through hacking, through lost or stolen computers, or even when people with legitimate access to data use it for inappropriate reasons. (Ask celebrities how strongly protected their health information is when they go to a hospital.) At one extreme, researchers make data anonymous, stripping all the identifiers from it and therefore, they argue, fully protecting privacy. But even this "anonymization" is a fiction. Give me date of birth, sex, height, weight, and place of birth of an anonymous person and I'll bet that, with a little work, I can usually give you his or her name. Give me all the information in a detailed electronic health record and it should often be easy.

This is not an entirely hypothetical boast. A few years ago, the then-governor of Massachusetts, William Weld, ordered all health records for state employees and Medicaid recipients to be made available to researchers. Don't worry, he said, no records will be identifiable. A couple of days later, a computer scientist named Latanya Sweeney visited the governor's office and handed his receptionist copies of the governor's medical records. She had identified them using only publicly available motor vehicle and voter registration databases.

Finally, biobanks will often contain important information about research participants that those people themselves don't know. Consider, for example, a brain scan that shows a tumor or a genetic analysis that reveals a very high health risk. Most people assume that researchers will tell them, or their doctors, about this important information. Most research projects, however, expressly promise not to return any information, no matter how medically significant. This is a scandal just waiting to happen, the first time the widow of a dead former research participant realizes that the researchers had information that could have saved her husband's life—if he had been told about it.

Now, in fact, many people—perhaps most people—are happy to participate in medical research. But they have reasonable expectations about taking part in research. They want to have their permission asked. They want to know what their samples and information will be used for. They want to know just how private their information will be. And they want to know that researchers, whose work—whose careers—the research participants make possible, will care enough about those participants to warn them of health risks they uncover.

The underlying problem in all these cases is the mismatch between what the research participants think is happening with their samples and data and what is actually happening. Whatever else happens, a systematic violation of the expectations of people whose tissue, blood, DNA, and personal health information is being used in research is just wrong. And, more practically, if—no, when—they find out that their expectations are being ignored, they will be upset. At least, they should be. And their unhappiness—in some cases, outrage—could hold back important research.

I have made these arguments before, over many years—in more detail in academic journals,[2,3] more broadly in a book chapter,[4] and with equal fervor in an editorial in an ethics journal.[5] The responses were, to be kind, underwhelming. I think it is time to move beyond careful analyses and impassioned editorials. It is time to challenge, in legal and political forums, the ways biobanks are created and run. It is time to get this right, to protect the rights and interests of research participants and to protect the future of important biomedical research. Who's with me?

11

Suspect Creatures: Balancing Ethics and Utility in Research Biobanks

A Conversation with George Annas, Robert Green, Patricia Roche, and Susan Wolf

Four scholars spoke to GeneWatch *in three separate interviews in 2011 about research biobanks.* **George Annas, J.D., MPH** *is Chair, Health Law, Bioethics & Human Rights at Boston University School of Public Health;* **Robert Green, M.D., MPH** *is Associate Director of the Harvard Partners Center for Personalized Genetic Medicine and a geneticist at Brigham and Women's Hospital;***Patricia Roche, J.D., M.Ed.,** *is Associate Professor, Health Law, Bioethics & Human Rights at Boston University School of Public Health; and* **Susan Wolf, J.D.,** *is McKnight Presidential Professor of Law, Medicine & Public Policy and the Faegre & Benson Professor of Law at the University of Minnesota.*

On populating biobanks

Susan Wolf: Part of the challenge is that biobanks come in all shapes and sizes. There is a tremendous variety.

Robert Green: There are thousands of different entities that you might call a biobank. So it may be helpful to consider whether we are talking about blood banks, tissue banks, clinic or research banks, tumor banks or actual DNA banks.

GeneWatch: Where do all of these samples come from?

Patricia Roche: A lot of projects have projected numbers, so in order to answer the questions they want to answer, they need a sizable group or the power of their results is not going to be there. Getting that number of samples is a challenge. If you stop me on the street and say, "I want you to give me a biological sample," I'm just going to say, "Go away." You need the opportunity to engage with someone to get the sample, or you have to get the samples from someone who has already collected them.

Robert Green: Every time a hospital pathology department takes a surgical sample in the course of a patient's care, they maintain that sample somewhere in the pathology department of the hospital. That's been going on for decades in the name of good clinical care. If pathology decides you have a certain kind of tumor, and you get treated on the basis of that decision, it's important to be able to come back years later and look at those slides. Holding onto stored tissue in hospitals is considered standard care, at least in certain circumstances, and there is a good reason for it.

George Annas: Last year researchers collected samples at the Minnesota State Fair from kids. The alternative was to collect them in a hospital, and they thought that might not be representative, and they wanted to study healthy kids. They plan to follow the kids at the fair each year; and we'll have to wait to see whether this strategy is successful. Of course, when the children turn 18 they should have the right to either continue or have their samples and information destroyed.

Robert Green: You might look at drug studies, too. Every drug that's on the market had to go through clinical trials, and every clinical trial had to take samples for blood levels of the blood, monitoring for ill effects. I don't know how long drug companies keep those samples, but they certainly store them while the study is ongoing, and they may analyze the genetics of those samples for pharmacogenetic features of the new drug as part of their study analysis.

GeneWatch: Once a sample is being stored, how long might it be kept there?

Patricia Roche: One thing that I think is overlooked is any kind of a standard for destruction of samples. It seems to me that rules for destruction of samples are often not built in. Sometimes this might be done intentionally, particularly in studies where they are building fishing expeditions to look for diseases (*i.e.*, trolling through large datasets for correlations between DNA sequences and diseases); they don't know how long it might take them to find something. But there are other instances where you could identify an end point at which the project will be completed, and without the endpoints being built in, what happens to a particular collection when whoever created it is done with it? It's up for grabs.

George Annas: There are no good rules on transferring samples if, say, the collection's owner goes to bankruptcy court. It's a big deal. People might be happy to give their DNA to Joe Shmoe's DNA bank for research on prostate cancer, but not to let it be sold to the highest bidder in a liquidation sale.

Patricia Roche: And what happens when he's done? Whose responsibility is it to make sure that either the samples are safeguarded and preserved somewhere, or that they are destroyed? It's up in the air.

George Annas: As a general rule, scientists don't like to throw out anything. And it's hard to blame them; you really don't know what the next research project will require.

Patricia Roche: Once I was using a consent form for a study as an example in my class. I asked the class what their understanding was after reading that form, and their understanding was that if they gave a sample for this study, after a period of time the sample would be destroyed. And I asked, "What do you think 'destroy' means?" There happened to be someone in the class who was actually involved in the study, and she said, "Oh, I suppose you'd think that would mean that we would flush it down the toilet or something, but we don't do that. We just put it on the shelf." I asked why they wouldn't destroy it, and she said, "Because we think maybe the person would change their mind, and we would have lost it. If we already have it, we don't have to get another one from them. . . . But now that I'm thinking about it, that's not really what we tell people, is it?" It's that mindset—"We don't want to destroy anything because it might be useful, we might need it, something else might happen."

GeneWatch: Are there any laws governing ownership of samples in biobanks or whether ownership can be transferred?

George Annas: Not really. There's not much doubt that you "own" your own DNA sample in the sense you can give it as a gift to a biobank; but just what other things you can (or can't) do with it are not as certain. There also aren't any gene bank transfer laws. You have to argue the legal status of DNA by analogy, and the analogies aren't very good here. Genetic information is unique, although most closely analogous to medical information; and the DNA sample itself could also be viewed as a medical information repository. And, of course, we will be able to derive more information from it in the future.

Patricia Roche: As we go forward and as we learn more and more about the human genome, that sample is going to get more valuable. If I gave you a sample for a particular purpose, do I really have any idea what I gave you?

GeneWatch: Is there liability for whoever collects or stores the samples to protect participants' privacy?

George Annas: DNA bankers and researchers need to take reasonable steps to protect privacy. To protect, not to *ensure*; nobody's guaranteeing that.

Patricia Roche: George Church's idea is that we need to get over treating DNA samples and genetic information like this is such sacred or harmful stuff. He theorizes that if volunteers in genomic studies provided unrestricted access to identifiable data from their DNA and medical records, nothing bad or harmful would happen to them. Consequently, the public would be less hesitant to take part in genetic research and researchers would have no need to make unrealistic guarantees that privacy will be maintained.

George Annas: It would be unreasonable to ask for a total guarantee of privacy. If you need a guarantee, don't give your sample.

GeneWatch: What do we mean when we refer to "de-identified" samples? Is there really such a thing?

Robert Green: "De-identified" means the sample is given a code, and the link between the code and the actual names and identifying features of the subject are kept in a locked drawer or a computer file somewhere that's considered confident.

George Annas: "De-identify" is an inherently ambiguous term that could permit, for example, a trusted intermediary to keep sole possession of a code used to identify the samples. A much better term is "unlinkable," even though you could argue that nothing's unlinkable, because someday we're going to be able to link everything.

Susan Wolf: De-identification has for a long time been of concern to researchers; but the technology for re-identifying people has advanced. That has really, I think, led to the recognition that we probably shouldn't use words like "anonymous" or "anonymize" anymore, because with a reference sample you really can reidentify it. People more commonly talk about degrees of de-identification. I think that has made us cautious about the potential for re-identification and modest about what we accomplish when we de-identify.

Patricia Roche: If I have this little bit of someone's sequence but I don't know the person's identity, the theory is that I could go into some public database, and when it hits on a match, I can figure out who it is. But I think all that would tell you is that whoever's sample it is that you started out with, they are also in this database. Depending upon what else I know about how folks were selected for inclusion in the database (*e.g.*, only individuals with a history of heart disease) or how large the pool is, I might draw some conclusions about the source of my sample. But whether I can also readily figure out who they are is less clear.

George Annas: So, can you identify someone from an "unlinkable" or "anonymized" sample? Theoretically, but the odds are slim.

GeneWatch: If researchers need to communicate with participants later, who does the re-identification?

Susan Wolf: The most common answer to that, in the past, has been that it should be the collecting site and the original researchers; but there is the question of whether the biobank or perhaps a trusted intermediary should hold the codes, so that the full burden of dealing with potential incidental findings from research results doesn't fall on the collecting research site—where the original researcher may be dead or retired or have run out of funding.

On incidental findings and returning results

Robert Green: Say you've given me your sample for some kind of genetic testing, and I find—either by accident, or because I happen to be studying breast cancer—that you have a very prominent risk of breast cancer. Do I have a moral responsibility to tell you? Do you even want to know? Could I even contact you if you did want to know? And is it my responsibility as a researcher, on a research budget, to somehow hire the staff to call you up and contact you? At the present time, this is a huge ethical mess.

Susan Wolf: This whole question of incidental results and return of findings has really exploded and become a very pressing question in the design of research biobanks, both prospectively and also for established biobanks, facing the question of what to do with the incidental findings and potentially actionable results they've already got in hand. There are people who argue that biobanks shouldn't be in the business of returning incidental findings and research results at all; that these are important, crucial resources for the conduct of genetic and genomic research, particularly large-scale research; that they have limited resources, they are not clinical care entities, and their resources shouldn't be used for the purposes of clinical care. Furthermore, many biobanks are constructed based on the receipt of de-identified samples and data, so they really are not very well set up to return information to participants. So there are a lot of practical barriers, resource barriers, and, some people would argue, ethical barriers to returning incidental findings at the biobank level. The pushback is that we have growing data to suggest that participants are interested and want clinically important and perhaps reproductively important information. In Europe, in particular, there has been a lot of movement toward suggesting that biobanks and researchers owe participants clinically important information as an expression of reciprocity, solidarity, and respect.

GeneWatch: Are there very many research biobanks currently that return findings to participants?

Susan Wolf: We've discovered that there are studies and biobanks that do, but I do not think that is the dominant approach right now. I think there is a great deal of concern about the cost of doing this and whether biobanks are set up to do it. There is not widespread consensus yet on exactly what incidental results and research findings are appropriate for return. I think the question of how to handle incidental findings and return of results is still very much up for grabs.

GeneWatch: How might the determination be made whether or not a particular result or incidental finding should be reported to the individual who provided the sample?

Susan Wolf: I think there is a trend in the ethics literature toward recognizing a duty to consider some results for return; for example, where there is high clinical importance and actionability. "Actionability" has become a term of art and is defined somewhat variously; but the basic idea is that if you return these results, does it have the potential to change the clinical course for this participant, either by opening up treatment options or by increasing surveillance to look for disease, for example the development of cancer. I think that there is a developing consensus that high clinical importance plus actionability suggests considering return.

Patricia Roche: When it comes to that issue of whether to give results back to participants, the first thing researchers have to ask themselves is: what's the reliability of the results to begin with? A lot of research is so preliminary that they really don't know whether they will find something that will have any utility. And if they were going to give the information back to people, they need to know how confident they are about the results—if not in terms of the clinical validity, at least in terms of analytical validity. I don't have a problem with somebody saying upfront, "We may find something, but until we do subsequent research and confirm either the test we're coming up with is a valid way of doing this or that the results have significance, we're not going to give you back information."

GeneWatch: Committing to return results seems to open up another ethical can of worms.

Patricia Roche: To say to people, "We're not going to give you back any information. Is that OK?" is one thing; it's another thing to know that the way you're designing it, you're likely to find something that might be significant to people, and not say to them upfront, "By the way, we're likely to find something that would be beneficial to you, but we're not going to tell you. Is that OK?"

George Annas: This really goes to the question of what the person giving the DNA sample thinks the deal is. Is that person simply giving a "gift" to science with no expectation of any individual knowledge or information coming back to him or her, or is the person making a healthcare agreement with the DNA bank in the expectation that any information found about my particular genome will be shared with the donor? This makes all the difference, because if it is the latter expectation, and if you don't contact me, am I supposed to think I'm OK? Should I think, "They didn't contact me, my genome must be clean?"

Susan Wolf: I should introduce the caveat that almost nobody talks about imposing this information on research participants. So "return of results" is really a little bit of a misnomer; it's really offering back results and incidental findings to participants who are interested in receiving them.

Robert Green: Say I perform a research study and I say to you explicitly that I am going to take and store your biological sample, and I would like your permission to give it to any researcher I want. It will stay de-identified—meaning it's theoretically possible to hook up to your identity, but we'll make every effort to keep it confidential—and you will never learn anything back from me. Is that enough? The alternative is almost unthinkable to many people who actually run biobanks, because it starts to turn the researcher into a population screening machine and to conflate the role of researcher with the role of the clinician.

Susan Wolf: Researchers conceive of themselves as researchers, not clinicians. So the idea that researchers owe some kind of duty to participants to find information of clinical importance, and then to offer that back to participants, kind of exists in tension with the research mission. It takes money, it takes resources, and it really is not what research is set up to do, at its core. It challenges the traditional line that we've drawn between research duties and clinical care duties. Clinical care duties are very robust and broad; research duties have really been conceived as primarily pursuing generalizable knowledge, rather than caring for individual research participants and returning information to them of their own particular clinical care or reproductive decisions.

GeneWatch: Research biobanks make a lot of very important research possible. How much is really at stake in how an individual's sample is handled?

Patricia Roche: I think the worst that could happen is for people to give away samples without knowing what they are signing up for. Firstly, they may be exposed to a risk that they wouldn't accept; but I also think that it's insulting and disrespectful not to inform participants. If I'm going to ask you for something, and you're not going to get any benefit from it, and I need it to do something that I think is going to be important, then it seems to me the very least that I can do is be honest with you, to tell you what it is I want to do.

George Annas: There's no excuse for a researcher not to tell you what they're doing. And not just for the DNA donor's sake. Anthropologists used to warn their colleagues: "You have to be nice to the group you're studying, for the next anthropologist who comes along. Don't irritate (or exploit) them too badly or nobody else will be able to study them again." You don't want to get the public to view researchers as suspect creatures who believe that participating in their research is a matter of civic duty.

PART II

Genetics and Medicine: Can Gene Therapy Cure Cancer and Other Diseases?

Artist: Sam Anderson

The Human Genome Project (HGP) was the largest international collaborative research project ever undertaken in biology, with the goal of sequencing the three billion bases of genetic information that reside in every human cell. It developed detailed genetic and physical maps of the human genome and made them accessible for further biological study. It was led by the Department of Energy and the National Institutes of Health in the US, soon joined by the Wellcome Trust Sanger Institute in the UK, with additional contributions coming from Japan, France, Germany, China, and other working groups. It began in 1990 and was completed in 2003. A parallel project was conducted privately by the Celera Corporation (or Celera Genomics), which was formally launched in 1998.

The completion of the Human Genome Project promised a new era of disease treatment and personalized medicine. Tremendous advances have been made; new tools and techniques are increasingly allowing researchers to dig ever deeper into the workings of biological processes. Scientists are discovering new disease pathways, identifying genes linked to Mendelian disorders and allowing for early intervention, making some advances in pharmacogenomics and therapeutic interventions with cancer patients, and offering parents a growing list of pre- and post-natal screenings for inherited abnormalities.

As the science progresses, however, there continues to be a large gap between basic research and clinical applications, and the "genetic revolution" has been much slower to unfold than predicted. Indeed, there is still much confusion between scientific fact and fiction and all too often the simplicity of the "DNA is everything" model, and the outside commercial and scientific incentives available for such a focus, prevails.

We are still very much at the beginning of a long process of discovering the subtle interplay between our genes, coded in our DNA, the proteins that are wrapped around or are external to the DNA (epigenome), our lifestyle, and the environment we live in. Incorporating genetics into medicine promises to be a long process, far from the "revolution" that was promised. As research inevitably progresses, we must work to set reasonable expectations for the utility of medical genetics in clinical care and create ethical frameworks for its inclusion.

From direct-to-consumer genomics to clinical care, the chapters that follow explore the current impact of genetic research on medicine, separating the myths from the reality.

12

The DNA Era

BY RICHARD C. LEWONTIN

Richard C. Lewontin *is Professor of Zoology in the Museum of Comparative Zoology, Emeritus at Harvard University. He is the author of numerous works on evolutionary theory and genetic determinism, including* The Genetic Basis of Evolutionary Change *and* Biology as Ideology, The Dialectical Biologist *(with Richard Levins) and* Not in Our Genes *(with Steven Rose and Leon Kamin). This has been adapted from an article that appeared in* GeneWatch, *volume 16, number 4, July-August 2003.*

No one who reads the newspapers or scientific journals can have missed the fact that this is the 50th anniversary of the publication of the correct three-dimensional structure of DNA. That structure, a double helix of two chains of nucleotides, has become a popular icon and the very phrase, "double helix" has been spoken and written so often as to become part of ordinary discourse.

The fact that genes were composed o f DNA had already been established nine years before the publication of Watson and Crick's paper on its structure, and the chemical, as opposed to the spatial, configuration of DNA was also well known before 1953. Yet, despite the obvious importance of DNA in understanding the molecular details of both heredity and development, it was not until after the publication of the proposed double helical structure that DNA started increasingly to occupy the interest of biologists and finally became the focus of the study of genetics and development. The last fifty years have seen the reorganization of most of biology around DNA as the central molecule of heredity, development, cell function and evolution. Nor is this reorganization only a reorientation of experiment. It informs the entire structure of explanation of living processes and has become the center of the general narrative of life and its evolution. An entire ideology has been created in which DNA is the "Secret of Life," the "Master Molecule," the "Holy Grail" of biology, a narrative in which we are "lumbering robots created, body and mind" by our DNA. This ideology has implications, not only for our understanding of biology but for our attempts to manipulate and control biological processes in the interests of human health and welfare, and for the situation of the rest of the living world.

The first step in building the claim for the dominance of DNA over all living processes has been the assignment of two special properties to DNA, properties that are asserted over and over again, not only in popular expositions but in textbooks. On the one hand, it is said that DNA is self-replicating; on the other, that DNA makes proteins, the molecular building blocks of cells. But both of these assertions are false—and what is so disturbing is that every biologist knows they are false.

First, DNA is not self-replicating. It is manufactured out of small molecular bits and pieces by an elaborate cell machinery made up of proteins. If DNA is put in the presence of all the pieces that will be assembled into new DNA, but without the protein machinery, nothing happens. What actually happens is that the already present DNA is copied by the cellular machinery so that new DNA strands are replicas of the old ones. The process is analogous to the production of copies of a document by an office copying machine, a process that would never be described as "self-replication." In fact, many errors are made in the DNA copying process: there is protein proofreading machinery devoted to comparing the newly manufactured strands to the old ones and correcting the errors. An office copier that made such mistakes would soon be discarded.

Second, DNA does not make anything, certainly not proteins. New proteins are made by a protein synthesis machinery that is itself made up of proteins. The role of the DNA is to provide a specification of the serial order of amino acids that are to be strung together by the synthetic machinery. But this string of amino acids is not yet a protein. To become a protein with physiological and structural functions, it must be folded into a three-dimensional configuration that is partly a function of the amino acid sequence but is also determined by the cellular environment and by special processing proteins that, among other things, may cut out parts of the amino acid chain and splice what remains back together again.

The other function of DNA is to provide a set of "on-off" switches that are responsive to cellular conditions so that different cells at different times will produce different proteins. When the conditions of the cell set a switch associated with a particular gene to the "on" position, then the protein manufacturing machinery of the cell will read that gene. Otherwise the cell will ignore it.

In this mechanical description of the relation of DNA to the rest of the cellular machinery there is no "master molecule," no "secret of life." The DNA is an archive of information about amino acid sequences to which the synthetic machinery of the cell needs to refer when a new protein molecule is to be produced. When and where in the organism that information is read depends on the physiological state of the cells. An organism cannot develop without its DNA, but it cannot develop without its already existing protein machinery (unless it is a parasite like a virus that has no synthetic power of its own but gets a free ride on its host's protein machinery).

The unjustified claim for special autonomous powers of DNA is the prelude to the next step in building a picture of a DNA-dominated world. This picture is simply the molecular version of a biological determinism that has dominated explanations of the properties of organisms, and especially of humans, since the nineteenth century. Differences in temperament, talents, social status, wealth, and power were all said to reside "in the blood." The physical manifestations of these claimed hereditary differences could be seen by criminal and racial anthropologists in the shapes of noses and heads and the color of skins. With the rise of Mendelian genetics, genes were substituted for blood in the explanations, but they remained, for the fifty years of genetics, merely formal entities with no concrete description beyond the fact that they were some bit of a chromosome. The discovery that DNA is the material of the gene, and the subsequent determination of the correspondence between nucleotide sequences of genes and amino acid sequences of proteins, then provided a concrete molecular basis for a total scheme of explanation of the organism. The fact that organisms are built primarily of proteins and that DNA carries the archive of information for the amino acid sequence of the proteins gave an immense weight to the conclusion that the organism as a whole is coded in its DNA. A manifestation of this view is the claim made, at a symposium in commemoration of the 100th anniversary of the death of Darwin, by a founder of the molecular biology of the gene: that if he were given the DNA sequence of an organism and a large enough computer, he could compute the organism. One is reminded of Archimedes' claim that, given a long enough lever and a place to stand, he could move the Earth. But while Archimedes may have at least been right in principle, the molecular biologist was not. An organism cannot be computed from its DNA because the organism does not compute itself from its own DNA.

It is a basic principle of biology, known to all biologists but ignored by most of them as inconvenient, that the development of an organism is the unique consequence of its genes and the temporal sequence of environments in which it develops. The current fascination of developmental genetics is with the way in which information from different genes enters into the formation of the major features of an organism. How does the front end of the animal become differentiated from the back end? Why does the egg of a horse develop into an animal with four legs while the egg of a bird produces an organism with two legs and two wings, and the egg of a butterfly results in an animal with six legs and two sets of wings? This concentration on the major differences and similarities between different species has resulted in a genetically determinist view of development that ignores the actual variation among individuals. There is an immense experimental literature on plants and animals showing that individuals of the same genetic constitution differ widely from each other in physical characteristics if they develop in different environments. Moreover, the relative ranking in some physical trait of individuals of different genotypes changes from environment to environment. Thus, a genetic type that is the fastest growing at one temperature may be the slowest at another. But even genes and environment together do not determine the organism. All "symmetrical" organisms show a fluctuating asymmetry between their two sides, and the variation between left and right sides is often as great as the difference between individuals. For example, the fingerprint pattern on the left and right hands of a human individual are not identical; on some fingers, they may be extremely dissimilar. This variation is the manifestation of random growth differences that arise from small differences in the local tissue and cell conditions in different parts of the body, and from the fact that there is random variation in the number of copies of particular molecules in different cells. A consequence is that two individuals with identical genes and identical environments will not develop identically. If we want to understand human variation, we need to ask far more subtle and complex questions than is the rule in DNA-dominated biology.

The other side of the movement of DNA to the center of attention in biology has been the development of tools for the automated reading of DNA sequences, for the laboratory replication and alteration of DNA sequences and for the insertion of pieces of DNA into an organism's genome. Taken together, these techniques provide the power to manipulate an organism's DNA to order. The three obvious implications of this power are in the detection and possible treatment of diseases, the use of organisms as productive machines for the manufacture of specific biological molecules, and the breeding of agricultural species with novel properties.

The Human Genome Project[1] has been largely justified by the promise that it will now be possible to locate genes that cause human disease by comparing the DNA sequences of affected and unaffected individuals. Once the nucleotide difference has been established, that difference can be used as a diagnostic criterion, as a predictor of a future onset of the disease, and as a basis for a cure by gene replacement therapy. It is undoubtedly true that some fraction of human ill health is a consequence of deleterious mutations. However, while family studies can strongly suggest that a disease is being inherited as a single Mendelian gene difference, the determination that it is a consequence of mutation of a particular gene is not a trivial problem. A blind search for a genetic difference that is common to all affected individuals is impractical given that, on the average, any two humans differ from each other at 3 million nucleotide sites. On the other hand, if the biochemistry of the disease is sufficiently well understood, it may be that a few candidate genes can be singled out for investigation. Alternatively, studies of the pattern of inheritance may show that the disorder is inherited and associated with a gene of known location in the genome, greatly narrowing down the search for the DNA variation implicated in the disease.

As with all other species, for any given gene, human mutations with deleterious effects almost always occur in low frequency. Hence, specific genetic diseases are rare. Even in the aggregate, genes do not account for most of human ill health. Given the cost and expenditure of energy that would be required to locate, diagnose and genetically repair any single disease, there is no realistic prospect of such genetic fixes as a general approach for this class of diseases. There are exceptions, such as sickle cell anemia and conditions associated with other abnormal hemoglobins, in which a non-negligible fraction of a population may be affected, so that these might be considered as candidates for gene therapy. But for most disease that represents a substantial fraction of ill health and for which some evidence of genetic influence has been found, the relation between disease and DNA is much more complex and ambiguous. Claims for the discovery of "genes for" schizophrenia and bipolar syndrome have repeatedly been made and retracted. It is generally accepted that cancer is a consequence of mutations in a variety of genes related to the control of cell division, but even in the strongest individual case, the breast cancer-inducing BRCA1 mutations, only about 5 percent of such cancers are linked to these specific mutations.

Up to the present we do not have a single case of a successful cure for a disease by means of gene therapy. All successful interventions, whether in genetically simple disorders like phenylketonuria (PKU) or in complex cases like diabetes, have been at the level of biochemistry and were in place well before anything was known about DNA. Of course, a successful gene therapy for some disease may be produced in the future, but the claim that the manipulation of DNA is the path to general health is unfounded. In fact, on a world scale, most ill-health and premature death is caused by a combination of infectious disease and undernourishment—factors that genetic manipulation will never solve.

The second implication, the possibility of using genetically transformed organisms as factories for the commercial production of biologically useful molecules, has been realized in practice. The most famous case, the mass production of human insulin by bacteria, is particularly instructive. Insulin for diabetics was originally extracted from cow and pig pancreases. This molecule, however, differed in a couple of amino acids from human insulin. Recently, the DNA coding sequence for human insulin has been inserted into bacteria, which are then grown in large fermenters; a protein with the amino acid sequence of human insulin is extracted from the liquid culture medium. But amino acid sequence does not determine the shape of a protein. The first proteins harvested through this process, though they possessed the correct amino acid sequence, were physiologically inactive. The bacterial cell had folded the protein incorrectly.

A physiologically active molecule was finally produced by unfolding the bacterially produced protein and refolding it under conditions that are a trade secret known only to the manufacturer, Eli Lilly. This success, however, has a severely negative consequence. For some diabetics this "human" insulin produces the symptoms of insulin shock, including loss of consciousness. Whether this effect is caused by a manufacturing impurity, or because the insulin is not folded in the same way as in the human pancreas, or because the molecule is simply too physiologically active to be taken in large discrete doses rather than internal, continuously released amounts calibrated by a normal metabolism, is unknown. The problem is that Eli Lilly, which holds the patent on the extraction of insulin from animal pancreases, no longer produces pig or cow insulin. Hypersensitive diabetics for whom Eli Lilly's standard treatment is dangerous no longer have an easily obtainable alternative supply.

All of the elements that characterize the era of DNA have in common an underlying simplistic view of living organisms. By concentrating in practice and in theory on the properties and functions of a single molecule, biologists, both in their professional work and in their public statements, reduce the extraordinary complexity of life processes to the structure and metabolism of DNA. This emphasis ignores the intricate and multiple ways in which organisms are built and function. The intricacy is a consequence of the structural and metabolic functions of proteins and the interactions of those proteins with each other, with other molecules, and with the environment in the course of development.

Moreover, for human life, no account at all is taken of the role of social and economic processes in determining health and life activities and molding the processes of industrial and agricultural production. We cannot understand our size, shape and internal functioning except by a detailed understanding of the extremely complex web of interactions among the various molecules that form the body in concert with influences exerted by our environments. We cannot understand the origin and development of our mental states except by an understanding of the map of nervous connections and how that map is influenced by experience. We cannot understand why agricultural technology develops in particular directions if we do not understand the social, political and economic interactions that drive technological innovation. The bottom line is that life in all its manifestations is complex and messy and cannot be understood or influenced by concentrating attention on a particular molecule of rather restricted function.

Autism and Genetics

BY CHLOE SILVERMAN AND MARTHA HERBERT

Chloe Silverman *is an assistant professor in the Science, Technology, and Society Program at Pennsylvania State University.* **Martha Herbert, M.D.,** *is a former Board member of the CRG and a pediatric neurologist and brain development researcher at Massachusetts General Hospital. She is also the author of* The Autism Revolution. *This article was originally published in* GeneWatch, *volume 16, number 1, January-February 2003.*

Autism, a devastating childhood neurodevelopmental disorder, was first characterized by Leo Kanner in 1943. Kanner described eleven children with odd and disturbed language, poor human connectedness, and repetitive, disturbed behavior.

Hitherto considered "genetic," autism is now being diagnosed in unprecedented numbers. Yet while parents, clinicians, schools, and a growing number of researchers confirm the flood of autistic children, this biological epidemic seems to have touched off not an epidemic of administrative or public concern but a near-pathological denial of both the fact of increasing rates and the role of extra genetic factors implicated in the upsurge. This denial threatens to precipitate an educational, financial, social and human disaster in coming years.

Epidemic denial

Diagnoses of autism and pervasive developmental disorders (PDDs) or autism spectrum disorders (ASDs) have increased since they were first observed, but the increase has accelerated over the last decade. Where rates were once 3 in 10,000, they are now, depending where you look, 1 in 500, 1 in 150, or even more—between a three- and a tenfold increase. A recent state-funded study conducted by the M.I.N.D. institute in Davis, California [and published in January 2009] confirmed a 273 percent increase in autism from 1987 through 1998, originally reported in a March 1999 study conducted by the California Department of Developmental Services, and argued that it could not be blamed on more aggressive diagnosis, improved ascertainment, or immigration. This increase is likely to be nationwide. Although no other state has kept comparable statistics, the US Department of Education has recorded a nationwide average increase of 544 percent in autistic students from 1992-93 to 2000-01, and comparable rates have been found in a number of local studies. A CDC study released at the end of December [2002] showed a tenfold increase over the last decade.

There are several ways in which these numbers considerably underestimate the magnitude of impact. For one thing, autism affects boys three to four times as often as girls; this means that if the overall rate is, say, 1 in 250, it will be closer to 1 in 100 in boys. For another, the figures in California count only the most severe cases; when mildly but still significantly impaired children are included, the figures can be considerably higher. And many people consider other disorders, such as Attention Deficit Hyperactivity Disorder (ADHD) and many learning disabilities, to be on a spectrum with autism. Considered in its broadest terms, this epidemic currently may affect from a few million to as many as 20 percent of U.S. children.

Such drastic increases imply the influence of non-inheritable or environmental factors, since, of course, there is no such thing as a "genetic epidemic." But research continues to focus almost exclusively on studies of brains, screening and genes, as well as on denying the increase or disproving the role of controversial environmental triggers, notably vaccines (see sidebar).

Thimerosal and Autism: An Emerging Scandal by Sally Bernard

The Homeland Security Act, signed into law by President Bush on November 25, contained a provision shielding a small group of pharmaceutical companies from liability for harm to children caused by thimerosal, a mercury-based vaccine preservative. The rider was inserted just prior to the full House vote, and few in Congress were aware of it.

The liability shield has nothing to do with homeland security. Thimerosal is not found in vaccines for smallpox or any other possible bioterror agent. Until recently, however, it was used in most routine infant vaccines—and many parents, along with a growing number of scientists, are linking it to childhood autism and related disorders.

The rider effectively ends hundreds of lawsuits filed, or about to be filed, by parents of autistic children. The main beneficiary is Eli Lilly & Company, the developer of thimerosal. From 1990 to 2002, Lilly made political contributions amounting to almost

$6 million—nearly a quarter of which was spent on November's mid-term elections—and has substantial ties to the Bush Administration. Mitchell Daniels, the White House budget director, is a former Eli Lilly executive. The company's current CEO, Sidney Taurel, is on the Homeland Security Advisory Council. George Bush Sr. was a member of Eli Lilly's Board of Directors during the 1970s.

The White House denies responsibility for the rider, but Representative Dan Burton (R-Indiana), who has a grandson with autism and has held hearings on autism and vaccines, lays the blame at their door. Bill Frist, the new Senate majority leader, has been linked to the provision.

Adding to the intrigue, the Department of Justice asked the special vaccine injury court to seal the records relating to more than one thousand autism-thimerosal cases now pending. This action would have effectively blocked public scrutiny of any vaccine manufacturer's documents to emerge from the proceedings. Although the record-sealing was eventually rescinded, parents of autistic children were left wondering exactly what in those documents needed to be hidden—and, ultimately, why Eli Lilly and the other manufacturers are so scared.

Apparently, there is plenty to worry about. Although autism was once thought to be a purely genetic disorder, the steep rise in the number of cases during the last decade shows that environmental causes are also at work. The huge increase in autism rates coincides with the addition, in the early 1990s, of two new thimerosal vaccines to the infant immunization schedule. In 1999, the FDA disclosed that the amount of mercury in vaccines exceeded EPA safety guidelines. Manufacturers were asked to remove thimerosal, although existing stocks were left on the shelves. Parent groups like Safe Minds demonstrated that the symptoms of mercury poisoning matched the abnormalities they saw in their children. Scientists are now showing that vaccine levels of thimerosal can cause neuronal apoptosis, immune imbalances, and autistic-like brain lesions and behaviors.

Documents obtained by lawyers show that Lilly and others knew about the dangers of thimerosal as far back as the 1940s. In fact, as the FDA has admitted, safety testing of thimerosal has never been conducted. A recent attempt by vaccine researchers to absolve thimerosal of toxicity through a poorly designed—but well promoted—study that used only 33 infants (Pichichero et al., The Lancet, November 2002) was derided by parent advocates as inadequate and overreaching.

It is not surprising that Lilly and the vaccine industry would fight so hard to dismiss concerns over thimerosal and, failing that, to have the concerns forcibly dismissed by law. Not only are lawsuits a threat, but the presence of an untested toxin in infant vaccines raises the ugly question of whether vaccines are being properly evaluated before being unleashed on the public. A lot is at stake. Vaccines are considered one of the few bright spots for pharmaceuticals in the future. Hundreds of vaccines are in various stages of development. Revenues are expected to reach into the hundreds of billions of dollars. Parents of autistic children are making substantial inroads, but the pharmaceutical giants are fighting back with a vengeance. Truth, however, has a way of becoming undeniable.

Sally Bernard *is Executive Director of Safe Minds (www.safeminds.org)*

How autism became a genetic disease

What does it mean to describe a condition as genetic? Every disease, including viral and infectious ones, involves vulnerabilities or resistances that relate in some way to genetic influences. But for many autism researchers, genetics is not about modulation or vulnerability, it is about "cause" and "determination." These researchers justify their hunt for autism genes by pointing to studies of twins: while anywhere from 36 to 96 percent of both identical twins have autistic features, this is true for only 0 to 33 percent of fraternal twins. The very spread of these figures shows that claims of full genetic causation are flimsy; and these same figures can equally be used to argue that environmental factors must play a role, since the concordance for identical twins is not 100 percent.

Nevertheless, this approach to autism as "genetic" has secured upwards of $60 million in research funding. To date there have been as many as eight genome scans and dozens of genetic studies. As in so many other gene hunts, at least a few candidate regions have been located on nearly every chromosome, but they have not led to specific genes and generally have not been replicated by later investigations. Meanwhile, other research approaches languish.

Autism as biological: paradigms break down

Many established autism researchers justifiably pride themselves on overthrowing the invidious "refrigerator mother" theory, which held that autism was a behavioral response of children to mothers who failed to display affection. Findings in the 1980s of abnormalities in autistic brains freed parents of this blame and shame, and opened the way to treating autism as a biological disorder. These brain abnormalities appeared to be of prenatal origin, and this seemed to fit with the evidence for genetic causes and the lifelong, apparently incurable impairment of people with autism. Researchers concluded that autism was determined by genes, was hard-wired before birth, and was treatable only by behavior modification. This has set the current research agenda, which is dominated by genetics, neuroscience and psychology.

But this picture is now unraveling. New evidence is emerging from both within and outside the dominant research domains that makes autism look more like an environmentally mediated illness. Many autistic children turn out to develop abnormally large brains—and do so after birth, in the first 2 to 3 years of life. Recent work suggests that other brain changes, previously thought to be prenatal, could occur postnatally. It also turns out that autistic children have substantial illness not only in their brains but in their bodies. While the mainstream research and clinical community considers physical symptoms to be "incidental" to the core autism, and pays little attention to them, subgroups of autistic children have common patterns of significant biomedical illness—notably immune system disturbances, disturbances in various biochemical pathways (including impaired detoxification, which may explain increased vulnerability to toxic exposure), and painful gastrointestinal disease. When treated biomedically, many autistic children have shown substantial behavioral improvement and improved receptiveness, suggesting that their behaviors aren't totally "wired in."

From the mainstream perspective, which defines autism in terms of unchangeable, genetically determined brain damage, it is incomprehensible that nutritional or metabolic interventions could have any effect. Therefore, such approaches are dismissed, usually without investigation, as "quackery." But for those who see autism as an environmentally mediated illness, it makes perfect sense that the body as well as the brain should be affected. After all, why would toxins affect us only from the neck up? A growing social movement of parents and doctors who take environmental influences seriously feel that the genetic approach has betrayed autistic children by assuming that biomedical treatment can't change anything. They argue that they look at their physically ill children and "believe what they see," while the genetic determinists "see what they believe."

Why the vehemence?

The mainstream's dismissal of new approaches to autism reveals a double standard about evidence. While parents and practitioners offering these methods are unable on their own to bear the staggering cost of double-blind controlled studies, neither have the accepted pharmacological and behavioral approaches been tested in this rigorous fashion. In truth, nobody in the field of autism treatment has much basis for invoking the legitimizing mantra of "Evidence-Based Medicine." But the deck is stacked. In October [2002], the FDA seized supplies of the amino acid taurine from Kirkman Labs, a nutrition company catering to autistic children. Their claim, that Kirkman Labs was making unsupported claims for this substance, came ironically within days of news stories exposing the agency's lowest-ever rate of taking action against misleading advertisements by pharmaceutical companies.

Because the environmental triggers underlying the autism epidemic are presumably injuring not just the brain but the rest of the body as well, we need to test and refine biomedical approaches, rather than just aiming for new psychopharmacological drugs, behavioral treatments and gene identification. But advances in biomedical treatments will only occur once we move beyond a strictly gene-brain paradigm and allocate serious funding to physiological and toxicological autism research. Unfortunately, pharmaceutical companies on their own are unlikely to investigate biomedical approaches to autism, since many of the most promising nutritional interventions have little potential for patentability and therefore profitability.

Understanding the source of these twin denials—the denial of increasing incidence and the denial of non-genetic biological and environmental factors—is key both to addressing autism and to understanding the ideological role of genes and genetics research in contemporary America.

What if we looked at environmental causes?

At a conference in October 2002 at Rutgers University, New Brunswick, entitled "Autism: Genes and the Environment," leading researchers of the genetic persuasion shared the stage with toxicologists in what seemed to herald the cautious beginnings of a new synthesis. But we do not yet see a concerted shift to a research model that incorporates genes and environment.

There are probably several reasons for this. Certainly there are significant economic forces that stand to gain from the current direction of research. The notion that identifying genes and brain circuitry will lead to targeted drug development not only benefits the pharmaceutical industry but also leads researchers down pathways that feel familiar and rational. In the belief system of many researchers, genes loom powerful, while environmental factors seem trivial and secondary—so that big effects just "naturally" are presumed to be "genetic." This new line of investigation calls for a knowledge of toxicology and biochemistry much more detailed than most current autism researchers possess.

Moreover, examining environmental causes opens up a Pandora's box of vulnerabilities. Some are personal: it is hard to think about toxic effects in autism without also wondering about one's own health and the health of one's family. That human actions, rather than genes, might be responsible for compromising the health of a significant proportion of a whole generation is so painful as to be, for many, unthinkable. And if there are environmental causes, then there may be liability and corporate accountability. If our carelessness with chemicals can harm children so profoundly, we may be called upon to fundamentally restructure the way we make things, the precautions we take, and the way we live—something industry and regulatory agencies go to great lengths to prevent.

Social costs

Meanwhile, the determination of many researchers, regulators, legislators, funders, and even some parent groups to explain away the increased incidence of autism will have grave social effects. School districts are feeling the burden of increased numbers of autistic children. These children are often unable to function within a mainstream classroom, whether because of violent or self-injurious behaviors, lack of toilet training, or inability to communicate their needs. The cost of providing them with individualized behavioral therapy, which requires up to forty hours a week for maximum effectiveness, can run from $30,000 to $60,000 per year, per child. As a result, already underfunded public school districts will do almost anything to avoid providing these services, and parents without ample financial means are left with few options for their children, since Medicaid will not cover the optimum amount of behavioral therapy. Community treatment centers are faced with too many disabled children to accommodate and are increasingly forced to turn away all but the most severe cases—although milder forms of pervasive developmental disorders, or high-functioning forms of autism such as Asperger's syndrome, are also potentially devastating conditions.

Furthermore, full-spectrum autism, even if treated early and intensively, continues to have a poor prognosis. As this generation of children ages and their parents are no longer able to provide full-time care for them, residential institutions will be unable to provide facilities for even a fraction of these autistic adults. Estimates of the lifetime costs of care for a child diagnosed with autism today range widely—from conservative predictions of $2 million to published figures as high as $12.4 million, depending on the extent of therapies, care, and support services figured into the equation. Over the next decade, the autism epidemic is likely to cost the U.S. economy hundreds of millions of dollars.

Conclusion

A children's epidemic by all rights ought to open floodgates of concern and set off an urgent search for creative responses. While discoveries that pathological events may be occurring after birth should provoke a search for environmental triggers, it also opens up hope for prevention and for treatment. Moreover, the improvements experienced by growing numbers of children from biomedical interventions should fuel an intensive study of the mechanisms by which these treatments work and a search for more finely targeted approaches.

Instead, we are treated all too often to methodological quibbles about epidemiological research; patronizing remarks by researchers about "hysterical parents" who "can't accept their child's genetic fate;" highly publicized, but methodologically weak, studies that claim to definitively refute any role for various vaccines in the increased rates of autism but raise no alarms about the increased rates themselves—and a press blackout on subsequent critiques and refutations. And we still get press conferences triumphantly announcing that "scientists are closing in" on the genetics of autism. The response of genetics researchers to the failure of their efforts so far is simply to increase the number of genes they expect to find—it's now up to twenty or more, where it used to be as few as three or four—rather than attempting, for instance, to discern increased vulnerability in small subgroups. What we don't get is a frank admission that a purely genetic model is inadequate.

To cling to a genetic explanation for autism, to insist that the epidemic is a consequence of methodological rather than toxicological effects, is thus a desperate attempt to maintain the illusion that one lives in a comfortable and rational world where all is basically well, new chemicals and technologies always mean progress, experts are always objective and thorough, and authorities can be trusted. This form of genetic reductionism is a roadblock to developing the forceful science and social policy called for by the epidemic, it sustains taboos within the scientific community against potentially controversial ideas about environmental factors and it distracts governments from addressing the financial and social demands that this epidemic creates. In short, it weakens our response to a disaster that has already begun.

14

Looking Upstream: Cancer Genomics Project Obscures Causes of Cancer

BY JONATHAN KING

Jonathan King, Ph.D., *is Professor of Molecular Biology at MIT and a founder of the Council for Responsible Genetics. This article originally appeared in* GeneWatch, *volume 19, number 2, March-April 2006.*

In December 2005, the National Institutes of Health (NIH) announced that it would be funding the Human Cancer Genome Atlas, beginning "with a pilot project to . . . systematically explore the universe of genomic changes involved in all types of human cancer." The initiative applies the newest gene-sequencing technologies to identify the different forms of gene and chromosome damage observed in human cancers. Such information is of great interest to the patient, medical and pharmaceutical communities because it might open new paths to developing therapeutic and anti-tumor agents specific for the altered cell functions in the diverse forms of cancer.

However the potential health significance is even greater, since if properly conceived and applied, the detailed characterization of the genetic damage could be used to identify the mutagens and carcinogens causing this damage. Rather than accepting the incidence of cancer as a fact and focusing solely on treatment, researchers could identify the etiological [disease-causing] agents of many human cancers, opening the possibility of the only true method of cancer prevention, the elimination of exposure to carcinogens.

Disappointingly, but predictably, signs suggest the project will be blind to such a preventive strategy. Singularly lacking from 1,200-word NIH announcement of its decision to fund this project, and from all of the follow-ups, was any mention of identifying actual etiological agents of cancer, the primarily chemical mutagens or carcinogens that cause the mutations and other damage. Also lacking was any mention of the existing *Atlas of United States Mortality by*

County, which captures some of the environmental and occupational origins of cancers in the US.[1]

The subtle message transmitted by the public presentation of the cancer genome project is that the cancer problem is located in our genes. Readers of *Gene Watch* are familiar with this tendency to blame our genes for our condition of life. According to NIH genetics chief Francis Collins, the project will "enumerate the complete list of genomic insurgents that lead to cancer." Articles in newspapers across the country described the program as intended to identify the "cancer genes" associated with most human cancers. The message transmitted to the reader is that cancer is due to tissues that carry altered genes—sometimes called oncogenes—and the NIH effort will identify these damaged genes in the various tumors that affect us. The effort to identify the genetic damage in cancer cells is indeed critical to progress in preventing cancer and in generating better treatments for those afflicted. However, the description of the genome atlas does not bode well for the use of this information for preventing cancer.

The distortion of the current representation is particularly obvious when we consider lung cancer. Clearly, any program that identifies the damaged genes in lung epithelial cells that are altered in lung cancer is valuable. However, to fail to point out that such genetic changes are due to exposure to carcinogens present in cigarette smoke masks the real problem—the carcinogens in cigarette smoke.

Why do we get cancer?

Do we get cancer because control of the proliferation of our cells depends on genes whose damage leads to cancer? Or do we get cancer because we are exposed to environmental carcinogens? The first model feeds into the current interests of the pharmaceutical industry and modern medicine, *i.e.*, to accept the occurrence of disease in the population and sell the patient some therapy, treatment or procedure to alleviate the condition. This also conforms to modern medical practice, which focuses on the illness of the individual and generally ignores the conditions that led to the disease in the population. The focus on the diseased individual further shifts liability and responsibility away from the corporations, institutions or processes that generated the carcinogens inducing the cancer.

An alternative formulation is to identify the agents that cause the disease, and either remove them from the human environment or help people protect themselves from exposure. However, this policy brings biomedical scientists and physicians into conflict with the social and economic processes that generate the carcinogens: the tobacco industry with respect to cigarettes, chemical manufacturers and the petrochemical industry with respect to industrial chemicals, pesticides and herbicides. Vivid examples of this resistance have been described in books like Paul Brodeur's *Expendable Americans,* Samuel Epstein's *The Politics of Cancer* and Robert Proctor's *Cancer Wars.* Even earlier, Walter Hueper, the principal scientist attempting to bring the relationship between synthetic aniline dyes and bladder cancer to medical attention and a leader in the development of the National Cancer Institute, was prevented from becoming its director due to opposition from the industry.

Molecular basis of tumor development

The NIH announcement and much National Cancer Institute (NCI) literature refer to cancer as a complex disease. There is some truth in this; for most tumors, multiple alterations are needed to transform the normal cells into cancer cells. Since the many regulatory processes controlling and inhibiting cell proliferation differ from tissue to tissue, the damage pathway to cancer is different for different tissues. However, one of the great contributions of NIH-sponsored research over the last three decades has been to reveal the underlying nature of cancer. It may be useful to review this, so as to demystify the disease.

Cancer represents the uncontrolled proliferation of a particular cell type. The actual pathology and course of the disease depends on the cells and the tissue of origin. Lung cancer has a different course than liver cancer, but they both represent cells that are dividing when and where they should not be. Malignancy generally represents the ability of the cancer cells to overcome the constraining signals imposed by normal contact inhibition by cells that surround it in the affected tissues and organs. (Contact inhibition is a property of healthy cells wherein a cell stops dividing once it comes in physical contact with a neighboring cell.)

Cancer is not transmissible from individual to individual, and the great majority of cases are not inherited. Rather, like neurological damage from mercury poisoning, anemia from lead poisoning, black lung disease from coal dust, or bysinnosis from cotton dust, cancer is due to cell damage from exposure of the cells to toxic agents. In the case of cancer, the ultimate targets of the damage are the genes and chromosomes of the affected cells. Thus, the progeny of tumor cells express similar properties as their altered parent cells, since they carry similar chromosomal damage.

A major breakthrough in basic research was the recognition that, in cancer cells, damage was to the genes and chromosomes of affected cells. Such damage includes single nucleotide mutations and more complex forms of DNA damage—deletions, translocations, and duplications—in genes controlling cell proliferation. These changes are transmitted to daughter cells, which continue to pass on the damaged genes, eventually generating a tumor. Since different sets and subsets of genes are controlling the proliferation of liver cells, as opposed to lung cells, the genetic targets for transforming cells from normal to malignant cells differ in different tissues. Over the past decade, cancer researchers have discovered that many of these genes involved in encoding proteins regulate the cell cycle, and thus cell division.

Another important step forward was the biochemical identification of the pathway to DNA damage. For many organic chemicals, the carcinogenic species are modified forms, activated by detoxifying enzymes of the liver or other tissues, as part of the body's effort to rid itself of these modified species. Transport of the activated species into the nucleus of a cell results in their attack on the nucleotides in DNA, which generates mutations. For a number of human carcinogens, including poly-aromatic hydrocarbons, aflatoxin, acridine dyes and some metals, the chemical character of the DNA modification reactions is known in considerable detail. For these and many other carcinogens, the tumor induction process has been extensively studied in rats, mice and other experimental animals, as well as in cultures of human cells and tissues.

For a number of colon, skin and breast cancers, the damaged genes leading to the malignancy have been identified. The tumor cells in some women with breast cancer exhibit damage to the essential genes, but the rest of the breast cells and other tissues in their bodies carry the normal, functional versions of these genes.

The relatively rare cases of inherited cancers have been important in solidifying this interpretation. In these cases the mutations occurred in the germ line of the parents, or earlier generations, and were inherited by the children. The mutations were therefore transmitted to all the cells of the offspring. Thus, for example, all the cells in women with an inherited propensity to develop breast cancer because of mutations in their BRCA1 and BRCA2 genes have the same genetic damage to the BRCA1 and BRCA2 genes in both their cancers and their other cells. These individuals represent a distinct, but small, fraction of those who currently have breast cancer. Examination of the graph of those with lung cancer makes it immediately clear that this change in cancer frequency within a single generation could not be due to inherited mutations, since the disease was coupled to the individuals' exposure to cigarette smoke, not their parents. It has been important to recognize that as long as the genetic damage is not in the germ line—sperm or eggs—damage will not be transmitted to subsequent generations.

When Robert A. Weinberg and coworkers first showed that the tumorigenic property of human bladder carcinoma cells was associated with mutations in a single gene, it was a dramatic step forward in establishing that cancer was due to somatic mutation. The genes carrying these mutations were named "oncogenes." The experimental work included clear evidence that the oncogenes were not inherited by the affected individuals but had occurred in the cells generating the tumors. That is, they were almost certainly the product of mutations occurring in the bladder cells of individuals who had contracted bladder cancer. Nonetheless, the dominant message in hundreds of subsequent papers has been that oncogenes themselves are the cause of cancer, rather than that damage to specific genes (which generates oncogenes) is the cause. The difference may seem subtle or semantic, but it is profound in terms of the formulation of policies for reducing—not treating—the incidence of human cancer. There is compelling evidence that many human bladder tumors are due to exposure to aromatic amines and related chemical carcinogens.

Known human carcinogens

The recent publicity given to inherited forms of cancer has failed to point out, in both the popular and scientific literature, that there is a high correlation of the incidence of cancer with where people live and work, and with the conditions and character of their workplace.

One of the earliest recorded identifications of a chemical carcinogen was the 1775 description by the London physician Percival Potts that the high incidence of scrotal cancer in young London chimney sweeps was due to their exposure to coal tar and coal combustion products. Thus began the long history of identification of cancers and corresponding carcinogens. Over the next two centuries, many human carcinogens were identified due to high incidences of particular cancers in a variety of workplaces and environments. These resulting cancers included bladder cancer from aniline dyes, liver cancer from aflatoxin produced by fungi, sarcomas in watch dial painters from exposure to radium, mesothelioma from asbestos in shipyard workers, and many other examples.

The medical community was relatively early in recognizing the cancer danger from exposure to radioactive substances. The first mines for radium and uranium were in Joachimstal & Schneeburg in the Black Forest region of Germany. Local physicians Harting and Hesse noted the high mortality among the miners and reported an unusually high frequency of deaths due to lung cancer. Experience with exposure to radioactive elements, together with the medical follow-up on the survivors of Hiroshima and Nagasaki, led to increased awareness of sources of radioactive exposure and cancer: uranium from mining and processing, strontium 90 from atomic bomb testing, plutonium from nuclear reactors and spent nuclear fuel.

Public understanding that cancer is induced by external agents developed out of the linking of lung cancer to cigarette smoking. Initially, this causation was denied, as cancer incidence was not well correlated with simultaneous cigarette smoking. As epidemiologists continued to collect data on the relationship of smoking and health, evidence of the 20- to 30-year lag between initial exposure and cancer began to appear, and it became clear that the lung cancer was closely associated with previous history of smoking. Though this correlation was resisted by the tobacco industry for decades, the scale of the tragedy and extent of the data eventually overwhelmed such resistance.

With the increasing strength of the environmental movement in the 1960s and 1970s, many studies that linked human cancers to specific carcinogens were brought to the public's attention, including Irving Selikoff's investigation of the relationship between asbestos exposure and mesothelioma and Thomas Mancuso's studies of radiation workers. The collection and publication of cancer incidence data in the US by geographical location was of great importance to the advancement of this understanding. Cancer incidence data revealed variations far too high to be accounted for by genetic variations in the US population. In most of these cases, the efforts to publicize the identity of the carcinogen were steadfastly resisted by the producers of these carcinogens and their allies.

Studies on people who had changed their environment were also important in advancing this understanding. Thus, for example, Japan has a much higher incidence of stomach and liver cancer than America. However, among Japanese people who migrate to the US, there is a lowered incidence. Their children, raised in the US, exhibit the same incidence as the general US population. Conversely, incidence of colon cancer is lower among Japanese males than American males, but it increases in Japanese men who have migrated to California.

Lung cancer is not the only human cancer strongly tied to exposure to distinctive chemical carcinogens. Striking data has long existed describing the relationship between bladder cancer and exposure to aromatic amines, which include aniline dyes. The effort to generate synthetic purple textile dyes to replace natural indigos was key to the launching of the German synthetic chemical industry in 1856. By 1900, workers in the industry and physicians observing them knew that those who worked in dyeing plants had a high probability of contracting cancer of the bladder.

British studies revealed that risk of bladder cancer in aniline dye industry workers was 33 times greater than that of a control population. Before 1930, one in every four workers in the industry contacted bladder cancer; by 1950 the incidence had been reduced, but was still a tragically high one in six. By the 1940s, experimental evidence emerged from work of Walter Hueper and colleagues at the newly established National Cancer Institute (which was the original NIH Institute) where they showed that treating dogs with aniline dyes induced bladder tumors.

After World War II, manufacture of aniline dyes and related aromatic amines was established in the US. In 1955, Melick *et al.* reported the first cases of bladder cancer in the US: 19 of 74 men exposed to 4-aminodiphenyl, used as an antioxidant in the rubber industry. By 1958, the incidence was up to 36 percent of those exposed. A later study of another cohort of 366 workers revealed bladder cancer in 26 percent.[2]

Indeed, if one examines the map of the bladder cancer incidence in the *Atlas of United States Mortality by County,* among the highest incidences are found in the region including Essex County, New Jersey. This was the site of the largest aromatic amine-producing plant in the US. There are also hot-spots in a number of midwestern cities historically associated with the manufacture of tires. One of the main consumers of aromatic amines was tire manufacturers, who used them as antioxidants in rubber processing and tire production.

The role of the medical and pharmaceutical industries

For the past 30 years, the cancer research establishment has moved steadily away from a focus on identifying carcinogens to a focus on the cancer patient and the development of therapies and cures. Of course, this latter component needs our full attention. Then again, preventing cancer from developing goes much further in preventing and reducing human suffering. Unfortunately, preventing cancer does not generate a market for consumables.

The pharmaceutical industry is currently the most profitable sector in the US economy. This is due in part to an increasing population that either is ill or is concerned over health issues, which provides a wide market.

The intensive interest in patenting human genes by companies such as Myriad, Incyte and Celera is driven by the disease model of "oncogenes cause cancer." The initial market is for diagnostics tools, in which the patent monopoly insures profitability. Their long-term business plans could involve gene replacement or gene therapy, where patent monopolies will also insure substantive profits. Of course, companies invested in such business strategies have no interest in research that focuses on the causes cancer and reducing the incidence of gene damage in the population.

However, if we want to protect ourselves from the scourge of cancer in the long run, we need to recognize that the largest social and health benefit will accrue from identifying the causative agent and removing it from the human environment.

15

Back to the Drawing Board?

By Carlos Sonnenschein and Ana M. Soto

Carlos Sonnenschein, M.D., *and* **Ana M. Soto, M.D.,** *are Professors of Anatomy and Cellular Biology at Tufts' Sackler School of Biomedical Sciences. This article was originally published in* GeneWatch, *volume 19, number 2, March-April 2006.*

In the December 5, 2005 edition of *The New York Times Magazine*, David Rieff described the last days of the life of his mother, Susan Sontag. She was an accomplished writer, to such a degree that we quote her in an epigram in our book, *The Society of Cells,* published in 1999. Mr. Rieff relates all the uncertainties to which a layperson is exposed by authoritative sources and his or her own shortcomings as a second-hand consumer of scientific news. It takes some effort to integrate this information into a somewhat coherent picture about the subject of cancer, mortality and the cost of survival. Of course, Mr. Rieff is presenting the compassionate views of a patient's relative. Those of us who play a professional role in this multi-layered drama called cancer often have a different perception of the problem. For one, we are not the ones who do the actual suffering. Even though we have been trained as medical doctors as well as researchers, we also want to know the reasons others suffer. In our case, we first wish to know why cancers develop and, based on this knowledge, how to prevent them from doing so. If this fails, we then wish to diminish the patient's suffering by providing meaningful therapeutic options. But then reality sets in.

By now, most readers of *GeneWatch* know that, despite a massive financial effort in cancer research, few benefits have been harvested. Truthfully, the last decades have provided opportunities in the biological sciences to better understand how cells function; however, many rightfully still wonder why it is that these discoveries have not benefited the cancer patient to the degree we all had hoped. The unraveling of DNA's structure, the discoveries about how gene expression takes place, the role of gene therapy and stem cell research represent just a few subjects that occupied the attention of science writers and scientists willing to insert hope for cancer patients into dry experimental bench accomplishments. Unfortunately, most of the wonderful promises made by these people never materialized, despite the rather generous timetable the prognosticators propose. "The ten-year promise" seems to be the most common variety.

Ten years is almost a lifetime in experimental biology. The funding period of a National Institutes of Health grant is much shorter than that. Our shared frustrations make us wonder whether a "frank" (as diplomats refer to when sharp opinions are exchanged in private) public debate will ever take place on the many aspects of cancer that are now out of bounds. Colleagues and "innocent bystanders" remind us that cancer is part of an "Industry of Hope" in which many sectors of society play both interested and naive roles.

So, once we have explained some of our own doubts on the subject, we will try to provide our own perception of the problem of cancer. For the most part, we will deal with our bench experience. However, being convinced that science is a social endeavor influenced by interactions comparable to those in any other human activity, we will editorialize when we feel it is justified.

Faulty assumptions

In *The Society of Cells,* we discussed our two decades of bench-based experience with control over cell proliferation and also speculated on the field of cancer research. Before writing the book, we had mistakenly assumed that by clarifying why cells proliferate, we could come to a reasonable understanding of how cancer starts. This is called carcinogenesis among cancer researchers. Though we are not the only ones who misread what cancer was about, it was somewhat comforting to realize that our misperception put us in the company of many others. Indeed, cancer research has been the intellectual graveyard of many distinguished scientists who plunged into this subject under the laudable aim of "curing cancer" or at least contributing to its understanding. Why is it that the "best of the brightest," regardless of their qualifications, have not succeeded in this task?

Cancer as a hierarchically complex entity

In our book, we analyzed cancer within four levels of hierarchical complexity, namely, a) a social level, b) an organismic level, c) an organ-tissue level and finally, d) a cellular/subcellular level. By social level, we meant the one where information is gathered mostly through epidemiological methods. In the organismic level, we considered the relationship between a patient and his or her physician from diagnosis to treatment. We introduced the organ/tissue level of complexity because the rigorous diagnosis of cancer has remained in the hands of the pathologist who, looking through a light microscope, makes the educated judgment of whether the tissue biopsy shows the presence of a cancer. Finally, there is the cellular/subcellular level of cancer complexity. This latter level has been the one where, in the last five decades, efforts have been concentrated to explain why cancer develops.

Lung cancer exemplifies the use of a hierarchical approach. Over a half-century ago, epidemiologists and public health officials were able to design and promote effective preventive campaigns for lung cancer. Reducing tobacco consumption in order to lower lung cancer incidence did not require those epidemiologists and public health professionals to know much about the lower levels in this hierarchy. These methods did not require deeper analysis between the subcellular level, where mainstream cancer researchers believe cancer originates, or of the tissue level at which cancer is diagnosed. However, when studying how cancer arises, it is crucial to establish whether this is a problem occurring at the cell or at the tissue level of biological organization.

How many varieties of cancer are there?

Over 90 percent of cancers are of the carcinoma or adenocarcinoma variety, namely, tumors arising in epithelial cells, *i.e.,* the cells lining the surfaces of organs. The overwhelming majority of these cancers develop mostly among adults and aged people; they are called "sporadic" and they represent over 95 percent of the cancers that come to medical attention. There are also heritable cancers that affect mostly children and youn g adults, which represent only 5 percent of clinical cancers.

Since 1914, when Theodor Boveri, a famed German biologist, formulated the first version of what became the somatic mutation theory (SMT), this most explored theory of carcinogenesis covered just one aspect of neoplasia, that is, the ability of cells to breed true to type. The rationale was worded along the lines that if normal cells beget normal cells, and cancer cells beget cancer cells, what causes normal cells to become cancerous? For almost one hundred years, the answer to this question was assumed to reside in the changes in the genetic material of a cell, *i.e.*, its DNA.

Following the successful reductionist strategy in chemistry and physics, many thought that this strategy would eventually succeed in the biological sciences as well. In the minds of these scientists, genetics and developmental biology could be successfully "reduced" to physics and chemistry. In fact, this strategy was instrumental in validating what is known today about cell biology and genetics. Indeed, the contributions of Franklin, Wilkins, Watson and Crick, who uncovered the molecular structure of DNA, and the introduction of other experimental tools helped immensely in looking deeper into the cell's components. In this regard, perhaps with his tongue in cheek, John Cairns, the noted British biologist, commented some years ago, "Biology and cancer research have developed together. Invariably, at each stage, the characteristics of the cancer cells have been ascribed to some defect in whatever branch of biology happens at the time to be fashionable and exciting."

Meanwhile, the notion that cancers are due to the accumulation of mutations in the DNA of a cell, and thus that cell eventually becomes the source of a tumor, gained influence. By inference, cancer cells were increasingly thought of as altered, or mutated, cells. Thus, it was not surprising that, when experimental data could not vindicate the notion that "sporadic" cancers could be explained by DNA mutations that caused the cells harboring them to proliferate excessively, as the SMT predicted, leading researchers suggested the involvement of epigenetic mechanisms related to gene transcription that would result in the same outcome. As in pre-Copernican astronomy, whereby epicycles were added to account for the lack of fit between theory and observation, this new epicycle added to the SMT preserved the theory. Research continued to focus on the interior of an epithelial cell that may eventually become a tumor due to altered control of cell proliferation.

The usage of the word mutation has changed since the early years of the twentieth century; today it means a change in the linear structure of DNA. But, is it necessary to invoke these genomic mutations to explain the different behavior of normal and cancer cells? Not necessarily. For instance, it has been acknowledged for some time now that all cells in the adult human organism contain the same genetic information. Hence, a change in the appearance and/or behavior of a cell (phenotype) does not require a change in the structure of its DNA (genotype). A liver cell and a kidney cell carry the same DNA sequence while they look different and do different things. Changing which genes are being expressed in each cell type, or epigenetics, could explain this fact equally well. Thus, the somatic mutation and epigenetic theories provide alternative, not necessarily complementary, ways of explaining the stability of the phenotype of cancer cells.

This is where the notion of hierarchical levels of inquiry becomes relevant. The somatic mutation theory places the problem of carcinogenesis at the cellular/subcellular level. It assumes that the main problem in cancer is altered cell proliferation patterns due to mutations in the genes that control this process, or in genes that, by affecting cell differentiation, would indirectly affect the proliferation characteristics of the mutated cells. However, the speed at which cells proliferate in cancers is not faster than that of cells in normal tissues and this fact challenges the former conclusions.

Moreover, the proliferation rate of cells in cancerous organs responsive to hormones is susceptible to the hormonal milieu in the host. For example, breast and prostate cancers regress when their specific trophic hormones (estrogens and androgens, respectively) are removed. This behavior cannot be readily explained by the accumulation of irreversible, stable mutations in those putative cancer genes. Thus, cancers are not necessarily autonomous entities that would place them beyond any "physiological" restrictive control.

Complicating components

About three decades ago, cell death (apoptosis) was first described to be equally important to the regulation of cell numbers in normal tissues as is the generation of new cells. By extension, lack of regulation of apoptosis was also assumed to play a role in the rate of tumor growth. This notion was further extended to carcinogenesis, *i.e.*, the process of initiation of cancer formation. Disabled genes involved in apoptosis would also have a role in causing cancer. This line of thought remains controversial today.

In this regard, the hypotheses positing that carcinogenesis is an error in cell proliferation, cell death or cell differentiation, or is a combination of all three processes, implicitly acknowledges again that the initiation of a cancer is a cellular/subcellular-based phenomenon.

Another aspect of the SMT is the idea that proliferative quiescence is the default state of cells in multicellular organisms. This means that in the presence of excess nutrients, cells will not proliferate unless they are stimulated by outside signals, generally called "growth factors." This is the opposite of what is widely thought to happen in unicellular organisms, like bacteria, yeast and others, and also in plants, since they are known to have proliferation as their default state. In other words, they reproduce as long as nutrients are available.

The bases for a change in perspective

In the late nineteenth century, pathologists began diagnosing cancer by describing the histological pattern of tumors while using a simple light microscope. These same types of microscopes are the ones that allow pathologists to diagnose cancer today, as will most likely be the case for the foreseeable future. This simple realization, however, itself suggests that tissue disorganization is at the core of carcinogenesis and cancer.

For the most part, cancer cells retain the distinctive structures that characterize their organ of origin, represented by its parenchyma (the distinctive cell types of an organ) and the adjacent stroma (the supportive framework of that organ usually composed of connective tissue). In fact, toward the end of the nineteenth century, observing that tissue disorganization was the most distinctive characteristic of cancer cells, some noted pathologists ventured to propose that carcinogenesis was a process akin to an altered development of organs. Later, from the 1930s to the 1950s, prestigious developmental biologists and pathologists reawakened the notion that carcinogenesis took place at the tissue level of biological complexity. Unfortunately, these ideas were then dismissed in favor of the cell-centered, reductionist explanation of carcinogenesis that has prevailed to this day.

Some insights have challenged this view. For instance, since 1975, several research groups have reported that, under rigorous experimental conditions, cells taken from some cancers gave rise to "differentiated" non-tumorigenic cells and tissues, *i.e.*, cells that looked and behaved normally. As a result, they postulated that the main defect in cancer cells resided in an altered control of their differentiation and showed that the assumption of SMT, "once a cancer cell, always a cancer cell," was not supported by data. Instead, these researchers considered that epigenetic, non-mutational, mechanisms were at work.

Based on these early insights and knowledge stemming from developmental biology, we proposed in our 1999 book a theory of carcinogenesis that claims that neoplasia is a defect of tissue organization, rather than a cell-based phenomenon. We call it the tissue organization field theory of carcinogenesis and neoplasia (TOFT). According to this theory, cancers are emergent phenomena resulting from a flawed interaction among tissues.

Let us expand a little on this subject. During embryogenesis, organ formation depends on the reciprocal interactions of cells within what is called a "morphogenetic field." A morphogenetic field is a collection of cells that eventually form the organs of an individual. In a more advanced stage of embryogenesis, the stroma (the framework cells) of an organ play a central role in the differentiation of the parenchyma (the interior cells). These interactions continue once the organ in question is formed, participating in the maintenance of their normal structure and function. Now, in contrast with SMT, which identifies DNA as what is being principally affected by carcinogens, our theory considers tissues as the level at which carcinogens create disorder. Depending on which is borne out by the evidence, a whole different strategy should be applied to resolve the origin of cancer. Hopefully, this will aid in devising a rational therapeutic approach for the eradication of cancer.

So, what are the ultimate targets of the carcinogens we are exposed to: a cell, a tissue or both? We maintain that the effects of carcinogens on the structures and components inside cells, while variably deleterious to each of them, are not directly responsible for the development of a cancer. In other words, according to the TOFT, carcinogens act initially on the stromal cells, but these cells do not ultimately become cancerous. Rather, the carcinogen-altered stromal cells affect the progenitors of would-be cancerous cells in the adjacent parenchyma, *i.e.*, the epithelial cells in their proximity. Carcinogenesis and neoplasia would then occur entirely through emergent phenomena once the signals that maintain normal tissue organization are disrupted by the carcinogen.

From what we know about the natural history of human tumors, which is not much, it is impossible to predict whether an individual initial lesion will progress, remain stable, or regress. However, it is clear that the probability of regression should decrease as the abnormal tissue interactions increasingly deviate from the normal state. In this regard, we should recall that the second postulate of TOFT claims that the default state of all cells is proliferation. Thus, as tissue disorganization increases, the proliferative restraint imposed upon cells by their neighbors becomes less effective and, hence, allows for neighboring cells to express their own proliferative potential and further disturb the tissue architecture.

Cancer realities and the media

We frequently hear fellow scientists say that data speaks for itself. Of course, depending on the premise one adopts, the same data can be interpreted differently. Given that a common, unified view of the level of biological complexity at which cancer originates does not exist, chances of real progress in understanding or curing cancer appear remote. Members of the media are presented with a choice of how to convey this state of affairs to their audiences. They can play a clarifying, educational role that is not incompatible with their desire to offer hope to patients and their relatives. But they seem to be content with the constant reruns of the traditional "ten-year promises," offered routinely by cancer researchers, eager to keep the patients' hopes high regardless of their justification. This approach, however, runs the risk of losing the public's trust.

At this time, the somatic mutation theory is rife with contradictions between theory and experimental data. A recent article by Gina Kolata published [December 27, 2005] in *The New York Times* states:

> At first, as scientists grew to appreciate the complexity of cancer genetics, they despaired. "If there are 100 genetic abnormalities, that's 100 things you need to fix to cure cancer," said Dr. Todd Golub, the director of the Cancer Program at the Broad Institute of Harvard and M.I.T. in Cambridge, Mass., and an oncologist at the Dana-Farber Cancer Institute in Boston. "That's a horrifying thought."
>
> Making matters more complicated, scientists discovered that the genetic changes in one patient's tumor were different from those in another patient with the same type of cancer. That led to new questioning. Was every patient going to be a unique case? Would researchers need to discover new drugs for every single patient? People said, "It's hopelessly intractable and too complicated a problem to ever figure out."

How are these contradictions dealt with? Our method is to go back to the drawing board and consider alternative theories to the one that is not getting results. However, another, more popular method of dealing with such contradictions is to put faith in the ability of new technology to overcome them. As cancer geneticists prepare to embark on the recently funded Cancer Genome Atlas project, applying "new" technology to an old, failed theory, we wonder when the tide will turn to a more comprehensive understanding of cancer, from carcinogenesis to therapy.

Where do we go from here?

Will a better understanding of how cancer starts come closer through the adoption of our tissue-based approach? Probably, yes. Have all questions relating to carcinogenesis been answered by this theory? Surely not. But, a consensus is starting to be built around this truly novel alternative. Five decades of intensive research commitments (both financially and in individual time and effort) have given cancer patients unfulfilled hopes, all based on a rationale that implies that there is something wrong with our genes. Epidemiological evidence is pointing instead to something being wrong with our environment.

Those who believe that faulty genes are responsible for the current increased incidence of "sporadic" cancer are embarking on The Cancer Genome Atlas project, a costly program that will allow the survival of a failed deterministic view for yet another decade. Meanwhile, a radical change is occurring: the mainstream no longer looks toward a "cure" but rather perceives cancer as a chronic disease, like AIDS. Consequently, drugs that will be administered for the remainder of a cancer patient's life are being proposed as treatment. Patients' relatives, like Mr. Rieff, as well as the public at large, might also start asking more probing questions of researchers, industry and media sources.

16

Direct-to-Consumer Genetic Testing: What's the Prognosis?

BY JORDAN P. LERNER-ELLIS, J. DAVID ELLIS, AND ROBERT GREEN

Jordan P. Lerner-Ellis, Ph.D., *is a Clinical Molecular Genetics Fellow at Harvard Medical School & Brigham and Women's Hospital and leads a genetic test development group at Partners Center for Personalized Genetic Medicine and Harvard's Laboratory for Molecular Medicine.* **J. David Ellis, Ph.D.,** *a consultant in public policy and regulatory affairs, teaches Communication Studies at York University in Toronto.* **Robert C. Green, M.D.,** *is a physician-scientist and a faculty member in the Division of Genetics, Department of Medicine at Brigham and Women's Hospital and Harvard Medical School where he is Associate Director for Research in the Partners Center for Personalized Genetic Medicine. He co-directed the NHGRI study in conjunction with Dr. Scott Roberts of the University of Michigan. He is a member of the CRG Board of Directors. This article was originally published in* GeneWatch, *volume 23, number 4, August-September 2010.*

Genetics has been making news lately, in large part because of the growing pains of a new and controversial industry: direct-to-consumer (DTC) genetic testing. DTC genetic testing raises questions involving privacy, how medical tests should be ordered and understood, who should regulate access to genomic information, and how individual consumers will understand and act upon such information.

DTC genetic testing has been around on a small scale for some years but began in a new form in November 2007 when three companies (23andMe, Navigenics and DeCodeMe) launched their genome-wide scan services within days of each other. Suddenly, individuals could send in a sample of their DNA and receive ancestry information or a wide variety of medical risk information based on the latest discoveries in genetics. In October 2008, TIME magazine recognized the 23andMe personal genome service as "Invention of the Year."

Celebrities turned over samples of their DNA at trendy and well publicized "spit parties."

Non-medical services, like ancestry testing, provoked few criticisms. The same was not true of medical risk reporting, which was immediately criticized on two counts. Firstly, the companies were reporting in most cases on DNA variants of common diseases, discovered through statistical comparisons in genome-wide association studies. While these associations were well-established for large populations, they typically accounted for only a tiny fraction of total disease risk. Genetic testing of this kind was hard to justify for an individual, since it provided no clearly useful information to either the patient or the health care provider.

Secondly, it seemed possible that other risks, like family history and lifestyle—currently much better predictors of common disease—might be de-emphasized to the detriment of the DTC genetic testing consumers. Thus, a customer who was obese and had a strong family history of type 2 diabetes might well receive a low genetic risk score for the disease. To be fair, the leading companies have taken care to be accurate on their websites as to the modest effects of genetic risk information and the importance of other risks.

By 2009, the DTC controversy revolved around conflicting visions of the future of personalized health care. To its supporters, DTC genetic testing offered private, scientifically supported and personalized information about the state of one's health, away from the intrusive gaze of insurance companies, freed from the paternalistic intermediation of harried and often uninformed clinicians seeking to preserve their economic advantage in an already dysfunctional health care system.

To its detractors, DTC genetic testing was exploiting widespread misunderstanding of genetic determinism to market common DNA risk variants that were poorly understood by the scientific community and provided little useful information to consumers. DTC genetic testing was simply the latest in an unending series of health-related pseudo-interventions, ranging from colonics to nutraceuticals, for the privileged who could afford the extra cost. And the major challenge was to keep conventional medical practitioners from taking it seriously, lest the cost of medical testing be driven up in response to dubious genetic "risks."

Fast forward to 2010 and the controversies have evolved, but by no means disappeared. For one thing, DTC companies have expanded their offerings to include the identification of variants associated with rarer, more highly penetrant diseases, as well as carrier states. Examples include BRCA1 [the breast cancer gene #1], cystic fibrosis, PKU [phenylketonuria] and Tay-Sachs. These disorders are more "fully penetrant" because if an individual carries mutations, he or she will either have the disease or be at high risk to develop the disease. Thus, the nature of the information being offered in the DTC genetic testing space is changing. Such changes have the potential to make the test results medically relevant for a small number of people. But this may also increase the potential for public misunderstanding, since companies will now be offering clinically meaningful rare DNA variant information alongside clinically less relevant, common DNA variant information.

Another disruptive development concerns the Food and Drug Administration's (FDA) regulatory actions. In May, 2010, DTC genetic testing almost went retail when the Walgreens drugstore chain announced it would stock $30 DNA collection kits for the DTC company, Pathway Genomics. The attempt to go to market was blocked at the last moment, and the controversy triggered new scrutiny and new revelations.

This summer the FDA decided to investigate the use of what it calls laboratory developed tests (LDTs), which had previously been unregulated.[1] The agency's main concern was focused on genetic tests intended for use without medical supervision. FDA scrutiny has grown for several reasons, among them greater complexity of genetic testing, the role of labs located far from the primary care setting, the involvement of profit-making firms and the focus on poorly understood genetic risks for common diseases. The FDA notes that patient risks include "missed diagnosis, wrong diagnosis, and failure to receive appropriate treatment."

The agency later announced it was holding public meetings to gather stakeholder views on LDTs. As shown through the diverse testimony presented at its July public meetings,[2] clinicians, researchers, advocates and business executives are far from united on the issue of whether or how to regulate genetic tests. During the July hearings, the General Accounting Office (GAO) made a surprise announcement that it had surreptitiously taped telephone conversations between investigators and representatives of the DTC testing companies. The GAO played the tapes on the record and exposed a number of inaccurate statements made by company representatives.

It is worth noting that the proposed oversight of marketplace behavior will not necessarily ensure that the services in question will improve in accuracy or prognostic value. These goals are not often achieved through regulation, and physicians are not necessarily the best gatekeepers to determine the pros and cons of ordering genetic tests. Moreover, although commercialization has brought these regulatory concerns to the forefront, the development of innovations in genetics is clearly benefiting from the energy and imagination of the biotechnology and DTC testing industries. It may be that the interplay of commercialization and scientific innovation that is represented in the DTC genetic testing industry will prove to have long-term value to society. The FDA is thus keeping a close eye on how innovation will be affected by its actions.

The Human Genome Project, completed in 2003, is rightly regarded as one of the great scientific achievements of our generation, well worth its $2.8 billion cost. But what scientists have achieved in the intervening years is every bit as significant: new technologies that have reduced the cost of DNA sequencing to one one-hundred-thousandth of what it was originally. It is considered inevitable that within the next 5 years, whole genome sequencing will be available to any individual for under $1,000! As it turns out, the big challenge for the future will not be the sequencing technologies but the cost and difficulty of interpreting the huge amounts of data they generate.

Therein lies the dilemma for scientists and regulators. On the one hand, the DTC testing companies may continue to be innovators in the interpretation of personal genetic data. On the other hand, the companies concerned will have to make a concerted effort to develop and refine precautionary measures covering a wide range of medical and ethical issues. Many unsettling results can turn up as part of an otherwise routine screening. A child might, for example, be found to have a variant associated with one disorder, say autism spectrum disorder, and the very same variant might later be determined to cause a separate neurodegenerative disorder. How will this development be reported and explained to the consumer? What procedures if any should be in place for tracking individual customers long after they've ceased doing business with the testing company? Examples abound of the challenges created by incidental and unexpected findings, and there may be no ready answers. While informed consent is the universal goal, making it work universally is not a simple matter.

Some experts suggest the crucial problem with DTC testing is lack of supervision by qualified medical personnel. Medical supervision may often be desirable, even essential. But there is another perspective here. Even the experts, including medical geneticists, continue to struggle with incomplete and incompatible genetic databases, poor risk models and disagreements over interpretation. Moreover, many primary care physicians are not well versed in genetics and may not know what kinds of tests their patients need. In other words, it is unrealistic to expect that medical supervision in and of itself will turn DTC tests—or any genetic tests—into accurate and reliable tools.

In summary, difficult questions face medical professionals and members of the public in their attempts to evaluate genetic tests. What does a set of genetic results actually reveal? How will they help promote medical care? And what difference will testing make to the individual's quality of life? These are not theoretical questions. Right now many in the testing business are raising the public's expectations by making suggestive statements like this one from a company web site: "Let your DNA help you plan for the important things in life."[3]

Given its mandate, the FDA will have to devote its attention to high-profile issues, especially patient safety, while encouraging innovation as best it can. Even then, the regulator can only do so much to manage public expectations, and rising consumer demand for inexpensive tests is certain to be a strong market mover as this debate unfolds. Over-regulation of a service that consumers want could simply drive such services offshore where they could operate over the Internet with impunity. In any case, no amount of market regulation can take the place of well-designed and well-funded research that will help geneticists and other scientists understand the complexities of the human genome in the service of better medical care.

Do customers of DTC genetic testing services really understand what they are purchasing? Do they understand the results? Do they consult their physicians about the information? Are unnecessary medical tests ordered or are valuable health lessons learned from the overall experience? At the present time, we simply do not have all the answers. The National Human Genome Research Institute (NHGRI) has recently [2010] funded a proposal to implement the first "before and after" survey of DTC genetic testing in order to better understand these questions.

17

Time to Reconsider?

BY MARCY DARNOVSKY AND STUART NEWMAN

Marcy Darnovsky, Ph.D., *is Executive Director at the Center for Genetics and Society, an Oakland, California-based public interest organization working to encourage the responsible use and effective governance of human biotechnologies.* **Stuart Newman, Ph.D.,** *is Professor of Cell Biology and Anatomy at New York Medical College and a founding member of the Council for Responsible Genetics. This article originally appeared in* GeneWatch, *volume 20, number 6, Winter 2007-2008.*

The Food and Drug Administration is still investigating whether the death last summer [2007] of a 36-year-old Illinois woman was directly caused by an experimental gene therapy procedure. Jolee Mohr was enrolled in a gene transfer clinical trial for arthritis, sponsored by Targeted Genetics Corporation of Seattle. She developed a fever of 105 degrees and multiple organ failure just after she received a second injection of genes carried by engineered viruses.[1]

The story is complicated and still unfolding. The Recombinant DNA Advisory Committee (RAC), which oversees gene transfer protocols for the National Institutes of Health (NIH), discussed Mohr's death in September 2007 and will return to it at its December meeting. The information publicly available from news reports and the RAC meeting strongly suggests that the gene transfer experiment was at the very least a significant contributing factor.

What do we know at this point about what went wrong? Is Mohr's death an isolated tragedy? Or is it a symptom of deeper problems with the way gene transfer experiments are being conducted?

In the 1980s and 1990s, gene therapy was widely touted—with virtually no clinical proof—as a medical miracle. Some called it a cure for cancer in an injection. Others predicted that it would enable the conquest of many others of our most serious diseases. The first experimental gene-based treatments were administered in 1990, with inconclusive results.

Since then, gene transfer researchers have conducted hundreds of clinical trials, received hundreds of patents, written thousands of articles, and launched scores of companies. In a 2005 article in *Human Gene Therapy*, Christine Crofts and Sheldon Krimsky commented that the size and strength of this infrastructure is "seemingly incommensurate with the demonstrated potential of the technique."[2]

More bluntly, the record of gene transfer has been worse than disappointing. In truth, gene therapy has harmed more people than it has helped.

Who should participate in gene therapy experiments?

Jolee Mohr had arthritis. While for some people arthritis is debilitating, it is not life-threatening. Mohr's husband Robb Mohr has said that she suffered from "occasional stiffness" and that she went boating the weekend before she was injected with genetically engineered viruses.

In 2000, after the gene transfer death of 18-year-old Jesse Gelsinger—in a clinical trial for a condition that he was able to control with a conventional treatment—one of us (Newman) appeared before the RAC to request that they limit enrollment in gene therapy protocols to people with serious conditions that cannot be treated by other methods.[3] This would have required a reformulation of the existing multi-phase clinical framework which, though useful for pharmaceutical drug evaluations, has severe limitations in other applications. Although one long-term committee member privately expressed his strong support for the proposal, the public discussion that day was entirely dismissive.

If the RAC had followed this recommendation, researchers could have spent the past seven years testing gene transfer in situations where the risks were far better justified. Mohr wouldn't have been a candidate for experimental gene transfer and probably would be alive today.

Do participants in clinical trials benefit from them?

The clinical trials in which Gelsinger and Mohr participated were both early-phase studies, which are designed to evaluate the safety of experimental treatments.[4] They are explicitly not meant to assess effectiveness. But Mohr was under the impression, her husband said, that the treatment would "make her knee better."

Hope for a cure strongly motivates a person to assume risks. That's why it is, or should be, a cornerstone of medical ethics to ensure that participants in early-phase clinical studies fully understand not to expect any therapeutic benefit from them.

What led Mohr to enter the study, then? The fact that her own doctor invited her to participate may well have contributed to her misunderstanding. Trust in our doctors to help us, or at least to "do no harm," is likely to outweigh the fine print in a release form. The consent agreement used in the Targeted Genetics arthritis study contained a single sentence saying that no medical benefits were anticipated, along with long descriptions of how the product might help.

Mohr's doctor, rheumatologist Robert Trapp of The Arthritis Treatment Center in Springfield, Massachusetts, may have had no reason to believe that the experimental treatment would harm her. But it turns out that his clinic received payments from Targeted Genetics for each subject he recruited. Studies show that doctors are influenced even by small gifts from drug companies. Should we allow them to receive direct financial payments to recruit their patients into trials that carry significant risks?

Who makes the rules?

The RAC, established to provide oversight for gene transfer, has always made its deliberations and the researchers' responses to them publicly available. But the FDA, to which actual authority for gene therapy experiments was transferred in the mid-1990s, considers most information about adverse reactions to be the property of the sponsoring company. Paul Gelsinger, Jesse's father, recalls an FDA representative telling indignant RAC members, "My superiors answer to industry."

In 2003, the RAC (now a purely advisory body) evaluated Targeted Genetics' arthritis study. Committee members questioned the justification for the trial, both because it involved patients who were not very ill and because evidence from animal studies didn't show much improvement. They worried that the viruses and the genes they carried could trigger dangerous immune system responses.[5] They criticized the informed consent document for not being clear enough about the fact that participants in the trial would be unlikely to get any benefit from it.

Targeted Genetics addressed some of these concerns in remarks at the 2003 meeting. But because resolution of issues discussed before the RAC was a matter for the FDA and thus no longer part of the public record, there is no way to know the extent to which the company was responsive.

Conflicts of interest?

The products of biomedical and biotech companies are often matters of life and death. They are also matters of profit and loss.

In the study in which Gelsinger was enlisted, both principal investigator James Wilson and his academic employer, the University of Pennsylvania, had financial stakes in a biotech company called Genovo, Inc. Wilson and the university stood to make money, perhaps a lot of it, if the experimental treatment could be commercialized. The investigations that followed Gelsinger's death disclosed dangerous shortcuts and clear-cut ethical violations in that study. These findings were strong enough to lead to extensive reforms in the financing of research at Penn and many other universities.

Were any similar dynamics at play in the arthritis study? In 2005, Targeted Genetics CEO H. Stewart Parker told a Seattle newspaper that her company's gene transfer system for arthritis was aimed at "a $7 billion market . . . by 2011" and that she was "looking at 15 to 40 percent of that opportunity." Targeted Genetics (which purchased Genovo in 2000 in a deal that gave James Wilson $13.5 million in stock for his 30 percent share) clearly wants to get its gene transfer product to market.[6] One important step in that process is enrolling subjects for its clinical trials as rapidly as possible. Could financial pressures created by expectations of a lucrative market have endangered study participants?

What now?

There are still many unanswered questions about this latest gene transfer death. But two lessons seem clear.

First, regulators should rethink the drug-related multi-phase clinical trial model for testing genetic agents. Moving toward a model in which subjects are chosen on a "compassionate use" basis, in cases where there are no good alternatives, may not ideally suit commercial prerogatives. But in linking testing to the possibility of actual therapy, it may be good medicine, and because it is subject-oriented, ultimately more ethical. There is no question that valid scientific data can be extracted from such studies, despite their complexities.

Second, we need much greater transparency in the conduct of clinical trials, whether they're conducted at universities and medical schools, or sponsored by commercial enterprises. In the present instance, the FDA should make public complete information about the Targeted Genetics arthritis gene transfer trials. This means not just officially confirming the cause of Mohr's death but also making public information about how Mohr (and other participants) came to be enrolled in the trial, whether they received independent medical advice and care, and releasing all records relevant to the issue of whether and how Targeted Genetics addressed the concerns raised by RAC members in 2003.

Clinical research is a major function of medical schools and collaborating university departments in basic and social sciences, relating to their roles in the production of knowledge and the education of physicians and scientists. It has also been a mainstay in the funding of such institutions since the end of World War II in the form of overhead from federal grants and, increasingly since the 1980s, in the form of commercial investment in anticipation of revenues from patentable products of the research. Distortions of academic functions are inevitable when the financial aspect looms so large and have been well documented (Washburn, 2005). Distortions of the medical function, with loss of life one of the consequences, are unfortunately no longer rare.[7]

The precedents set by the FDA's handling of the Mohr tragedy will have ramifications beyond gene transfer experiments. The level of protection the agency offers to human subjects will affect the safety of experiments with new drugs, new medical devices, and novel medical approaches like stem cell treatments.

Prospective patients and the public need to keep in mind that clinical trials often involve risks to participants—and significant financial stakes for researchers, companies, and sometimes their own doctors.

Anyone who puts profit before patient safety must be held to account. And government regulators must answer not just to industry but, above all, to patients and the public.

18

Where Are the Genes?

BY RICHARD LEWONTIN

Richard Lewontin, Ph.D., *is Professor of Zoology in the Museum of Comparative Zoology, Emeritus at Harvard University. He is the author of numerous works on evolutionary theory and genetic determinism, including* The Genetic Basis of Evolutionary Change *and* Biology as Ideology, The Dialectical Biologist *(with Richard Levins) and* Not in Our Genes *(with Steven Rose and Leon Kamin). This article first appeared in* GeneWatch, *volume 22, number 2, April-May 2009.*

The announcement in February 2001 that researchers had sequenced the entire human genome sparked immense publicity that was without precedent in the biological sciences. The public attention was a consequence of the putative chief motivation in personalized medicine for the sequencing efforts. While it is true that the human genome sequence is of great interest to evolutionists trying to reconstruct human ancestry and to biologists whose attention is on understanding the processes of development and of molecular interaction, the promise of benefits for human health has been the overwhelming justification for that immensely expensive effort.

The underlying claim is one of extreme genetic determinism. The assumption is that all-important variations in basic physiological and developmental processes are the direct result of genetic variation, so that pathological states are reflective of "abnormal" gene function due to mutations producing variant nucleotide sequences. This goes beyond the claim that systemic disorders like cancer, stroke and cardiac disease would eventually be treatable and even preventable using either gene therapy, or other interventions based on our knowledge of how genetic variation causes pathological states. While the race to sequence the genome was under way, William Haseltine, as CEO and Chairman of Human Genome Sciences, assured us that "Death is a series of preventable diseases." Knowledge of our DNA apparently assures us of not only better health but immortality as well.

There is a long history, predating genomic studies, of the discovery of genetic mutations that are responsible for some human disorders. The classic example is phenylketonuria (PKU), in which the affected individual is homozygous for a single mutation. In PKU the enzymatic pathway that normally breaks down the amino acid phenylalanine is blocked, with the result that lethal concentrations of the amino acid accumulate in the body. The disease is rare, however, as might be expected for a simply inherited lethal disorder. We do not expect to find single gene mutations of large effect that explain the prevalence of diseases, such as cancer, stroke, and heart disease, that are the common direct causes of mortality in populations not suffering from severe malnutrition or epidemic infections. Even in the famous case of the BRCA1 and BRCA2 mutations, where the presence of the mutation results in a very high probability of contracting breast cancer, only about 15 percent of all breast cancer sufferers carry the mutation. The chief sources of breast cancer remain to be found.

The belief that genes determine the characteristics of individuals, together with the lack of evidence for simple single-gene defects as the cause of the major sources of disease and mortality, has led to a deterministic model of genetic causation of disease and a new approach to searching for genetic causation. This approach has been made possible by the availability of a complete DNA sequence of the human genome and of tools for detecting nucleotide differences between the genome sequences of individuals. Humans are genetically polymorphic: no two individuals (except for identical twins) will have identical nucleotide sequences. At any nucleotide position, some fraction of individuals will carry a different nucleotide than the common one, a phenomenon called Single Nucleotide Polymorphism (SNP). One technique for searching for genetic causation of diseases is to scan the genomes of a sample of both diseased and healthy individuals for positions in the genome where statistical differences exist. Ideally, all of the diseased individuals would have a different nucleotide at a particular position compared to the healthy individuals, but in practice there is only a difference in the proportion of the four nucleotides, except in cases of well-known simple genetic disorders like PKU.

The screening of whole genomes for variant SNP's has been a major industry in both academic and commercial biomedical laboratories during the last decade, and reports of newly discovered genetic differences between healthy and diseased individuals have been a weekly phenomenon in medical journals, in general scientific publications like *Nature* and *Science*, and in the science sections of major newspapers. Then, suddenly, it was revealed that the whole enterprise had failed to produce useful results. On the front page of *The New York Times* on April 18, 2009, there appeared over the byline of one of the greatest boosters of genetic determinism, Nicholas Wade, an article whose headline read, "Study of Genes and Diseases at an Impasse." In the same week, the News of the Week section in the April 24 issue of *Science* reported on the "relatively low impact" of the SNP studies done so far. Both of these reports were instigated by an article appearing in the April 23 issue of *The New England Journal of Medicine* reporting on the search for "genes underlying the risk of stroke in the general population" and several commentaries on the approach to finding such genes. The general consensus of all of these reports is that the search for genetic causes underlying major causes of mortality has so far been a great disappointment.

The facts certainly bear out their pessimism. The usual measure of a specific genetic difference's importance is to calculate a risk ratio, asking: What is the risk of persons with this genotype for contracting the disorder relative to the risk in persons with a different genotype? Another form of risk calculation is the sibling risk ratio, which asks how much more likely it is that two siblings will be affected than two unrelated persons, taking into account the genes shared by siblings. In the study on stroke, two candidate SNP's were found on the same chromosome with risk ratios of 1.3 (or a 30 percent relative risk increase). This hardly represents a major increase in risk, but it is actually higher than the usual outcome of such studies. In a study of type 2 diabetes, for example, seven gene variants have been identified; the one with the strongest effect had a sibling relative risk of only 1.02 and the remaining six had ratios between 1.005 and 1.01.

The various commentators in *The New England Journal of Medicine* do not dispute these results, and one might suppose that they would begin to doubt the assumption of genetic causation—and that Wade's article in *The New York Times* might reflect that doubt. Yet, in actuality, their underlying assumption of genetic determination is unshaken. In reporting on the disappointing results of genome wide association studies (GWAS), Wade writes that the method "has turned out to explain surprisingly little of the genetic links to most diseases." Moreover, he states flatly that "common diseases like cancer and diabetes are caused by a set of several genetic variations in each person."

Like Wade, none of the authors of the articles in *The New England Journal of Medicine* has the slightest doubt that genetic differences really underlie these common diseases. What they disagree about is the best methodology for finding them. The standard GWAS method for screening for SNP's relevant to diseases is to use chips that contain about 500,000 of the approximately 3.5 billion nucleotides in the human genome sequence. These 500,000 nucleotides are those that are known to have common variants (that is, variants in frequencies of 1 percent or greater). The supporters of this technique point out that over 200 spots in the genome have already been shown to be associated with diseases, and as chips are improved many more such nucleotide positions will be identified. For example, 35 spots in the genome have already been associated with Crohn's disease, a form of bowel inflammation. It is these researchers' contention that the large number of variants of small effect that have been discovered reveal the truth about the genetics of disease, namely that "many, rather than few, variant risk alleles are responsible for the majority of the inherited risk of each common disease."[1] They point out that new sites associated with a given disease are constantly being discovered as more samples are probed. They claim that having some information is always useful, and the fact that complete information about the genetic causes of a disease is not available at a given time does not make the method useless. However, their analysis does not make clear what preventive or curative action should be taken if scores or hundreds of individual nucleotide substitutions, each of vanishingly small effect, constitute the collective cause of common diseases.

Some of those who question this method of looking for nucleotide variants point out that the very fact that variants are in medium-to-high frequency is evidence that that they cannot be of major effect on the disease. Indeed, the integrated physiology of organisms makes it likely that all kinds of variation in genes whose developmental and physiological effect is far removed from the primary disease pathway will have minor effects on the disease condition. One suggested alternative method is to carry out intensive complete sequence studies of the entire genome on a small number of affected individuals. The purpose would be to find genetic variants of major effect that are at low frequency in the population and would not be detected by the chip technology. It is to be expected, after all, that genetic variants of large causal force on disease will be in very low frequency since natural selection will have been effective in reducing their frequency in the population. It is the hope of this school that a study of such rare variants of large effect will suggest therapeutic directions that are not apparent when small effect variation is studied.[2]

Both sides in the struggle over how to study the genetics of common disease make deep assumptions which we know not to be true. The first is their reliance upon genetic determinism. This approach finds all diseases that are not the result of infectious agents to be the consequence of faulty genes. The failure to find the gene defects that cause a disease must therefore be the result of faulty technique; nowhere are environmental effects taken into account. At the purely methodological level, the very concept of "relative sibling risk" assumes that similarity between siblings in disease pattern must be a result of genes in common. What about the common environment within families? Is there no evidence that heart disease, cancer, and hypertension leading to stroke are induced by environmental stresses? Is it not possible that genetic effects are minor in comparison?

Secondly, there is the issue of gene-environment interaction. The methods of population sampling for genetic studies take no account of the fact that different genotypes have different sensitivities to environmental effects. Moreover, there is no reason to suppose that genetic and environmental effects are additive or even simply related. Genotype A may be more likely to lead to a disease state than genotype B in one environment, but less likely in a different environment. This is a common observation in experimental outcomes of varying genotypes and environments, yet the population sampling that is carried out for genomic disease studies takes no account of such interactions.

Thirdly, there are complex interactions between physiological and developmental pathways within organisms. Some of these interactions are of a homeostatic nature, so that the effect of large perturbations to one pathway may be dampened by reactions in peripherally related pathways, yet felt in those peripheral reactions. Just because I have a headache doesn't mean that the real problem isn't in my stomach.

The doctrine that we are the product of our DNA leads to the fantasy that by manipulating our DNA we could avoid or cure all disease and even escape eventual death. That is indeed a fantasy. All flesh is mortal.

19

The Broken Clock: Accuracy and Utility of Direct-to-Consumer Tests

Interview with James Evans

James Evans, M.D., Ph.D., *directs the Clinical Cancer Genetics Services at the University of North Carolina, where he is the Bryson Distinguished Professor of Genetics and Medicine. He delivered testimony at the July 2010 congressional hearings focusing on a Government Accountability Office (GAO) investigation that criticized DTC genetic tests as "misleading and of little or no practical use to consumers." This interview originally appeared in* GeneWatch, *volume 23, number 4, August-September 2010.*

GeneWatch: How did you get involved in the GAO report hearings?

Evans: They called me a year or two ago. They told me that they were concerned about some of these direct-to-consumer testing companies' offerings, and I have some concerns as well. I don't think that the sky is falling because of the existence of these things, but I have some concerns. My biggest concerns have to do with the false claims that are made by these companies and the fact that we don't really know how to interpret this kind of information.

So the GAO was looking to design a way to investigate some of these concerns, and I think that the strategy they ultimately took was a good way of investigating and illustrating that we are not ready to interpret this information with any degree of reliability. Then there is a second question: Even if we were able to interpret these results reliably, would it tell us anything of any real significance?

They investigated that first question and eloquently and elegantly demonstrated that we are not ready to interpret much of this data. There's just no way of reconciling claims that it's usable information with the fact that reputable companies conduct analyses on the same DNA and come up with radically different interpretations. There's no way to reconcile that information with claims that it's ready for prime time.

What the GAO did not investigate—and really could not investigate, I think—was the other issue. The question that remains is: Even if there were consistency, and we learned to interpret it, would it provide any utility to patients? And the answer to that, I think, is largely a resounding no.

GeneWatch: Industry critics have claimed that the GAO's methods were not scientific. Do you share any of those concerns?

Evans: It's an interesting criticism, because they actually did an elegant experiment; it was entirely scientific. They had, really, the ultimate control. They took the same sample to different companies and simply presented the results. For the same exact sample, one company says the individual has an increased risk of prostate cancer, one says he has an average risk, and one says he has a low risk. One of them is right, but it's a little like the broken clock that's right twice a day.

We have no idea, as that experiment readily demonstrated, how to interpret some of this genetic information. So the idea that the report was not scientific is, I think, a rather silly accusation. They did an experiment, and the results speak for themselves.

GeneWatch: In your testimony you took issue with the marketing claims made by DTC testing companies. Are there any specific claims that you see commonly made that you find especially egregious?

Evans: Yes. The three big players in this field, the top strata of these companies, are doing a fine job of telling you reliably which nucleotide you have at a given position, but all of these companies make implicit and explicit claims that the information will improve your health. All you have to do is look at their home pages on their websites, look at their advertising, and they all make some claim along the lines of "Understanding your genes will be a roadmap to better health" or "Take control of your future with genetic analysis." They are all making explicit or implicit claims that knowing your genetic information will improve your health, and, frankly, there is no evidence that this is the case.

They spout platitudes that, for example, people will be motivated to lose weight or live a healthier lifestyle. Firstly, there is little evidence that this is the case; secondly, even if this is the case, if somehow genetic information has some magical properties that make it particularly motivating, then we have a bigger problem: it is arithmetically guaranteed that for everyone who has increased risk of a condition, there are an equal number with decreased risk. If this information is actually so motivating, we run the significant risk of altering people's behavior for the worse. And frankly, the magnitudes of the risk shift that they are giving people are practically meaningless. Finding out that you are at a twofold or a fifty percent risk of heart disease over the general population is essentially meaningless since these are common diseases that we remain at significant absolute risk for whether or not we are at some relatively decreased genetic risk.

GeneWatch: So you are concerned not only about the possibility of false reassurance, but even reassurance that isn't necessarily false.

Evans: Exactly. It may not even be false! In other words, I might tell you that you're at a 50 percent risk of heart disease over the general population, but that relative risk is rather meaningless. You are still at a high risk for heart disease simply because it is a very common malady. Millions of people out there who are at relatively "low risk" for heart disease end up dying of it!

GeneWatch: Because of environmental factors?

Evans: Yes, and because it's simply a common disease. But what you're getting at is exactly right: genetics is only one small part of our risk for most of these diseases. Therefore, even if we understood completely the genetic risk for diabetes and heart disease and cancer, we still would be left with a huge amount of uncertainty because the causation of these maladies is multifactorial.

My other gripe is that in the results they send to the consumer, some of these companies mix pure entertainment like, "Do you have thick earwax?"—with a tiny subset of information that is very medically meaningful. A small percentage of the information they give, like BRCA1 and 2 (relating to risk of breast cancer), and LRK 2 status (relating to risk of Parkinson's disease), are very predictive, and in the right circumstances have important medical implications. And yet they're being dumped into this big pot with all kinds of tests that are purely of entertainment value and some tests that are misrepresented as being medically useful when they are not.

I don't have a problem with the public gaining access to information about their own genes. I'm not so paternalistic as to say you can't have the information in your genome. What I do feel strongly about is that people shouldn't be lied to about the significance of that information, and that people should be able to be assured that the claims that are made are accurate and that their privacy will be protected.

GeneWatch: Do you think that the whole concept of the way these tests are marketed clings to the old concept that your genome can tell you everything about yourself?

Evans: I think what's happened is that there's an understandable impatience to apply all of this wonderful, cool genetic technology to medical care. There's a seductive appeal to thinking that because we understand some things about the genome, we now understand a lot about its role in health and disease. The difficult and the sober reality, however, is that we don't have a very good grasp of precisely how to relate your genetic information to your health. That's going to be the work of many years. What we need to do before we just start willy-nilly selling this idea to people is to find evidence of what's real and what's not, what works and what doesn't. All I ask is that we have data that back up the things that we introduce into the realm of patient care.

And these companies want to have it both ways. They implicitly and explicitly make claims about the health value of this information, and yet on every page of their results, they say, "This isn't medical advice." Pick one or the other: it's either medically important or it isn't. And I would say it is not, demonstrably, as the GAO report really pointed out.

GeneWatch: Since there are plenty of traits tested where there is not a lot of utility, do you think there are traits that are useful for consumers to know about?

Evans: You bet. And I think that there may well be reasons that a patient might want to pursue various tests in what we consider a non-traditional environment.

For example, DNA Direct for many years has offered BRCA testing. I've never had any problem with what they do. They don't misrepresent what they offer, and, importantly, they have genetic counselors available to talk to customers about the meaning of their results. I haven't looked at their website in a while, but for many years they offered tests that arguably were medically valid tests, with real medically important results—and they didn't conflate entertainment with medical information.

I think there are ways of doing this that are reasonable and responsible, but I don't think companies like 23andMe are doing so.

GeneWatch: Going back to different companies' radically different interpretations of the same DNA sample in the GAO report, do you know of a reason that the interpretations would be so different?

Evans: I think there are several reasons. Now, the reason that the companies will give you is only part of the truth. They will tell you the problem is that one company did nine variants and another company did fourteen variants. The reason that explanation isn't the whole story, and the reason that settling on everyone testing the same twelve variants isn't a valid response, is the following: We still don't understand how to aggregate these independent risk factors into a net risk score. Genes interact with each other in ways that we are only dimly beginning to glimpse, and genes interact with the environment. It is entirely possible that you might have variant A, B and C and have an increased risk for a condition, but a person who has variant A, B, C and D is actually at decreased risk because of the interaction of variant D with variant B.

Secondly, genes interact with the environment. It could well be that an individual who has variant A, B and C, who should be at increased risk, is actually at average risk in the right environment—because those genes and our physiology interact with our environment.

So the reason that they come up with different results is that we don't know how to come up with a single net risk estimator from these variants. We just simply don't know enough at this point.

One of the things that was left hanging after the testimony was an idea that the industry would love to have you believe: "If we could all just get together and agree on standards, it would be fine." No, what would happen is you would all agree on standards and we would see the same risk prediction from all the companies . . . but that doesn't mean it would be correct!

GeneWatch: How much of this do you think could be solved just by having a genetic counselor involved throughout the process?

Evans: That's part of the solution, that the individual who avails himself of these tests would have a first responder who could give him information. That probably isn't all of it, though. As we learned from the GAO report, people can get told all kinds of different things, especially when there are conflicts of interest that are swaying people to give certain kinds of advice.

If 23andMe would just have a line that people could call, that would go a long way toward alleviating some of my concerns. But it doesn't relieve all of them, because I still think the claims that are being made in their advertising are simply wrong. And that seems to me something that doesn't necessarily require further regulation; it requires the FTC to enforce truth in advertising.

Can I say one other thing that I didn't get a chance to say during my congressional testimony? One of the things you will hear from these companies at first sounds quite convincing. Cholesterol and blood pressure confer only subtle relative risks for heart disease, so they'll say, "High cholesterol only confers a 1.4 relative risk on somebody for a heart attack and this is similar to the degree of risk conferred by genetic variants." And that is true. But what they fail to discuss is that your doctor doesn't check your cholesterol because he is primarily seeking predictive information. Your doctor checks your cholesterol because he can change your cholesterol. He isn't doing it just so he can say, "Oh, you're at increased risk for a heart attack. Have a good day." He's checking it because he can do something about it. And that puts it in an entirely different category than these direct-to-consumer genetic tests.

20

Consumer Genomics and the Empowered Patient

By Paul Billings

Paul Billings, M.D., Ph.D., *is Vice Chair of the Board of Directors of the Council for Responsible Genetics and Chief Medical Officer of Life Technologies, Corp. This article originally appeared in* GeneWatch, *volume 23, number 4, August-September 2010.*

My medical school classmates, Harvard physicians Jerome Groopman and Pamela Hartzband (who are married to each other), have railed recently against too much "cookbook" medicine.[1] They have cogently argued that care of patients is an art; it may not be adequately represented in computer algorithms, treatment guidelines or glib quality assessment reports. They have supported the import that the professionalism, experience and expertise of physicians can bring to interactions with patients. This art in the practice of medicine yields many—and, Groopman and Hartzband would argue, necessary—variations.

But sadly this necessary art can account for a multitude of sins. From frank medical paternalism, like withholding key facts from dying patients, to unacceptable variance in interpretations of pathology or radiology data, to even the provision of unnecessary or harmful treatments, variability or artfulness in medical practice can be wasteful and deadly. How do we improve care and get the most out of the trust sick people place in their doctors?

One way, it seems to me, is to make patients and healthy citizens more knowing and independent actors in the health care system. Professional norms demanding real-time honesty with patients, medical malpractice suits, and the Internet have all improved the power of patients in the doctor/patient relationship—and, I would posit, the quality of medical care provided. Measuring this change using outcomes like change in clinical outcome status could be very difficult at present, even if I am right!

If one's health and the management of illnesses is in fact one of the more basic personal responsibilities, then allowing people to pursue issues of prevention or disease treatment in whatever way they see fit could be reasonable. In such a model, government's role is to ensure honesty in the representation and claims of safety or effectiveness, while physicians act as solicited providers of expert opinion. The empowered and knowledge-armed citizen is a key agent for the improvement of all aspects of health care and illness prevention.

That is why I founded in 1999 one of the first direct-to-consumer (DTC) genetic information companies (GeneSage Inc.) and why I generally support efforts to expand this knowledge and service channel now. I fundamentally believe that we all have the right to know and test things about our bodies, unhindered by physicians gate-keeping those activities. Naturally, if someone else is paying for those services (a government, employer or private insurer, for instance), they may have some input in to or influence on what they pay for; but the underlying "right to health" (as Franklin D. Roosevelt put it) lies with each of our fellow citizens and should allow for a wide range of affordable activities.

Because there is so little known about exactly how most of the human genome impacts traits and participates in disease, many of the early entrants into DTC genomics have made overly optimistic, exaggerated or false claims. This is, of course, true as well for all the physicians who have proffered misinformed advice for decades in the field. Some of the practices that are offered (DTC or otherwise) or that were noted in the recent Government Accounting Office investigation [July 22, 2010] are bogus and should be curtailed. For instance, except in certain rare circumstances (for example, people with PKU should avoid phenylalanine), there is no genomic test for the right diet or exercise program that will yield better personal health. We may never see the day when attaching our DNA profiles to the grocery list as we head to the supermarket is a good idea!

But if someone wants to obtain a test that *may* identify a risk for disease or a susceptibility to a drug reaction, they should be able to do so without a physician's intercession. Those ordering such a test in person or from a website ought to have access to reliable and accurate data about what information the test may convey. They need to have a reasonable expectation of the quality of the lab that will conduct the test and access to easily understood lab performance data. Governments, professional societies, and good market practices should ensure those conditions are present. In fact, several states now allow consumers to order medical tests without a doctor's prescription.

23andMe and Navigenics are both companies I know and have advised. The tests they offer are pretty much "state of the art" in terms of assessing the genes of the human genome. Soon these companies or others will offer affordable sequencing assays of all the DNA in the human genome. It will be possible for many people to know exactly what nucleic acid sequences reside in the nuclei of most of their cells. Of course, the meaning of the huge preponderance of this data for an individual's health or disease development will take much longer to establish and will ultimately be individualized, primarily by each of us. But as that knowledge becomes available—unevenly, influenced in many ways, and with continued debate—some people will want that information, and so-called expert debate should inform but not stop them.

In general, the prominent commercial entities in the DTC genomics marketplace provide understandable information, are dedicated to support and participate in important research to make the field better, and engage in good clinical practices. Others do not and should be identified and shunned by consumers as well as scrutinized by legally designated regulators from HHS and FTC (but not congressional witch hunts). In my view, more knowledge and more highly empowered, well-informed independently acting consumers are needed to push improvements in genomic medicine and in all aspects of health care. DTC genomics is part of that. How medical care is ultimately personalized should be a topic that individuals acting as citizens, consumers, patients and in other roles actively control.

Genetic Counselors: Don't Get Tested without One

Interview with Elizabeth Kearney

Elizabeth Kearney *is a former president of the National Society of Genetic Counselors. This interview originally appeared in* GeneWatch, *volume 23, number 4, August-September 2010.*

GeneWatch: Is there a brief definition that you give people when they ask "what is a genetic counselor?"

Kearney: If I'm meeting somebody and they ask me what genetic counselors do, I tell them that genetic counselors work with families or individuals who are either at risk for or have a genetic condition. We take their family and medical histories, and we help assess what their chances are for that condition; we go over whether there are tests available and what is good and bad about that testing; and if there is a diagnosis, we explain what it means and connect them with the support that they need, whether it be medical professionals or a support group.

GeneWatch: Are there certain scenarios when you would tell someone not to get a test?

Kearney: If someone doesn't have a history that would predispose them to a condition, we really want to understand their reasoning for getting a test for it. Part of it is a matter of spending our health care dollars wisely. Obviously if someone wants to pay for it out of pocket, understanding that this information may not be as impactful as they were hoping, that's one thing. The more typical scenario genetic counselors deal with is when we are billing insurance, so we have to be careful to consider whether someone really has a risk that justifies using health care dollars to assess it.

The second part of it is the psychosocial element. If somebody really wants to have testing done, we want to ask them some questions about why, and why now? I had a patient who wanted to be tested for Huntington's disease six months before she was getting married. It was totally medically appropriate—her father had been affected—but she was six months from getting married, so I asked her: Why now? Would this change anything for you? And after that discussion, she thought about it and decided that now wasn't the time.

GeneWatch: A good deal of attention has been drawn to direct-to-consumer genetic testing recently on Capitol Hill, from the GAO report to FDA and congressional hearings. Do you think the attention is steering the conversation in the right direction?

Kearney: I think the positive thing that's coming out of all of this is that it's engaging people. I believe that more consumers and more physicians are aware of the availability of genetic testing, and I hope that they are learning about some of the possible benefits and drawbacks of obtaining genetic information. So I see that as a positive outcome. Genetic counselors work primarily with patients and obviously we care a lot about people having access to genetic information, so a real benefit that has come out of this for patients is that they probably are more aware and might be more likely to inquire about genetic testing to help them.

From the NSGC's perspective, I think it is most important for people to know that they have the opportunity to meet with a genetic counselor before they have testing, to determine if testing is right for them, to find an appropriate test, and to have support interpreting the results if they decide to have testing. So the benefit of all this is that it started a conversation, and I see that as fundamentally a good thing.

GeneWatch: Do you find that many customers of direct-to-consumer genetic tests are coming to genetic counselors first?

Kearney: There has definitely been an increase in recent inquiries, but I don't know whether it has been more frequently before or after the test. I certainly know of situations where people have contacted genetic counselors after the fact.

One example is a woman who'd had carrier testing for a number of genetic disorders and was found to be a carrier of Alpha-1 antitrypsin deficiency, a condition that results in early lung problems and basically causes emphysema even if the person is not a smoker. You have to have two copies of the gene in order to be affected, and this individual had only one copy. She had not had any genetic counseling beforehand, and she called the genetic counselor in a panic and thought that she was at risk for the condition.

GeneWatch: So the report wasn't clear about the difference between being a carrier and actually being at risk for the condition?

Kearney: Exactly, so she thought that being a carrier would mean she could be affected by those symptoms. It's an example of the value of meeting with a genetic counselor beforehand. A genetic counselor will ask why you want to have the test, go over which tests are right for you, and explain what you can learn and what you won't learn from it. If you still want to go ahead, that's fine, and you already have a relationship with that counselor and can call them up right away and go through the results and not have to go through that period of panic.

GeneWatch: Would you be concerned about a conflict of interest if customers go to a counselor on the DTC testing company's payroll instead of an independent genetic counselor?

Kearney: I think it's obvious there's some inherent conflict of interest. That doesn't mean that someone who is a board certified genetic counselor who works for a company cannot provide good care to a patient, but it's important to look at the incentives and how those counselors are evaluated . . . but I think it's fair to say there is some potential inherent conflict of interest, and patients could avoid all of that if they contact a genetic counselor who is not affiliated with the company.

GeneWatch: Is there a best practice scenario you can point to where genetic counselors are working together with test providers to reach the best outcomes for patients? Is there a model already in place?

Kearney: If you look at more classic genetic testing—testing for single gene disorders like cystic fibrosis, sickle cell, and Tay-Sachs disease—a lot of laboratories work with the requirement of having a provider involved and they have close relationships with genetic centers. For example, academically based labs will often have a genetic counselor based in the lab, primarily to get in touch with a provider if something doesn't look right. For instance, there may be a question as to whether the patient is really ordering the right test, and those genetic counselors who work in the lab might get back in touch with whoever ordered the test and advise them about whether this is the test that they really want. In this model the genetic counselor is in a sense the gatekeeper for t he appropriateness of testing.

GeneWatch: Is there any particular trait or set of traits being routinely oversold in terms of utility for patients? Put differently, is there a test or area of testing where you think a genetic counselor is most needed?

Kearney: One of the areas of concern is when someone is ordering the wrong test. One problem is simply that ordering the wrong test is a waste of money; but the more significant concern might be around not integrating information from genetic testing with medical and family history.

I would use diabetes as an example. Suppose somebody has a family history of diabetes and they are wondering about their own risk of developing diabetes, and they have a test result that shows they have decreased risk over the general population—but they have a family history of diabetes, and maybe it's even a woman who has had gestational diabetes during a pregnancy. A genetic counselor would look at all of that and integrate it and tell that patient that while the test result was reassuring, most likely the genetic factors responsible for the diabetes in your family are not those that were tested in this particular test. That's an example of when a test is misinterpreted as being sufficient information on its own, when you really want to integrate it with family and medical history.

GeneWatch: Has the profession of genetic counseling changed as more direct-to-consumer genetic tests have been introduced?

Kearney: Actually, I don't think that the practice of genetic counseling has changed that dramatically. The model for how we care for a patient is the same whether we're testing for single gene disorders or whether a patient is coming to a genetic counselor with a report from a direct-to-consumer lab. I also think that it's a pretty small percentage of the population that's pursuing direct-to-consumer testing without the provider involved. So I really don't think that, as of yet, it has influenced the practice of genetic counseling very much.

What Genomic Research Will Give Us

Interview with Eric Green

Dr. Eric Green *is the director of the National Institutes of Health's National Human Genome Research Institute. This interview originally appeared in* GeneWatch, *volume 25, number 1-2, January-February 2012.*

The pace of progress

One thing I've heard said repeatedly about genomics in the twenty-two years I've been involved with it is that we tend to overpredict where we'll get in the short term, say three to five years, and we tend to underpredict where we'll get in longer intervals, like ten years. I think that phenomenon has been described by someone else in another field, but it really applies to genomics. It seems that over and over again, we are way overly optimistic about what's going to happen in three to five years, and yet every time we look back at what we've done in the last ten years, we're shocked by how far we've come. I think that's absolutely the case now, especially in terms of data generation and DNA sequencing technologies. There's no evidence I can see that it's going to slow down; I don't think genomics is going to hit the wall. I think it has as much momentum now as it did a decade ago, and I would contend that ten and twenty years from now, we will be even more surprised than we thought we would be. So I guess one of my overarching comments is that I see no reason to think that the pace at which we are developing new technologies, understanding our genome, and figuring out how it's going to be medically relevant, will slow down.

Genome sequencing and analysis

One thing I would predict is that the technologies for generating data will create a situation where data generation is trivial and analyzing the data becomes the overwhelming challenge. I think genomics is going to become more and more an information science and less a technology science, and I think the great challenges are going to be in how we analyze and interpret data in creative and powerful ways; and every time we need to generate more data, that will be the least expensive part of the equation.

Once upon a time, the Human Genome Project was all about data generation; now we already find ourselves in a situation where we have data abundance but an analysis restriction. That disparity between the amount of effort to generate data and the amount of effort to analyze it will only grow with time. I can imagine twenty years from now it might cost $500 or $100 to generate a genome sequence, but to fully interpret it might cost more than the sequencing costs.

Discovering how DNA works

The second prediction I have—a very bold prediction—is that twenty years from now we will still be discovering basic ways that DNA confers function. I do not believe twenty years from now we will have figured out every last way that DNA encodes biological information; I still think there are major surprises out there to be found. I think there are major mechanisms still to be discovered, and with that will be a continued need for strategic interpretation.

I think there's a lot of biological information encoded in DNA that we will still be discovering. I'm even saying that we're going to be discovering basic mechanisms in ten or twenty years. Even if we say that we think we know all the promoters in the genome, I'm sure twenty years from now, we'll still be discovering new promoters acting in ways that we didn't know about.

I always say that the human genome sequence is like a great novel. We'll be spending dozens and dozens, maybe hundreds of years interpreting and re-interpreting it, just like a great historic novel. It's naïve to think that even in ten years or twenty years we'll have a complete catalog of every functional sequence and any deep understanding of how it works.

A revolution in evolutionary biology

My third prediction under the general research area is that we will see a completely new way of studying evolutionary biology that will be fully computational. I think twenty years from now, probably before then, we will have genome sequences of thousands and thousands of animal species. A 10th grade biology student's laboratory exercise will not be confined to dissecting frogs or looking at a fossil; he'll be sitting at a computer and will have tools in front of him to look at genome sequences of tens of thousands of different vertebrates, and the laboratory exercise will be to figure out how DNA changes have led to biological innovation.

There will be almost an entirely new field, a subcomponent of evolutionary biology, that will be dominated by computational analyses. Yes, we'll still be digging up fossils, we'll still be doing imaging and biometrics, but we will also have in front of us a database of tens of thousands of genome sequences from all different kinds of critters that walk and swim and fly on this Earth. Just imagine the experiment where you can look at a given stretch of a genome and trace the evolutionary history of every little piece through tens of thousands of vertebrate genomes. It's incredible, but it's absolutely do-able twenty years from now.

Genomics in medicine

I believe that certainly twenty years from now the use of genomic information about individual patients will be a standard of care. I think when it comes to cancer, it will be pervasive; I think genomic-based analyses of cancers will become a standard of care for many different kinds of cancer probably well before twenty years. For pharmacogenomics, it will be a standard of care for dozens, if not hundreds, of different conditions for which we will use genomic information on patients as a guide for selecting and dosing medications. And I'm very confident that we will use genome sequencing as standard of care for diagnosing rare single-gene genetic diseases.

Hand-in-hand with that, I can believe that the routine will be that you'll have a genetic sequence of every patient. Now, we can start wondering what it will look like, whether that genome sequence is obtained as part of newborn screening shortly after birth . . . I realize there are still many complicated issues, but I think one can certainly envision that whole-genome sequences might be generated as part of newborn screening.

I can't believe that electronic health records won't be standard of care in hopefully most places in the world; and I can't believe that genomic information wouldn't just flow into those electronic records. But, again, that is another area where there are lots of complexities and questions, and we're doing research in that area to clarify things.

Where I'm less certain is what the role of genomic information will be for truly understanding the genetic basis of common complex diseases in terms of individual patients. I don't know whether we'll get to the point in twenty years where we can look at a hundred different loci and say, "You are at a 42 percent greater likelihood of getting coronary artery disease, and this is what you should do." I think the jury is still out on what that's going to look like, and I wouldn't want to overstate that part. I think that's going to be a question mark for now.

Understanding gene-environment interactions

I believe we will gain a much more sophisticated view and understanding of gene-environment interactions. On the genomics side of that equation, the technology surge has really happened in the last decade and will probably continue over the next decade; but I think that we're getting to the point where it's going to become trivial to gather data about the genome. I think the surge to anticipate over the next ten or twenty years will be technologies for doing environmental monitoring. I think that one of the reasons we're ignorant in understanding the environmental basis of disease is that we just don't have technologies for doing fine-scale measurements of environmental exposures. I'm not sure my field is going to have anything to do with it—I don't think it's genomics, I think it's environmental science—but I think technologies are coming; and with that would come much more powerful studies to capture data on the environmental side that's just as powerful as the data we're getting on the genomics side.

Ethical, legal and social issues

Finally, I firmly believe that the societal issues that we are already starting to grapple with around genomics—the ethical, legal, and social issues—will continue to require significant attention, significant research, and significant debate. I don't think these ethical issues are going to go away. With technological advances and increasing knowledge will come a continued need to wrestle with very hard questions. I don't think the questions are going to become simpler; if anything, I think they will become more complex. We shouldn't fool ourselves into thinking that we'll eventually figure all this stuff out and the ethical dilemmas will go away. I just don't think that's true. It's not that I'm pessimistic; I think we can deal with them—we just can't ignore them.

Toxicology in the Genome

BY SHELDON KRIMSKY

Sheldon Krimsky, Ph.D., *is Chair of the Board of Directors and a founder of the Council for Responsible Genetics. He is the Lenore Stern Professor of Humanities and Social Sciences at Tufts University.* *This article originally appeared in* GeneWatch, *volume 25, number 1-2, January-February 2012.*

There was a time shortly before the human genome was sequenced that many believed genetic science was on the cusp of a medical revolution. Our sequenced DNA was thought to hold the key to understanding the onset of disease. Why are some children afflicted with autism? Why do some adults stop producing enough insulin? Why do some otherwise healthy individuals who reach their senior years lose mental functions and memory? Ten years after the human genome was sequenced, biomedical scientists have become ever more cautious in their optimism about how DNA sequencing will change medicine by revealing the existence of a disease years before its onset or by introducing new therapies with the tools of molecular medicine and stem cells.

The terms "gene-environment interaction" and "epigenetics" are now recognized as the clue to many disease conditions. The switches that turn genes on and off may be more important in understanding clinical pathology than mutations in coding sequences of DNA. These switches, which may stop or modify gene expression, are in the form of protein complexes that overlay the DNA code, such as histones or methyl groups, or the RNA interference molecules that reside in the genome.

On the website of the National Institute of General Medical Sciences we find the following statement: "A good part of who we are is 'written in our genes,' inherited from Mom and Dad. Many traits, like red or brown hair, body shape and even some personality quirks, are passed on from parent to offspring. But genes are not the whole story. Where we live, how much we exercise, what we eat: These and many other environmental factors can all affect how our genes get expressed."

Despite the growing awareness that environmental factors interact with and affect the human genome, most of the research remains focused on the mechanisms operating at the molecular level. Thus, there is much discussion about sequencing the epigenome to gain an understanding of the genetic switches or to probe deeply into non-coding DNA for discovery of RNA sequences that interfere or modulate gene expression.

Meanwhile, we know that around 100,000 people die from adverse drug reactions. Some people are highly sensitive to chemicals in perfumes or outgassing from carpets or plastic. Have you ever passed a nail salon gasping from the chemicals seeping through the open door, while a dozen women patrons and their handlers are breathing in those same chemicals without a trace of discomfort? The detoxification mechanisms of people vary widely. Without a sufficient quantity of enzyme production, our bodies cannot break down certain chemicals fast enough before experiencing harm. If we expect to make any major inroads into preventing the many environmentally-induced diseases, each of which may affect a small percentage of the population, then we must use the human genome and the epigenome to acquire an understanding of why some people are more adversely affected by environmental agents. What we need is a massive effort to unravel the "gene-environment" interaction in disease causation. We have over 100,000 chemicals in current industrial use. Many of these chemicals were introduced into commerce without much toxicological evaluation. It takes between twenty-five and fifty years to regulate or ban a chemical that has been shown to be harmful to humans. The United States has only banned about a half-dozen chemicals over half a century. In part this is because the regulatory system is geared toward industrial interests. The government requires very minimal safety studies to permit a chemical into industrial use but demands an extraordinary body of replicable scientific studies and cost-benefit analyses before a chemical is removed from the marketplace.

The one area where there have been major contributions in deciphering the gene-environment interaction is in the study of the genetic effects of ionizing radiation. Perhaps radiation effects on health is the low-hanging fruit because of the mutations the radiation produces, although the effects of low-level radiation remain highly controversial.

How can we learn what chemicals are adversely affecting the healthy human genome and what chemicals have differential effects on different genomes? How can we detect the detoxification potential of each individual toward a chemical? The differences in people's ability to detoxify a chemical may be the result of shorter genes coding for the relevant detoxifying enzymes, or the enzyme-producing gene is switched off.

Most of the commercial interest in the sequenced human genome has been focused on risk factors for certain diseases that are read from the individual's DNA. There is nothing in direct-to-consumer testing kits that reveals the cause of any disease other than what is encrypted in the code itself. And there are only a small number of illnesses where there is a one-to-one correlation between having a particular form of a gene and a disease, such as cystic fibrosis, sickle cell anemia and Canavan's disease.

Of course, a massive effort to determine how chemicals interact with the human genome may have unintended consequences. Instead of banning the chemical, it may result in a genetic classification of people—those hypersensitive to chemical X, those with peanut allergies, etc.—putting the onus on them about how to navigate through life. Many people have figured out they are hypersensitive to new carpets, latex or perfumes and learn how to keep away. But if we had a mechanism that showed us these people were not psychologically challenged but rather had a normal genome with less capacity to metabolize chemical toxins, we would have a new regulatory mechanism for removing the chemicals from the environment.

Biomedical scientists have been able to titrate chemotherapy agents to individuals based on genomic information. Gene expression patterns associated with sensitivity and/or resistance to chemotherapy may be used to help provide more effective treatment.

Scientists have used genetic testing to identify patients at high risk of bleeding from the drug warfarin. Two genes account for most of the risk. Recently, genetic variants in the gene encoding Cytochrome P450 enzyme CYP2C9, which metabolizes warfarin, and the Vitamin K epoxide reductase gene (VKORC1), has enabled more accurate dosing that takes account of the genome of an individual. Genotyping variants in genes encoding Cytochrome P450 enzymes (CYP2D6, CYP2C19, and CYP2C9), which metabolize antipsychotic medications, have been used to improve drug response and reduce side effects. Pain killers like codeine affect people differently. Tests for certain enzymes (P450-2D6) can determine whether someone will be an ultra-rapid metabolizer of codeine, which could induce life-threatening toxicity.

If we can understand through certain enzyme pathways that individuals react differently to drugs and that some of us cannot efficiently metabolize certain chemicals, why couldn't we do the same for industrial toxins within the next twenty years? Once we learn that many people cannot de-toxify a chemical that bioaccumulates in their bodies, it provides new grounds for finding a substitute for that chemical rather than waiting a quarter-century to complete hundreds of studies with mixed results.

Personalized Genetic Medicine: Present Reality, Future Prospects

BY DONNA DICKENSON

Donna Dickenson, M.Sc., Ph.D., *is a fellow of the Ethox Centre in Oxford, Emeritus Professor of Medical Ethics and Humanities at the University of London, and honorary senior research fellow at the Centre for Ethics in Medicine at the University of Bristol.*

We are in a new era of the life sciences, but in no area of research is the promise greater than in personalized medicine.— Barack Obama, as a Senator introducing the bill that became the Genomics and Personalized Medicine Act 2007.

The soaring promises made by personalized genetic medicine advocates are probably loftier than those in any other medical or scientific realm today. Francis Collins, former co-director of the Human Genome Project, has written that: "We are on the leading edge of a true revolution in medicine, one that promises to transform the traditional 'one size fits all' approach into a much more powerful strategy that considers each individual as unique and as having special characteristics that should guide an approach to staying healthy. . . You have to be ready to embrace this new world."[1] Certainly, vast sums are pouring into personalized medicine; plans to spend $416 million on a four-year plan were announced in December 2011 by the National Institutes of Health, and private sector interest is also intense.

But does the science bear out the claim that there's a genuine paradigm shift toward personalized genetic medicine? It has been said that ten years after the completion of the Human Genome Project, geneticists are almost back to square one in knowing where to look for the roots of common disease.[2] As of March 2012, current genetic tests and molecular diagnostics have been applied to about 2 percent of the US population.[3] A Harris poll of 2,760 patients and physicians in January and February 2012 indicated that doctors had recommended personal genetic tests for only 4 percent of their patients; hardly the stuff of a paradigm shift—at least not yet.

It has been asserted that a baby could have her genome fully sequenced at birth, revealing her susceptibility to particular diseases. She could then enjoy the benefits of made-to-order diagnostic tools and drugs throughout her lifetime. That really is the "Holy Grail" of personalized genetic medicine, but it makes huge and currently unfounded assumptions about how much we are actually able to predict. Most major diseases are caused by the interplay of many genes rather than one, and they arise from both environmental and genetic causes.

The most recent policy update from the American Society of Clinical Oncology accepts that genetic testing for personal cancer susceptibility is now a routine part of clinical care, especially for high-penetrance mutations like the alleles (variants) of the BRCA1 and BRCA2 genes implicated in some breast and ovarian cancers. However, the Society also notes that such cancers are comparatively uncommon. The Society believes that there is little clinical value in testing for the hundred or more relatively common single nucleotide polymorphisms (SNPs) linked to parts of the genome that are associated with cancer in a yet undetermined way, because the risk from each individual SNP variation is generally too small to serve as the basis for clinical decision-making. By contrast, a family history of breast and ovarian cancer could alert a clinician to order a direct and specific test for the BRCA1 and BRCA2 genes implicated in some such tumors. But BRCA 1 and 2 testing may be restricted by monopoly patent protection on those genes, leading to prices of up to $3500 for the diagnostic tests. Although these patents were challenged in a recent court case,[4] they still stand at present.

In *pharmacogenetics* or *pharmacogenomics,* clinical genetic typing is used to determine a patient's probable response to drugs such as cancer treatments and to tailor the pharmaceutical regime personally. It might be possible, for example, to identify patients who are genetically programmed to respond more quickly to chemotherapy and to give them lighter dosages, so as to avoid the worst side-effects. Pharmacogenetics is not confined to oncology, but there the goal is also to adjust treatment to the sequenced genome of the cancer, which differs from the patient's normal cells. This double approach is crucial because cancer is so heterogeneous, even in patients with the same diagnosis. After sequencing the entire genomes of fifty patients' breast cancers, researchers found that only ten percent of the tumors had more than three mutations in common.[5]

Outside oncology, there has also been progress in pharmacogenetics. For example, the drug warfarin is an oral anticoagulant commonly used to prevent or manage venous thrombosis. It is sometimes difficult to determine the correct dosage for an individual patient; thinning the blood excessively can be an unwanted side-effect, carrying its own risks. But now warfarin dosage can be tailored to identify particular patients at increased risk of bleeding, by sequencing two genes that account for most of the variation in how people react to the drug. In public health, a major study—the five-year "Human Heredity and Health in Africa" (H3) study, jointly funded by the National Institutes of Health and the Wellcome Trust—aims to apply genome scanning and sequencing techniques to major communicable diseases like HIV/AIDS, tuberculosis and malaria, as well as to non-communicable conditions like cancer, stroke, heart disease and diabetes. The hope is that the project will finally bring some of the benefits of advanced genetics research to the world's poorest continent.

These and other developments give reason to be hopeful about pharmacogenetics, certainly more so than about direct-to-consumer retail genetic testing. However, a genome-wide analysis of biopsies done on four kidney cancer patients showed that a single tumor can have many different genetic mutations at various locations. Two-thirds of the genetic faults identified were not repeated in the same tumor, let alone in any other metastasized tumors in the body.[6] That is quite discouraging, because if a pharmacogenetic drug targets one mutation in the tumor, it may not work on other mutations.

The former head of the American Society of Clinical Oncology, George Sledge, has gone so far as to declare that the only cancers that have been outwitted so far by pharmacogenetics are the "stupid" ones—the minority of cancers caused by mutations in only one or two genes. "One danger of stupid cancer is that it makes us feel smarter than we are," Sledge concedes ruefully.[7] That overconfidence is obvious in many of the more exaggerated paeans to personalized medicine.

Trials in cancer pharmacogenetics additionally have to contend with an inherent paradox of personalization: the more unique or specific the proposed drug is to particular genetic sub-groups of patients, the harder it becomes to find enough patients for statistically significant results. This profound problem makes some commentators skeptical that individualized drug therapy will be possible for most conditions any time in the foreseeable future.

The continuous discoveries of new surprises about the genome call into question the claim that personalized medicine is almost here, or that individualized drug therapy will soon be a reality. In fact, it probably never will be, or at least not by DNA testing alone, because most genotype-phenotype associated studies are hampered by limited size and therefore decrease in statistical power.[8]

If the scientific evidence alone fails to bear out the bigger claims for personalized medicine, why is there such great interest? We need to look to social and economic factors as well as scientific ones. For a pharmaceutical industry facing the expiry of patent protection on many of its best-selling drugs, new markets have to be found. By breaking an existing medication down into different "size ranges," and by persuading customers that they cannot simply rely on a "one-size-fits-all product," pharmaceutical companies can create new niche markets.

It would be even more advantageous for the pharmaceutical industry if the individual patient could be persuaded to pay for genetic typing out of her own pocket, so that she would then know which of the niche pharmaceuticals is her "size." Now that the $1,000 whole-genome test has become a reality, retail genetics may well extend its reach from subsets of SNPs to offering whole-genome mapping. Customers could thus have all their personalized genetic information ready for access when needed, so that prescribing on a pharmacogenetic model could become much more commonplace. In that event, diagnostic costs would be transferred from the public health system or insurers to the private individual, while some individuals might find themselves excluded from coverage on the basis of their genetic profile.

Patients' enthusiasm for pharmacogenetics would take quite a dent if they saw it as a rationale for denying them therapy, but in an era of cost-cutting, that could well happen. Cancers driven by a number of different genetic pathways may require different regimes of drug combinations for different patients. With drugs required by smaller-size patient groups, it may not be economical for drug companies to produce every drug required for the regimen of any particular patient. From the drug companies' point of view, big blockbuster drugs have traditionally been the money-spinners. Unless a stratified patient group is large (or wealthy) enough to constitute a niche market, it would not necessarily be in drug companies' interests to tailor medicines too narrowly. This is the largely ignored economic reality of personalized genetic medicine: the more personalized it becomes, the narrower the range of customers grows and therefore the less incentive there is for firms to produce the drugs.

Alternatively, pharmaceutical firms might pursue a strategy of high price increases for personalized cancer drugs. The pricing of a group of oral oncolytic (anti-tumor) drugs, including Gleevec, has gone up by over 76 percent since 2006.[9] The drug Xalkori, which was developed with a small group of patients whose lung cancers had a particular mutation, is being made available at a price of $9,600 per month.[10] This high price is driven by the small size of the potential market; the total target population for the drug is expected to be fewer than 10,000 patients.[11]

Against the trend of genetic personalized medicine, some of the most promising research in cancer prevention actually comes not from the complexities and costs of individually tailored drugs but from simple, cheap and comparatively safe "one size fits all" drugs, even for genetically caused conditions. In October 2011, a UK team found that a daily 600 mg dose of aspirin resulted in a 63 percent reduction in the number of colorectal cancers in patients with a hereditary disease called Lynch syndrome. This genetic condition increases the risk of colorectal and uterine cancer in about 2 percent to 7 percent of the population, by affecting genes responsible for detecting and repairing DNA damage.[12] Every one of the 861 people with this syndrome in the trial got the same dosage of the same simple drug against the same threat. It worked.

Déjà Vu All Over Again? Incidental Findings in Whole Genome Sequencing

BY MARC WILLIAMS

Marc Williams, M.D., *is Director of Geisinger Health System's Genomic Medicine Institute.*

In response to the precipitous drop in the costs of whole genome and exome sequencing, groups are now implementing this technology in a variety of research and clinical settings. Whole genome and exome sequencing (WGS) has remarkable power to detect causal genetic disease variants in populations of patients who had eluded diagnosis despite extensive diagnostic odysseys, particularly when one considers the challenges of interpreting the vast amounts of data coming off of the next generation sequencing machines.

A consequence of doing WGS is that one ascertains all of the variants from the portion of the genome that is being analyzed. Some of these variants will occur in genes. The vast majority of these variants will have no effect, but a small subset will be deleterious. In some cases the mutation will affect one allele in a gene associated with an autosomal recessive condition (CFTR and cystic fibrosis), in which case the individual would be a carrier but would not have any manifestations of the disease. More infrequently a deleterious mutation in a gene could predispose the individual to disease. Examples could include autosomal dominant cancer predisposition genes (*e.g.,* BRCA1/2); autosomal recessive genes, where an individual is found to have a mutation in each allele (HFE and hemochromatosis); and X-linked genes, where male or female hemizygous carriers could be identified (PRPS1 and Charcot-Marie-Tooth X-linked recessive 5 or GLA and Fabry cardiomyopathy). In some cases these variants will be identified as being causally related to the indication for performing WGS, but there is also the possibility that these could be found in situations where WGS was done for other reasons. In the latter case the findings would be considered incidental or secondary. The recognition that clinically significant variants would be discovered in the course of WGS has generated much discussion from a variety of groups involved in the performance and interpretation of WGS. For this article I will focus on one aspect of the debate: The contention that incidental information in the context of WGS represents a novel problem in medicine and, as such, consent to testing should allow the patient and/or provider the choice to opt out of receiving clinically important variant information that is unrelated to the indication for testing in order to protect patients from unneeded or unwanted stress.

Is this new? Is this different?

Based on thirty years of medical practice I can say unequivocally no—incidental findings are an inherent part of medical practice. Most agree that whether uncovered as part of a history and physical examination or as a consequence of a diagnostic test, this is a part of medicine. Perhaps the more relevant question is: Are incidental findings from WGS different from those found in traditional medical practice? Here there is much more disagreement. A non-exhaustive list of potential differences include: impact on other family members; identification of risk for adult-onset conditions in children; lack of information about the risk of specific variants (or findings for non-genetic situations) in particular information about the penetrance and variability of expression of the identified variant; and cost to the system of confirming incidental findings. While these all merit consideration, I've experienced each of these issues in my general pediatric practice.

Example 1

In a well-child examination, I identified a heart murmur that I thought was not an innocent childhood murmur. Echocardiography identified a hypertrophic cardiomyopathy that ultimately required heart transplantation for the infant. Most of the cardiomyopathies are genetic with the majority being autosomal-dominant (**impact on other family members**). Recommendation was made for echoes to be done on the parents and siblings of the patient. While these turned out to be normal, the **variability of expression and penetrance** for these conditions in families presented uncertainty in counseling. Should serial echoes be done over time and, if so, in both parents and children? At what age are we certain that a cardiomyopathy would be present on an echo? Clearly these issues affect not only the clinical issues but also impact the **cost of care** to the system related to the incidental finding. This case occurred long before genetic testing, but application of genetic testing in the form of a cardiomyopathy gene panel could have helped with some of the questions. If a causal variant was identified in the child, is it present in one of the parents? If not, this is likely a new mutation and other family members would not be at increased risk. If so, the questions of variability of expression and penetrance would still need to be addressed, but at least the number of family members that would be offered screening is restricted to those who carry the variant, lowering the impact on the cost of care.

Example 2

I performed a well-child examination on a three-year-old asymptomatic child of missionaries who were temporarily back in the US. They informed me that their mission was in an area where tuberculosis is endemic, so their child had received a BCG vaccine. Because the vaccine rendered the recommended screening test (intradermal PPD) uninterpretable, I performed a screening chest X-ray to check for occult disease, as the child would be attending day care while in the US. The X-ray revealed a 10 cm paraspinous mass which, at surgery, turned out to be a cystic neuroblastoma. In reviewing the X-ray finding, none of the consulting physicians could definitively define the **risk of the specific finding** and thus the need for surgery. Cystic neuroblastoma is a very rare tumor type, so even with a pathologic diagnosis, it was unclear if additional treatment with chemotherapy and radiation was necessary, as some of these tumors do not progress and occasionally resolve spontaneously. Ultimately no additional treatment was given and the child did well on interval visits. Several questions occur: If the X-ray had not been done, would the mass have ever led to problems? Would it have spontaneously regressed? Should additional treatment have been given beyond surgery? What are the risks and benefits of such treatment?

Example 3

At the time of a sports physical on a thirteen- year-old girl, a family history was obtained that was clearly consistent with a breast/ovarian cancer syndrome in the mother's family. No genetic testing had been done in the family, but segregation analysis showed that the patient's mother was at a 50 percent risk, thus the patient was at a 25 percent chance of carrying a highly penetrant mutation in a cancer predisposition gene (most likely BRCA1/2). In the context of obtaining a medical history, the incidental finding of a family history of breast/ovarian cancer identified high risk for an **adult onset condition** in an adolescent. The mother was referred for genetic counseling. If testing confirms a BRCA mutation in the patient's mother, should the child be tested? If so, when? If testing is deferred to adulthood to preserve autonomy, who is responsible for communicating the information to the child when she reaches adulthood? Has her ability to choose already been compromised? When should surveillance begin? When should chemoprevention and/or prophylactic surgery be offered? These questions reflect uncertainty not only in the **risk of the specific finding, variable expression and penetrance** but they also introduce a host of questions about choice and autonomy.

The questions raised by these examples illustrate that while the application of WGS in the clinic may impact the number of incidental findings quantitatively compared to any other single test, the questions raised by the incidental findings are not qualitatively different.

An instructive analogy?

One "advantage" of having been in practice for several decades is that it is easier to understand historical perspectives one has personally experienced. In thinking about clinical WGS and incidental findings, the analogy of chemistry panels seems useful, if somewhat imperfect. The primary motivation for the introduction of chemistry panels into practice was economies of scale. If one wanted to order a test covering sodium, potassium, calcium and phosphorus, it was cheaper to run these tests on a panel of 20 (or 36 or more) than it was to run two or more of the tests individually (a similar argument to WGS over gene-by-gene testing). When chemistry panels moved into clinical practice, with very few exceptions the entire panel w as ordered and reported, as this was more convenient. The challenge is that if you are doing twenty tests where the normal range is defined by 95 percent confidence intervals derived from populations, there is a near certainty that one of the twenty tests will be out of range (*i.e.,* "abnormal"). Clinician responses to these out-of-range results varied.

An aspartate aminotransferase (AST) a couple of points out of range in a context where liver disease was not suspected (*i.e.,* a low prior probability of disease for the specific test) could reasonably be ignored and was by many if not most clinicians. However, more compulsive (or less secure) clinicians would pursue more specific liver function tests, or even imaging studies despite the low prior probability of disease, leading to higher likelihood of subsequent false-positive out-of-range tests and added cost to the system without additional benefit and, in some cases, creation of harm. An extreme example of this is whole-body CT scanning as a screening test, where 37 percent of scans detect an "abnormal" finding the vast majority of which are inconsequential given that the testing is being done outside a clinical indication, therefore the prior probability of disease is low. Indiscriminate use of incidental findings from WGS has the same potential to lead to cascades of evaluation, particularly if the finding is not contextualized through the use of other information like family history. In chemistry testing, the trend has been to require clinicians to order the specific tests they are interested in based on the clinical context. For economy's sake, the tests are still run as part of a larger panel; however, only the results of the requested tests are reported to the clinician. This eliminates the requirement to follow-up on other tests that are statistically out of range but are unlikely to be of clinical significance. The exception is a value that is so out of range that it must be assumed to be clinically significant. Findings such as these are categorized as "panic values" and the clinician will be contacted even though they did not order the test.

The challenge as I see it over the near future for WGS is to develop a list of genetic 'panic values' based on an accepted level of certainty of pathogenicity for the variant and known clinical interventions so that if a given variant is found that is not relevant to the indication for the test it will still be reported. An example could be a known deleterious mutation in the MLH1 gene that causes Lynch syndrome discovered through WGS in an individual being evaluated for adult onset deafness. Disclosure of this information could have important implications for the patient, in that earlier and more frequent colonoscopy has the ability to identify and remove pre-cancerous adenomatous polyps, thereby dramatically reducing the risk of developing colon cancer. At present several organizations, including the American College of Medical Genetics and Genomics, the National Human Genome Research Institute and the Evaluation of Genomic Applications in Prevention and Practice as well as many private groups are examining the content and ramifications of such a list. Implicit in this work is the requirement for a regularly updated centralized repository of well-annotated deleterious variants that can be accessed by laboratories and clinicians to aid in the interpretation of variants identified through WGS.

Finally, there is one additional issue relevant to this topic that should be addressed. Some have called for consent for WGS to allow patients and/or providers to 'opt out' from receiving incidental findings. If this were to be implemented it would be true genetic exceptionalism; in no other area of medical care do we allow patients to 'opt out' of receiving clinically significant incidental findings. Take the examples presented above and imagine the conversation with the parents: "I'm going to do a physical examination and I may find a heart murmur, but I want you to tell me now if you want me to disclose if I find something since it may not be information you want to deal with." While meant to elicit a smile, the reality is that this is not the standard of medical practice nor does it reduce the clinician's liability. What it does require is agreeing on an acceptable level of certainty about the clinical impact of the variant that can inform the decision to disclose, as has been done with incidental findings in all other areas of medicine. As such, the aforementioned proactive attempts to address incidental findings from WGS are most welcome.

PART III

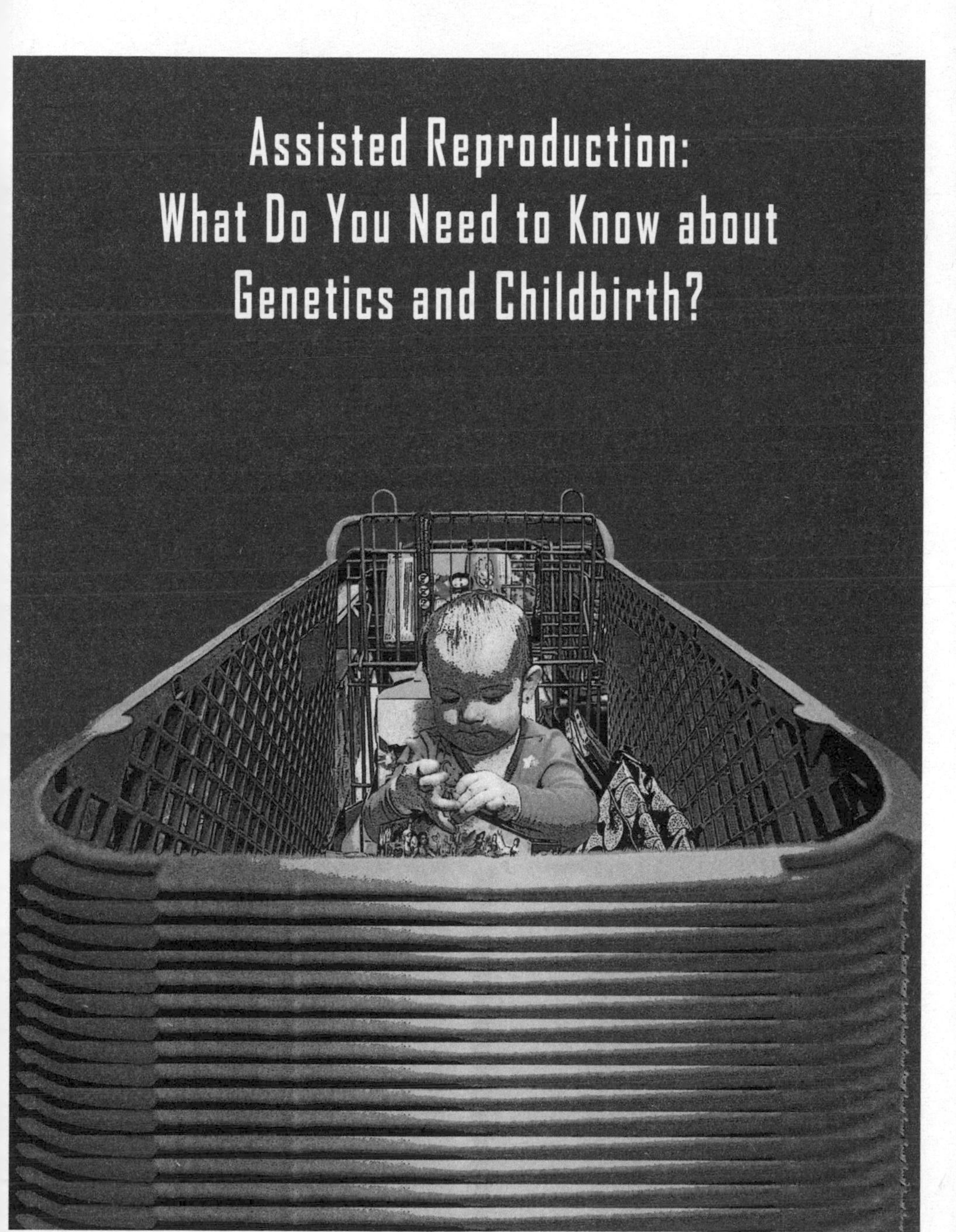

Artist: Sam Anderson

Today, women are faced with a rapidly expanding array of reproductive technologies. Indeed, the realm of assisted reproduction has become a multi-billion-dollar industry, visible in the increasing availability of in vitro fertilization, surrogacy, prenatal genetic diagnosis, and chemical and chromosomal testing of the fetus.

So far, these procedures have not been closely regulated. In part this is because the reach of federal oversight extends exclusively to publicly funded research, leaving private sector activity largely unregulated. Many interest groups are resistant to any regulation over the fertility industry due to the continuing political controversy over the abortion debate. Women in the United States who waged a successful struggle to have courts recognize a woman's right to choose are wary of opening the door to government regulation of anything to do with women's eggs, embryos or fetuses. The word "choice" has powerful connotations, and progressive women's groups themselves are not eager to restrict its power. These new options are often presented to women as an issue of "choice"—providing them greater control over the process and outcome of their pregnancies.

Yet the world of assisted reproductive technologies is far more complex than it is often presented as being. The subjection of human eggs and embryos to experimental manipulations and their use as products of commerce have raised concerns both for eugenic applications as well as for health risks to women and their offspring. Commercial surrogacy has drawn criticism over the lack of rights and health and safety concerns for the disadvantaged women who often assume these roles. There are very active debates over the use of genetic technologies, the scope and limits of assisted reproduction and the control of women's bodies.

Popular knowledge of in vitro fertilization and multiple embryo transfers may begin and end at the case of Nadya Suleman, aka "Octomom." There's no shame in admitting that you first learned of commercial surrogacy through *Baby Mama*, the 2008 comedy starring Tina Fey and Amy Poehler, or that up to this point you thought egg donation was the exclusive domain of the Easter Bunny. Education and discussion of these issues are key. The unique role of women in reproduction puts them on the front line of bio-technological experimentation, and, as such, women have the potential to play a leading role in determining the direction and scope of these developments.

From in vitro fertilization to human egg donation, from commercial surrogacy to prenatal genetic testing, these chapters explore childbearing in the age of biotechnology.

Childbearing in the Age of Biotechnology

BY RUTH HUBBARD

Ruth Hubbard *is a founding board member of the Council for Responsible Genetics and Professor Emerita of Biology, Harvard University. This article originally appeared in* GeneWatch, *volume 14, number 4, July 2001.*

The women's health movement of the 1960s and 1970s rallied around the call of "Taking our Bodies Back." In doing this, we were responding to our sense that, over the previous century, the male medical establishment had taken control of women's bodily functions by interpreting them as disease. Medical men had established not just one, but two medical specialties dealing with women's bodily functions—obstetrics and gynecology—within which menstruation, pregnancy, birth, lactation, and menopause all were looked on as pathologies in need of medical intervention.

One way of taking charge of our health led women to assert control over our experiences of pregnancy and birth. We advocated for unmedicated childbirth and a return to midwifery and home birth, and not just for women who could not afford medical help, but for women who wanted to experience their pregnancies and births as normal, social occurrences rather than as medical phenomena, unless there were specific reasons to go the medical route.

This independence from medical interventions proved to be short-lived because soon a series of technical innovations began to make it possible to monitor fetal development by means of chemical tests and ultrasound monitoring. As a result, many women began to feel that, while it is all very well to reject medical interventions on our own behalf, of course we want the best for our children. And, in our society, that tends to mean whatever interventions and reassurances science and medicine can offer.

In 1973, when American women acquired the legal right to terminate unwanted pregnancies, that "best" came to include an increasing number of prenatal tests that enable women, and families, to use health indices to decide whether to bring a pregnancy to term. Let me say at once that I fully support a woman's right to terminate her pregnancy, whatever her reasons. But, the fact is that the new predictive medical tests have opened up a range of supposed "reasons" to terminate pregnancies that raise hard questions for women, families and society.

For one thing, supposedly predictive prenatal tests have created what the sociologist Barbara Katz Rothman has aptly called "tentative" pregnancies. In the pre-test days, once a woman had embarked on a wanted pregnancy, she and her family could begin to ready themselves emotionally and socially for the fact that a new, entirely unknown person would shortly be joining them. Once supposedly predictive prenatal tests became available, two new things happened. One was that a woman who decided to have her fetus tested could no longer be sure the decision to have this baby would be final if the test came up with the "wrong" answer.

The other, quite subtle, result of prenatal testing is this: after a child is born, if she or he turns out to have a disability, that condition is just part of who that child is and the family integrates it into the total experience of living with her or him. But, once a test predicts some particular trait—even though tests usually cannot predict what that will mean for either that child or the family—the test result constricts the family's imagination. From that moment on, when the woman imagines her child, all she sees is cystic fibrosis or spina bifida, or whatever condition the child has been predicted to have. The child is no longer a person with desired abilities and traits; it has become its "predicted" diagnosis. The existence of supposedly predictive prenatal tests thus profoundly alters the experience of childbearing. What's more, once a test exists, only a very determined woman is likely to resist the medical and societal pressures to have that test.

Physicians and genetic counselors tend to strongly believe in the benefits of prenatal testing and therefore subtly or openly urge it. They are also under considerable legal pressure to not only make women aware of the existence of prenatal tests, but to be sure that women and families understand what they are doing if they refuse to have their embryo or fetus tested. Families have sued health care providers who didn't alert them to the existence of a prenatal test for a condition their child was born with, or even for not advising them strongly enough to have the test.

The existence of prenatal tests in itself, therefore, necessarily impacts women's experience of childbearing. What is more, with ultrasound imaging, what used to be private has become a very public experience: friends and relatives want to see "the baby," meaning the ultrasound image, and people open web sites in the name of the fetus, whose sex is usually revealed by ultrasound. This new approach to childbearing has recast an experience, which for most women is a normal, healthy function, into a risky process, requiring constant medical supervision.

Entire new industries have been developed that produce technically sophisticated equipment and offer a wide range of prenatal tests to monitor the course of the pregnancy and "predict" the potential existence of one or another, relatively rare health condition. It is very easy to make expectant mothers worry about the future health of the baby they are gestating; it is also very profitable. But the result, as one young woman recently told me, is that "It is very hard to be pregnant these days."

Let's stop and think about Down syndrome, the most frequent condition for which women are tested prenatally. As with just about all inborn conditions, the expression of Down syndrome is quite variable. Now that a great deal has been learned about how to successfully integrate children with Down syndrome into family life and schools, more and more people with Down syndrome are being educated, holding jobs, and living more or less ordinary lives. (See Michael Berube's recent book, *Life As We Know It*, about his family's life with his son Jamie.)

Some people's situation may be such that they do not feel they can take on the extra expense or work that having an unusual child is likely to involve. But, having children is always unpredictable and the most "usual" child can turn out to be "unusual" in ways we as parents find difficult to accept. Future parents are attracted to predictive tests because, once we become worried, we want to be reassured. And because human beings are, on the whole, a pretty sturdy lot, most tests turn out to be negative and so provide that reassurance.

As the range of human variation and difference becomes more and more medicalized and geneticized, and gets diagnosed in terms of variations in DNA, we become increasingly tempted to try to predict what the different variations will mean for people. This is driven, in part, by a curiosity to foresee the future. But, it is also driven by the fact that genetic tests constitute just about the only health-related products that are generating profits for the biotech industry. Other than tests, the industry has so far offered mostly promises. So there is lots of pressure to market tests, and therefore lots of hype about their presumed benefits.

The overall problem is that we Americans have come to look at our health not as a continuous, unfolding process, with variations and ups and downs, but in terms of "risk factors." And we have a very distorted notion of what, in fact, produces "risks," which risks to look out for, and what, if anything, to do about them.

In fact, the principal risks to newborns' and babies' health are low and very low birth weight and these usually are the result of premature birth. And what is responsible for those? For poor women, it is primarily poverty and malnutrition. The fact that *The New York Times* earlier this year [2011] reported that infant mortality is on the increase in specific parts of New York City has nothing to do with "bad genes." It has to do with what is euphemistically called welfare reform, which has left large numbers of poor women and families without the resources they need for a healthful existence.

At the other end of the income spectrum, increasing numbers of well-to-do, so-called older women are using fertility drugs and IVF. Both these interventions increase the likelihood of having twins, triplets, or higher-order multiple pregnancies and births. Multiple pregnancies are likely to result in the babies being born prematurely and thus with low birth weights. These procedures therefore also increase the incidence of developmental disabilities.

We need to reexamine the way we have come to think about our lives. In our continuous struggle to ward off "risk factors" from before birth to old age, we place more and more constraints on what constitutes an acceptable baby, an acceptable body, and acceptable health. The most effective ways to promote health require societal and public health policies that make for more healthful living and working conditions.

On Order

BY BARBARA KATZ ROTHMAN

Barbara Katz Rothman *is Professor of Sociology at the City University of New York.* *This article was excerpted from* Genetic Maps and Human Imaginations: The Limits of Science in Understanding Who We Are *by Barbara Katz Rothman.* *It also appeared in* GeneWatch, *volume 14, number 4, July 2001.*

The technologies of procreation are about control: from the simplest of contraceptives, through in vitro fertilization, amniocentesis, embryo biopsy and pre-implantation diagnosis, all the way to the newest twist, cloning. Cloning is quite markedly about control. It's about trying to introduce predictability and order into the wildly unpredictable crapshoot that is life. If normal procreation is the roll of 30,000 dice, a random dip in the gene pool, cloning is a carefully placed order. And that's where it gets interesting: it is order both in the sense of predictability and control; and it is order in the sense of the market, a (human) being on order.

In a perfect world, we could think about the value of the first form of order, the value of predictability and control in procreation, without thinking about the second form of order, the power of the market. But in our world, the two are hopelessly, endlessly entangled.

We've already moved in the direction of the marketing of babies, even without the technological sophistication of cloning. Of course, there is the adoption market, which often compares unfavorably in terms of decency and honesty in human relations with the marketing of used cars. But marketing is also operative in the technologies of procreation. If you want to purchase sperm or eggs, you can get listings of the available merchandise. The characteristics of the donors/sellers/producers of the genetic material are provided. Some of the traits listed are fairly direct predictors of likely genetic attributes: height, hair color. Some are in the far more questionable areas: intelligence, personality traits, talents. Some are downright strange, like religion and hobbies. Think there are genes for Methodists or stamp collectors?

But sperm and eggs are a long way from people. They're even a long way from the genetic attributes of the people that might grow from those gametes. Someone who has any given gene might not pass that gene on in the particular gamete the purchaser ends up with. People who use sperm from the "genius sperm bank" have to understand that their kid might inherit the genius' nose, not his brains. And a gene that is passed on may or may not express itself when in combination with the other gamete with which it pairs.

Cloning moves control up one more level. There are two things that are sometimes called cloning. One is what scientists mean by true cloning: an adult cell is used as the nucleus of an egg. The being grown from that egg is a genetic twin of the adult being from whom the cell was taken. That's the form of cloning that made the news with Dolly, the sheep cloned in Scotland. That's the form of cloning people have in mind when they talk about creating another Einstein or another Hitler out of their body cells.

Sometimes the word "cloning" is used when what is meant is really "twinning," in which embryos are split and can be artificially split over and over again, producing vast arrays of genetic twins. That's what they commonly do with cows all of the time: embryos produced by expensive cows are twinned in quantity, and then placed in the uterus of cheaper, more disposable cows, to be bred. Given enough time there's not much difference between the two techniques. If a sample of the embryos is grown to adulthood while the rest are frozen, you can still end up with genetic twins of vastly different ages. From a marketing point of view, that means you can show people the grown merchandise before they purchase the kit. When selling to dairy farmers, you can show them grown cows with known levels of milk production, and then sell the frozen embryos that are their twins. I think of Julia Child as she's appeared on television, putting the bread in one oven and walking across the kitchen to show us the finished product coming out of the other.

We already know that the two loaves, even in Julia Child's capable hands, will never be identical. Predictability and control are forever slipping out of our hands. When you're working with small herds of sheep cloned to produce some expensive exotic protein in their milk for medicinal purposes, as Dolly was, a certain percentage of error is to be expected, and accounted for. It is accounted for in two senses: it is part of the expectations, the narrative of what happens.

And it is accounted for in the ledger books, an anticipated expense. Most of the expected errors will not matter; only a few will affect the single purpose for which the sheep were cloned: the production of that protein in their milk.

With people, the accounting is a lot more complicated, in both senses. Errors are not to be written off, and our expectations are rarely so narrowly confined.

Genuine human cloning, cloning from adult cells, is probably not the direction in which we are likely to go. People do find it distasteful. It might well serve as a convenient boundary line, defining what we don't do, which will make what we do more palatable.

With twinning, producing sets of identical embryos, parents could choose an embryo from a variety of models. With the addition of the genetic technologies that permit DNA splicing, specific characteristics could be added, permitting and maybe even encouraging some customizing. (I'll take that one, but with red hair and no dimples.)

Could such a thing ever happen? We know, those of us who have been around the track a few times, how this latest "advance" in human procreative technology is going to be the solution to some problem. There's going to be a very good reason to do it the first time: some baby conceived through in vitro fertilization, with an embryo twin still in the freezer, is going to need a bone marrow transplant. Or maybe it will have to do with "isolates of historic interest," communities of people dying out. Or maybe.... But I don't want to give anybody any ideas. Do I think we could end up with catalogues of people the way dairy farmers have catalogues of cows? Why yes, I do.

How could we get from a set of technologies of procreation that were designed to end pregnancies where the fetus would suffer and die, to a set of technologies that turns babies into consumer objects? Was it the slippery slope or the camel's nose? Or was it, as my mother taught me, that one thing leads you to another?

Once when I was a kid I was supposed to do a quick straighten, dust, and vacuum before my mother came home from work and company came to dinner. Just get the shoes, socks and toys out of the living room, dishes out of the sink, make the house "presentable." My little brother's muddy fingerprints were on a closet door in a tiny hallway. I took a cleaning rag and some Spic & Span and gave it a wipe. Wow. A big white spot. I got the mud off, but also the year's accumulation of dust, cooking grease, whatever else yellows walls between paint jobs. I cleaned off that whole door. And the frame. The other doors in the hallway looked awful. I did them. The rag brushed the ceiling over the doorframes. My mother came home. The house was a mess, and what was I doing on a ladder cleaning the ceiling of the little hall?

Well, one thing leads you to another.

This way of thinking reminds us that people aren't quite so passive. Yes, it is a slippery slope indeed, and it surely is an aggressive camel. But I'm the damn fool with a rag on top of the ladder.

Still, I'm too much the sociologist, and far too affected by the women I've talked to who have used technologies of procreation, to frame these issues entirely in terms of individual choice. Women making decisions about prenatal testing often use the phrase, "my only choice," to mean the choice that they felt forced into, a no-choice choice. As individuals we are often without real choice. Even the scientists themselves who develop and introduce the various technologies are helpless to control it. Someone artificially twins cows or clones sheep and says, "Oh, we never meant for this to be used on people." But slopes and camels being what they are, one thing does lead you to another.

How do you come to find yourself on top of a ladder cleaning a ceiling? You let yourself move from one thing to the next without remembering what the larger project is.

In the area of reproductive technologies, some of us are hampered by not being entirely sure what the larger project is. It has to do with the meaning of life. And while I don't know what that might be, I'm sure it won't be served by turning the creation of people over to the forces of the market.

We began by selecting against very specific characteristics, dreadful diseases, horrific possibilities. And, some would argue, some not-so-dreadful conditions, too. But in so doing, we started splintering the self, started crystallizing the potential person into component parts. Once we've done that, removing the parts we don't want may not be enough. We might begin to sort through for the characteristics we do want.

In earlier times, the distinction was made between negative eugenics, the sterilization or even murder of "genetic weaklings" and positive eugenics, encouraging the "genetically fit" to breed. Once we start reading the code, we come to see that we are all made up of some genes we do want to perpetuate and others we do not. This micro eugenics is no longer about groups of people: it is now about specific stretches of code.

I've said that we shouldn't think of these stretches of code as "blueprints" for the construction of a person. I've said to think of them rather as bread recipes, with all of the individual diversity and variation that growth and time introduce. But what happens when we turn such a process over to the forces of the market? Wonder Bread: a nearly perfectly predictable bread.

I cannot afford to do what my great-great-grandmother did out of necessity, bake my own bread, any more than I could afford her uncontrolled fertility. I can't afford the time. Home-baked loaves of bread, like eight children, are luxuries well beyond the likes of me. I almost always buy my bread, baking only for a treat, a holiday, for the occasional pleasure of it. I can afford to buy a more customized bread than Wonder Bread: I buy it at the bakery, at the farmers' market. I buy the services of the baker, considerably more industrialized than what I could do at home, somewhat less industrialized than the factory supplying my supermarket. And sometimes, when I'm rushed or broke, I grab those plastic-wrapped loaves off the shelf.

Could people ever be mass-produced like supermarket loaves? I don't actually think so, but the image frightens. The factories toss out all the errors—the loaves that come out misshapen, that failed to rise evenly, that burned a bit or were underdone. It happens. They take that into account. It is part of their quality control program.

The current technologies of procreation introduced first quantity control and more recently, and increasingly so, quality control. Prenatal screening and testing with selective abortion are a form of quantity control, avoiding the production of the children we claim we can no longer afford to raise. The choice of quantity control, choices offered to us by (relatively) safe and (relatively) effective contraception, eventually lost us the choice of not controlling the quantity of our children. Who now feels they could really afford the eight children my great grandmother bore? And so it is with quality control: the introduction of that choice may ultimately cost us the choice not to control the quality, the choice of taking our chances in life's great, glorious and terrifying roll of the 30,000 dice.

One thing does lead you into another. Cloning, or embryo twinning, splicing specific characteristics in and out, may be eventually offered to us as a way of avoiding the tragedy of prenatal diagnosis and selective abortion. Cloning may eventually—and eventually isn't as long a time as it used to be—be offered to us as a way of inserting a measure of predictability and control earlier in the process. Placing order in procreation: placing our orders.

In the hands of the market, the "book of life" becomes a catalog.

What Human Genetic Modification Means for Women

BY JUDITH LEVINE

Judith Levine *is a journalist, essayist, and author who has written about sex, gender and families for three decades. An activist for free speech and sex education, Levine is a founder of the feminist group No More Nice Girls and the National Writers Union. "What Human Genetic Modification Means for Women" originally appeared in* WorldWatch, *Volume 15, Number 4, July-August 2002.*

Seduced by the medical promises of genetic science or fearful of losing reproductive autonomy, many feminists have been slow to oppose human genetic engineering. But GE is a threat to women, and in the broadest sense a feminist issue. Here's why. If anyone should be wary of medical techniques to "improve" ordinary reproduction, as GE purports to do, it's women. History is full of such "progress," and its grave results. When limbless babies were born to mothers who took thalidomide, the drug was recalled. But the deadly results of another "pregnancy-enhancing" drug, DES, showed up only years later, as cancer in the daughters of DES mothers. The high-estrogen birth control pill was tested first on uninformed Puerto Rican mothers, some of whom may have died from it.

Today's fertility industry takes in $4 billion a year, even though in-vitro fertilization (IVF) succeeds in only 3 of 10 cases. Virtually unregulated and highly competitive, fertility doctors often undertake experimental treatments. Recently [2002], the Institute for Reproductive Medicine and Science at New Jersey's St. Barnabas Medical Center announced the success of a new fertility "therapy" called cytoplasmic transfer, in which some of the cellular material outside the nucleus of one woman's egg is transferred into the egg of another woman who is having difficulty sustaining embryo survival. The transferred cytoplasm contains mitochondria (organelles that produce energy for the cell), which have a small number of their own genes. So the embryo produced with cytoplasmic transfer can end up with two genetic mothers. This mixing, called "mitochondrial heteroplasmy," can cause life-threatening symptoms that don't show up until later in life. When the Public Broadcasting Service's *Nova* enthusiastically reported on the procedure, complete with footage of its cute outcome, Katy, it mentioned no risks.

Didn't these women give informed consent? Yes and no. Most read warnings and signed their names. But with genetic therapies there's no such thing as "informed," says Judy Norsigian of the Boston Women's Health Book Collective, "because the risks can't be known." Adds biologist Ruth Hubbard, the risks associated with DES were identified "only because it showed itself in an otherwise very rare condition. If the effects [of human genetic engineering] are delayed, and if they are not associated with a particularly unusual pathology, it could take quite a long time to find out." Indeed, "we might never know."

"Perfecting" human genetic modification would require experimentation on women and children

Scottish biologist Ian Wilmut, the "father" of the famous first cloned sheep, Dolly, provided these statistics in 2001: Of the 31,007 sheep, mice, pig, and other mammalian eggs that had undergone somatic cell nuclear transfer (cloning), 9,391 viable embryos resulted. From those embryos came 267 live-born offspring. In these animals, *The New York Times* reported, "random errors" were ubiquitous—including fatal heart and lung defects, malfunctioning immune systems, and grotesque obesity. In all, "fewer than 3 percent of all cloning efforts succeed." Dolly may be a victim of accelerated aging, another problem in cloned animals. In January, it was reported that she has arthritis at the unusually early age of five-and-a-half. Mothers of clones are endangered too, since their bodies have trouble supporting the abnormally large fetuses that cloning often produces.

Perhaps scientists will get better at cloning, but some of the problems are intrinsic to the process. As health activists quip, if it works on a mouse, they will try it on a woman. The problem, warns Stuart Newman, a cell biologist at New York Medical College in Valhalla, is that if it works on a mouse, it is likely not to work on a woman: "Every species presents a new set of problems." How might the process be perfected in humans? In clinical trials?

"The degree of risk to be taken should never exceed that determined by the humanitarian importance of the problem to be solved by the experiment," reads the Nuremberg Code, drawn up after World War II to forbid future experiments of the horrific sort that Nazi "scientists" inflicted on concentration camp inmates. The science to find "safe" means of human GE, says Newman, would constitute "an entirely experimental enterprise with little justification." In other words, "We can't get there from here."

We are not our genes

When the Human Genome Project finished sequencing our DNA, its press releases called it the "blueprint" of humanity, the very Book of Life. The newspapers had already been filling up with reports of the discovery of a "gene for" breast cancer, and a "gene for" gayness. Many people had begun to believe our genes determine who we become.

This line of thinking should sound familiar to women. Not long ago, we were told that hormones, not sexism, explained why there has never been a U.S. female president (she might start a nuclear war in a fit of PMS). A decade after that came the notion that gender is "hard-wired" into the brain. Not incidentally, these claims were made just when social movements were proving Simone de Beauvoir's adage that women are not born but made. Now the old determinism is raising its ugly head once again, with genetics. As "non-traditional" families finally bring legitimacy to social parenting, proponents of inheritable genetic modification tell us not only that we can predetermine the natures of our children, but that cloning is the only means by which gays and lesbians can become real parents. "Real" parental ties, they imply, are biological, genetic.

"Genetic determinism" is not biologically accurate. "It is very unlikely that a simple and directly causal link between genes and most common diseases will ever be found," writes Richard Horton, editor of the British medical journal, *The Lancet*. If this is true of disease, it is even more true of musicality, optimism, or sexual orientation. The more complex a trait, the less genetics can explain it. Hubbard writes, "The lens of genetics really is one of the narrowest foci to define our biology, not to mention what our social being is about."

Genetic modification is not a reproductive "choice"

For feminists, one of the most galling aspects of the debate about human genetic manipulation is the way its proponents have hijacked the language of "choice" to sell its products. IVF clinics and biotech research shouldn't be regulated, say the companies that run them, because that would impinge on "choice" (for the paying customers, if not for their unsuspecting offspring). The Book of Life is becoming a "'catalogue' of consumer eugenics," says sociologist Barbara Katz Rothman.

Some ethicists, too, have posited a reproductive "right" to prenatal baby design. People decide whether or not to reproduce based on an expected "package of experiences," wrote law professor John Robertson, an influential bioethicist, in 1998. "Since the makeup of the packet will determine whether or not they reproduce . . . some right to choose characteristics, either by negative exclusion or positive selection, should follow as well." Already, selective abortion is widely accepted after prenatal genetic screening uncovers an "anomaly." Although some commentators (notably disability rights activists) critique such "negative eugenics," many people accept this practice for serious medical conditions. In any case, selecting from among a small number of embryos is a far cry from attempting to rearrange the DNA of a future child to achieve some preferred traits.

What feminists mean by "choice"—the ability to control fertility with safe and legal birth control and abortion—is far more concrete. It confers existential equality on the female half of the human race, which is why women worldwide have sought it for centuries. But genetic engineering creates inequality: it will artificially confer heritable advantages only on those who can afford to buy them. Performed prenatally, moreover, it affects the new person without that person's prior consent and possibly to her physical or emotional detriment. "Ending an unwanted pregnancy is apples, and mucking around with genes is oranges," says Marcy Darnovsky of the Center for Genetics and Society. "We support abortion rights because we support a right to not have a child or to have one. But we don't support a woman's right to do anything to that child once it's alive, like abuse it or kill it." Ironically, as Lisa Handwerker of the National Women's Health Network has pointed out, anti-choice, anti-GE forces share with GE's proponents an obsessive focus on the embryo as an independent entity, while they both virtually ignore the pregnant woman and the child she may bear.

Bans on dangerous genetic technologies do not give fetuses "rights"

Some choice advocates fear that any perceived concern about embryos will cede territory to abortion opponents, who want full legal protection of embryos and fetuses. U.S. Congressman Henry Waxman reflected this confusion when he said at a congressional hearing, "I do not believe that the Congress should prohibit potentially life-saving research on genetic cell replication because it accords a cell—a special cell, but only a cell—the same rights and protections as a person."

But pro-choice opponents of cloning do not propose to give cells rights. Rather, we worry that cloned embryos might be implanted by unscrupulous fertility entrepreneurs into desperate women, where they'll grow into cloned humans. And from cloning, it is not a big step to designing children.

For legal, political, and philosophical reasons, University of Chicago medical ethicist Mary Mahowald proposes clarifying the pro-choice position. "It does feminist support for abortion no good to confuse life with personhood," she told me. "We can admit that the embryo is life and therefore afford it respect," the respect, for instance, of not exchanging its genes with those of another cell. But respecting life is not the same as granting rights. Rights are reserved for living persons."

Individual freedom must be balanced with social justice

"We're against bans," said a member of a coalition of mainstream reproductive rights groups, explaining why the coalition was reluctant to join a campaign against human cloning. This reaction is not surprising in the United States, where defense of personal freedom can often trump the public interest.

Women's liberation means more than personal freedom, though. Rooted in the left, feminism is a critique of all kinds of domination and therefore a vision of an egalitarian world—racially and economically, as well as sexually.
In the case of species-altering procedures, social justice must prevail over individual "choice." Arguing for an international ban on reproductive cloning and regulation of related research, Patricia Baird, chair of Canada's Royal Commission on New Reproductive Technologies, put it this way: "The framework of individual autonomy and reproductive choice is dangerously incomplete, because it leaves out the effects on others and on social systems, and the effects on the child and future generations." The good news is that good public policy protects individuals too. Baird offered the example of overfishing, which might benefit the fisherman in the short run but deplete the fishery for everyone, including that fisherman, in the long run. Regulation sustains his and his children's livelihoods. "We all have a stake in the kind of community we live in," Baird said.

Feminists can work alongside anti-abortion conservatives against species-altering procedures

"We are repelled by the prospect of cloning human beings because we intuit and we feel, immediately and without argument, the violation of things that we rightfully hold dear," wrote Leon Kass, conservative social critic and chair of President Bush's committee to investigate stem-cell research.

Not every feminist holds dear what Kass holds dear: the "sanctity" of the family based in God-given, "natural" forms of reproduction. Still, Kass sat beside Judy Norsigian and Stuart Newman to testify before the U.S. Congress against cloning.

The genetic engineering debate has made strange bedfellows. But it has also rearranged the political definitions that made those bedfellows strangers. "Social conservatives believe [genetic engineering] is playing God and therefore unethical, while anti-biotech activists (of the Left) see it as the first step into a brave new world divided by biological castes," writes social critic Jeremy Rifkin. "Both oppose the emergence of a commercial eugenics civilization." Others suggest that the new political landscape divides differently, between libertarians and communitarians. Whether of the left or the right, the former would support an individual right to choose just about any intervention on one's own body or one's offspring, whereas the latter esteem public health and social equality and would reject those interventions, including GE, that endanger them.

Choice activists may at first be surprised when they find that their anti-cloning and anti-eugenics sentiments are shared by opponents of reproductive rights. But passionate arguments for the same position from historically sworn enemies can only make a legislator, or any citizen, listen up. Feminists need sacrifice no part of the defense of women's reproductive autonomy when we champion health and social justice for the future human community.

Reproductive Trafficking

BY HEDVA EYAL

Hedva Eyal *is Women & Medical Technologies Project Coordinator with Isha L'Isha (Woman to Woman) Haifa Feminist Center, Israel. This article originally appeared in* GeneWatch, *volume 23, number 5-6, October-December, 2010. The editorial assistant was* **Kathleen Sloan**, *former Program Coordinator, CRG.*

Women's inferior status throughout the world renders them more exposed than men to poverty, gender-based exploitation and all forms of trafficking. One of the newest forms of trafficking has emerged as a result of exponential growth in the worldwide use of new reproductive technologies. The international community faces a reality in which medical knowledge and technologies that were developed for healing purposes are used not for saving lives but rather for allowing people with economic means to fulfill a very specific kind of parenthood that often exploits the economic distress and risks the health of certain other women. This situation causes great concern among feminists, as transnational surrogacy and the trade in human eggs have become pervasive international phenomena.

Mediating agencies use various international trade routes to connect doctors, parents-to-be, and women willing to sell their eggs or rent their wombs.

Ova trade

Would-be parents often travel from western countries like the U.S., Germany, the U.K and Israel to Eastern Europe, India, Cyprus and other countries where payment for ova is cheaper or not legally regulated as it is in their home countries. In countries where ova sale is prohibited, hormones may be administered in the home country to the women who are then flown to a second country where their eggs are harvested, fertilized and inserted in the receiving woman's uterus. Many significant short-term health risks to women are incurred in the process of extracting abnormally large numbers of eggs from their ovaries per a single cycle. Long-term effects are less well understood but are known to include infertility and possibly reproductive and other cancers. Inadequate long-term research has been conducted on these health risks.

Surrogacy trade

Fertilized eggs often travel across international borders, sometimes arriving from one country while the sperm arrives from another country. The fertilized eggs may be passed to a third country where the woman who will become the surrogate mother resides and where the procedure of the insertion into the uterus is performed. Great concern regarding human rights violations resulting from the exploitation of low income and poor women, along with the absence of truly informed consent, is growing among advocates and many fear that the situation resembles trafficking in such organs as kidneys.

These procedures frequently involve the administration of synthetic hormones injected to produce abnormal numbers of eggs. As a result, when a pregnancy is successfully produced, it often consists of multiple embryos. The surrogate is then subjected to much higher levels of risk than a normal single pregnancy involves. Post-natal health care costs for treatment of complications are often not covered. Furthermore, the costs of health care provided to infants born with congenital problems may not be claimed by the intended parents. A typical situation is the case of the surrogate mother who was pressured to have an abortion after the fetus she was carrying failed to meet the quality specifications of those who hired her when they were informed that the fetus had Down syndrome.[1] Surrogate transactions are often facilitated by entrepreneurs operating without supervision or monitoring of women's health, and with little apparent concern about protecting surrogates from exploitation or criminal abuse.

In 2009, what is known as the "Romanian scandal" was exposed. Israeli doctors were involved in the trafficking of eggs of young poor women at SABYC clinic in Romania, some of whom were only fifteen years old, with little understanding of the health risks involved. In one example, a sixteen-year-old factory worker was left in critical condition after the procedure. The arrest of the doctors and agency operators by Romanian police revealed the ways the international reproductive industry is working as a free zone with no ethics or responsibility for respecting human rights or human dignity.[2]

Information on the magnitude of international trafficking in reproductive organs is scarce and partial, rendering many questions unanswered. Of particular interest is information regarding the health and social repercussions of this industry on women, and the economic gains of the various players in this field.

While there is legislation of varying degrees among different countries, ranging from full prohibition of egg donation and surrogacy to weak attempts at regulating aspects of this reproductive trade, the gaps are enormous and the industry metastasizes largely without any accountability. Such laws as exist are implemented solely at the national level.

The phenomenon of trafficking is not limited to countries that lack regulation. Black markets are known to develop alongside the publicly authorized services, offering leniencies that give them advantages from the buyers' point of view. Within this market, many grey areas remain. For example, in Israel, young women may receive hormonal treatments to stimulate egg production legally, while the harvesting takes place illegally outside the country.

The transnational nature of reproductive trafficking means that activism at the local level to reduce it is not enough. In light of these realities, women's health advocates have begun an international initiative to address the increasing problem of reproductive exploitation and commodification. Over the past year, action has begun to draft a call for a UN declaration on human rights abuses in reproduction.

Documents that should inform such a declaration include the Nuremberg Code (1947)[3] on human experimentation; the World Health Organization's Draft Guiding Principles on Human Organ Transplantation (1991) and its Commentaries;[4] the European Convention on Human Rights and Biomedicine (1997)[5] and its Additional Protocol on Transplantation of Organs and Tissues of Human Origin (2002);[6] and the Helsinki Declaration (Sixth Revision 2008) on human experimentation.[7] Although these documents more specifically pertain to human experimentation and organ transplantation, in some cases even deliberately excluding reproductive tissues including ova, they can be applied to reproductive organs and tissues as in the case of the Draft Guiding Principles on Human Organ Transplantation of 1991.[8]

Universal legal agreements on the trafficking of ova and surrogacy should protect the basic rights and interests of women, which form part of the set of legally binding obligations on countries that have agreed to these treaties. An international regulation might offer a basis for determining when eggs are being trafficked. It could also determine what precautions every country should take within its boundaries in order to protect all women concerned.[9] For example, the European Committee's 2004 resolution condemns any trade in human tissues and places crucial importance on encouraging EU members to incorporate in their laws the principal of voluntary donation (see clauses 14 (2) 12(1)). Furthermore, the resolution mentions the need for creating a monitoring mechanism on import and export that will take into account both the protection of donors and the quality of organs and tissues (clause 9).[10]

The practices of reproductive organ, tissue and cell trafficking and trade, particularly ova sale and surrogacy, infringe upon several basic human rights under international law and are violations of international agreements on health and medical standards. Trade in reproductive surrogacy, organs and tissue must be recognized as a unique kind of human exploitation.

The absence of an international stance on reproductive trade allows the free market to set the standard for using and selling human organs, including reproductive organs, cells and tissue. Medical technologies become a tool for people who, in the name of creating or saving life, are using women as objects of human organ stock.

It is our belief that taking a feminist stance on this issue will be beneficial for women throughout the world. There must be an international platform for discussion. The field of reproductive technologies raises complicated issues regarding autonomy and commodification of women's bodies, the medicalization of women, trafficking, children's rights, and a plethora of other issues. Alongside these questions is another set of issues regarding the right course of action. Could international regulation protect women or simply achieve legitimization of reproductive trade? Although there is awareness of these questions, recognition of the legitimacy of the fertility-industrial complex and its practices affects women's rights and health negatively. Recognizing the problem and bringing these issues to the forefront of the international agenda is the crucial first step. The need to convene an international meeting of women's health and human rights advocates to develop a universal feminist stance on the issue of reproductive trade is urgent.

The opportunities that these medical technologies provide also require that we face the question of who truly is paying the price. The obligation of each one of us is to acknowledge the irrelevance of national borders concerning this issue and to focus on the abuses occurring and to develop protective measures that will ensure the health and safety of women throughout the world.

Several elements provide a framework for the creation of an international declaration on human rights abuses in human reproduction that protects the rights and health of women and the children to whom they give birth:

- The commercial use of women's reproductive capabilities both within and across national borders should be prohibited.
- Surrogates and ova donors should be recognized as human participants in a complex birth-giving process, rather than as biological resources. Consequently, practices that bar human contact between surrogates, egg providers, and the children born through these processes should be strictly prohibited.
- Surrogacy and egg donation should be permitted only under circumstances allowing for the viable possibility of a prolonged relationship between and/or among surrogate, gamete donor, child, and growing family.
- All medical procedures must be conducted within the country of origin of the intended parent(s) by legally authorized fertility experts in licensed hospitals and/or clinics.
- Recipients of fertility treatment hormones of any kind must be informed that past uses of synthetic hormones have led to significant increases in cancer rates among women to whom they were prescribed and that the long-term medical risks of hormones currently used in fertility treatment (often unapproved for this purpose) are unknown due to a dearth of long-term studies of the effects of these drugs on recipients.

The hope is that these elements will be a basis for international dialogue between women and organizations that addresses the increasing problem of reproductive organ trade and trafficking. We have seen the effects of these phenomena on women's lives, and we are asking the international community to take social responsibility and actively defend human rights.

The Booming Baby Business

Interview with Debora Spar

Debora Spar, Ph.D., *is president of Barnard College and author of* The Baby Business: How Money, Science, and Politics Drive the Commerce of Conception. *This interview originally appeared in* GeneWatch, *June-July 2011.*

GeneWatch: The big question is: how did we end up with an essentially unregulated multi-billion-dollar industry?

Debora Spar: I think that we wound up with this unregulated industry for reasons that are actually quite understandable, if not good. We wound up here because, to begin with, the United States has a deep aversion to regulation of any sort. Compared to many other countries, particularly Western Europe, we don't like regulation. Virtually all of our markets, whether it is banking or Internet or education, are less regulated than they are in many other places. So right off the bat, it's not surprising that we also have a lack of regulation in this sphere. What makes it so pronounced in this particular area, though, is that we also have a deep aversion to having any kind of political conversation around reproduction, and particularly around the status of the embryo. It's almost impossible to have those conversations without getting deeply involved in the abortion debate, which is essentially political suicide in this country. So I find it not in any way surprising that there is virtually no politician out there who wants to tackle this issue.

GeneWatch: From either side of the aisle....

Spar: Absolutely. It's poison on both sides. Poison across the spectrum.

GeneWatch: That explains why it's not being addressed in a legislative way, but does it seem to carry over into agency rules and state and local policies?

Spar: Yes, because these are not regulations that would ever sneak under the radar. This is too politically charged. It's politically charged across the political spectrum in ways that can create very strange bedfellows, so you also don't want as many folks who tend to cluster on the left, particularly people in favor of same sex marriage or same sex relationships. They don't want the government touching the arena of assisted reproductive technologies because they understandably fear that it allows for government intervention in the question of who can have children. You have those on the far left who don't want to touch this, and then those on the right who don't want to touch it because of their concerns about abortion, so there is no political constituency in favor of reform.

I would add one more piece. There are also the parents who become involved in this area; they don't want regulation because when they're in the reproductive center, all they want is babies. They don't see the government as a way to get there. So if the parents in the process don't want regulation, the industry doesn't want regulation, and there is no political constituency pushing for regulation, there's really no voice for regulation . . . with one exception. We are starting to see flickers of a movement made up of the grown children of assisted reproduction, and I suspect that it will grow much, much louder. In particular, we're seeing that children who were born of sperm donors are increasingly fighting to have some sort of way of identifying their genetic fathers. And I think that's just the first wave: this is the first generation of grown-ups who were born through assisted reproduction.

GeneWatch: As you've pointed out, this is a multi-billion-dollar industry. What sort of sense do you get about where the money is ending up? Who is getting rich off of this, if anyone?

Spar: Reproductive doctors. The doctors and their intermediaries. There is an utterly unregulated group of egg donors, egg brokers, surrogacy brokers; they are truly a new sector of the market. They're making money, the clinics are making money, and the companies that make the drugs are making money. None of which, let me underscore, is necessarily bad. People make money doing all kinds of things, like creating cancer drugs. I don't think the fact that people are making money is bad at all. I do think it is bad that there is less regulation in this area than in other fields of medicine where people also make money.

GeneWatch: *The Baby Business* came out in 2006. Do you have a sense of how much has changed since then?

Spar: My sense is that there have been two major trends. When we hit the economic downturn, it appears that there were fewer people who could afford or were willing to pay the expenses to go into the assisted reproductive market, although I suspect that may be ending now. By the same token, though, there were also reportedly increased numbers of women interested in donating their eggs for the same reason. And then the other main trend, of course, is that as same sex marriage becomes more socially and legally accepted, I think we're going to start seeing the same sex part of this market steadily increasing. It's already increasing very, very rapidly, but obviously, for gay men, if they want to have a child together, they almost certainly have to go into the reproductive market to some extent—especially since it's become harder and harder for them to adopt children.

GeneWatch: So would surrogacy be the most common way to go about this?

Spar: It's really the only way. And adoption, but as I said, that's being cut off. And surrogacy is almost most certainly the most expensive way to acquire a child.

GeneWatch: In terms of cost, how does surrogacy compare to other options, like egg donation or adoption?

Spar: It's purely additive. If you're doing surrogacy as a gay couple, you have to buy eggs because you don't have your own. And you have to go through IVF because you have to put those eggs together with sperm. And you have to hire the surrogate. So you're doing every single piece of it, whereas if it is a heterosexual couple, they don't have to do all of the pieces, depending on what the circumstances are. But gay couples have to basically buy everything—aside from sperm, which is the cheapest part of the equation.

GeneWatch: How much does an egg go for?

Spar: Anyone who sits on a college campus as I do will see ads in the back of newspapers that cluster around $25,000 to $30,000, particularly at the elite colleges—although the ASRM (American Society for Reproductive Medicine) guidelines are much lower.

GeneWatch: Do you mean that is how much someone gets paid to donate an egg or how much it costs to give the egg?

Spar: That's how much the donor will get paid per harvest. So the price of that egg to the purchaser will be higher because it will include that cost, obviously, plus whatever the egg broker is making, plus the cost of whatever you do subsequently with that egg, whether it is then frozen, whether it is made into an embryo. That payment to the donor is only your, if you will, input cost.

GeneWatch: You also write about when efforts to get a baby can approach actually selling babies. Is there a place where you draw that line? How does that happen?

Spar: I think that this is a semantic distinction, but it is an important one. Adoption isn't baby selling. The laws here are quite good. As compared to assisted reproduction, adoption is very heavily regulated. It is illegal to sell a baby, and if people are caught, they go to jail or they pay huge fines. So people aren't selling babies. What they are doing is they are paying intermediaries to help them acquire a baby. So the great irony here—and I'm not sure if it's a bad thing, but it is somewhat ironic—is that what is illegal is paying the mother for the child. If you are never paying the mother for the child, you are not buying a baby. What you are doing is paying all of the intermediaries in the process.

GeneWatch: Is a lot of that international?

Spar: Yes. Adoption is basically regulated at every level. In the United States, it's mostly at the state level if you're adopting domestically, but if you're adopting internationally, one has to go through state level regulations, and national regulations, and the regulations of the target country, and The Hague, which is the international overseeing body.

GeneWatch: Out of all of the different things you looked at, what are some of the most egregious practices that you came across or that are actually happening?

Spar: Well, I think that the most egregious umbrella, if you will, is that what gets lost in all of this is a consistent focus on the health of the mother and children. There are not, in my mind, sufficient studies of what the long-term implications may be of assisted reproduction on the health of the mothers, the donors, and the children. There may be trust that everybody is fine, but we don't have the data yet to confirm that. And that, in my mind, is really the most egregious set of practices, because it includes things like transferring way too many embryos at once. There is no cap on the number of times women can go through massive hormonal stimulation. And I think that those are the very egregious practices—not the money side.

GeneWatch: How do the customers, so to speak, the couples or the single parents who desperately want a child, but can't have their own . . . how do they find their way to these sorts of places?

Spar: Like everything else, these days; it's all online. It's online, and then there are support groups, and list servs, and all of these wonderful things we've discovered and created.

GeneWatch: What do you think the government or other actors need to do to address some of these bigger problems? What kinds of regulations would you like to see and what can be done aside from regulation?

Spar: I say this with the caveat that I don't think it is particularly realistic, quite frankly, but I think that we could have a minimal set of regulations at either the state or federal level—I think that the state is more likely—that would really bring some of these practices in line with current regulations that we have in the general medical area; in the case of egg donors, bringing it in line with the kind of regulation that we have for other people who donate organs, blood, other body parts, even sperm. Sperm donation is much more heavily regulated than eggs.

The other thing I would really push for if I had the magic wand would be just to begin a basic process of data collection. There was one piece of legislation some years ago that does provide for collecting data from the fertility clinics.

That's a good thing, and I think that if we had a similar set of data on the children who were born through these practices and the women who undergo them, I think that would go a long way toward at least giving us the information to start to understand whether or not there are any long-term medical implications. And it would also give us the ability to enable children born from these practices to find their genetic parents if they choose to as they get older.

Commercial Surrogacy and the Cost of Reproductive "Freedom"

BY MARSHA DARLING

Marsha Darling, Ph.D., *is Professor of History and Interdisciplinary Studies and Director of the African American & Ethnic Studies Program at Adelphi University. This article originally appeared in* GeneWatch, *June-July 2011.*

The act of women birthing children for other women is a very old social behavior dating to biblical times. Commercial gestational surrogacy is recent and of compelling importance because it invokes sharp ethical dilemmas regarding the use of invasive and risky protocols on women's bodies, and the well-being of children born into arrangements where the genetic, gestational, nurturing and even institutional (as in when birthed surrogate infants are placed in orphanages staffed by women) mothers of infants may not be the same women. Commercial surrogacy also portends challenges for our advocacy as stakeholders, who in the United States participate in a democratic social order that promotes reproductive freedom "choice," and at the same time social justice.

Exactly what does "choice" mean in the context of participation in surrogacy arrangements when one lives below the poverty line in the United States or especially in a "low resource" country like India or Thailand? This question bears on us now as present-day gestational surrogacy has become a transnational commercial enterprise, linking multiple individuals and their interests in an interconnected global web: mostly poor women's labor, gametes, infants, medical services, surrogate hostels, legal contracts, state and national policy makers, sometimes human traffickers, international trade institutions [such as the World Trade Organization] and, of course, lots of money. This article identifies some of the challenges associated with women's work in the commercial gestational surrogacy industry, and argues that we should move beyond "just wait and see what happens" to the complex work of legislating responsible governance of the new biotechnologies that reach so very deeply into our bodies.

Few people might have imagined the exponential expansion of gestational surrogacy taking place within and across national borders in recent decades. Used primarily to intervene in infertility, contract gestational surrogacy is the centerpiece of a fertility production complex with a set of contractual arrangements that mark: a) the convergence of women's reproductive labor (whether paid or uncompensated), b) female and male gametes (purchased or donated), c) written contracts or just verbal agreements offered by commissioning individuals or couples, d) fertility brokers and doctors and medical staff often working in private hospitals specializing in fertility services in "low resource" countries, e) fertility hostels as places where contract surrogate mothers-to-be are often housed especially in "low resource" countries, f) some regulatory policy makers often intent upon allowing citizens some measure of autonomy in utilizing surrogacy in response to infertility and g) other policy makers who are intent on taking a place in the global economy by offering medical and reproductive services and encouraging citizen participation, especially in "low resource" countries.[1]

In vitro reproductive interventions are ushering forth an era where for an increasing number of persons, mostly in western societies, reproduction is shifting from conception as a possible outcome of coital sex to laboratory managed medical intervention and control of conception. Social identity rights based movements and the sexual revolution begun in the 1970s have altered definitions of "family" in much of the global North, creating new alternatives to traditional familial relationships. Fertility interventions enable greater agency on behalf of reproductive freedom, especially for heterosexual couples challenged by infertility, genetically transmitted disorders, or just wanting to opt out of the inconvenience of childbearing. However, persons who are disabled or differently able who seek to reproduce, same sex couples, single persons, and at some point transgender persons who seek to enable their reproductive choices, often encounter a refusal of fertility and surrogacy services.

The revolution in reproductive biotechnologies has made parenthood possible for thousands of people around the world for whom this would otherwise have been impossible, and their use is steadily growing. The number of babies born to gestational surrogates in the United States grew 89 percent from 2004 to 2008.[2] The commercial fertility industry which is nearly entirely unregulated in much of the world is a multi-billion-dollar annual business, generating enormous profits for fertility specialists, clinics, and brokers. While in the main these powerful stakeholders seek to minimize health and economic risks, they rigorously fight regulation. In many countries, including the United States, the industry self-reports, and as such sometimes covers over errors of judgment or adverse outcomes—preferring to focus on successes. Recognizing multiple dangers and ethical issues, policy makers in a number of countries have banned or severely limited fertility and surrogacy practices, turning countries with no regulation into "free market" zones. Many are concerned about negative impacts on women's health and infants in the fertility industry.[3]

The challenges to protecting surrogate women's health and well-being everywhere deepen, as women who labor as surrogates are almost always economically vulnerable and are lured by the promised payment that is a desperately needed income supplement. In "low resource" countries like India, where a large segment of poor women workers are illiterate, they are vulnerable in a system of intermediaries, all of whom have more power and authority than impoverished women. Needless to say, fraud and exploitation of women fertility workers are very real issues. At many clinics in India, surrogates recruited from among the poor are warehoused assembly-line style in long rows of beds where they are confined for the duration of their pregnancies. In the United States, it is estimated that nearly 50 percent of gestational surrogates are "military wives," a much sought-after group for agencies and brokers. Many "military wives" are low-income ($16,000 to $28,000 USD in annual income), and tend to marry and have their children at young ages, so the prospect of doubling their income by serving as surrogates can be irresistibly attractive. These women have few legal or regulatory protections, however, making them ripe for exploitation and fraud.[4]

Egg donors are a significant factor in the commercialized reproduction business. Egg suppliers, largely young college women struggling to finance their educations, are drawn by ubiquitous ads in campus newspapers and social media. One typical ad running in campus newspapers in the United States offers "$25,000 for a donor who is one hundred percent-Jewish with high SAT scores, attractive, at healthy body weight and free of genetic diseases." Many of the young, child-bearing-age women who enter ova selling networks, or the "embryo trade" as it is quickly becoming known, enter the industry to donate or sell their ova for the infertility industry. More recently, much of the quest to further stem cell research initiatives rests on recruiting young women to undergo ovarian stimulation so as to produce multiple embryos for research. This industry is also expanding rapidly. Much of the discourse about the arrangements undergirding these practices reduces them to "market arrangements." Yet, the well-being and health of many of the young women is at risk as they enter ova trade networks.[5]

The potential for serious health risks from these practices is very real. Synthetic hormones and other drugs utilized in IVF can have harmful short- and long-term health consequences for the women who work as egg suppliers and surrogates, as well as for the women undergoing these procedures to conceive and give birth to children of their own. Short-term health risks associated with these procedures include ovarian hyperstimulation syndrome (OHS), which ranges from mild to life-threatening. Nausea, vomiting and abdominal swelling are common; more serious complications may include kidney and other organ failure, strokes, and even death. Long-term risks include increased rates of uterine, breast and other cancers and infertility.[6] Despite the seriousness of these potential health risks for women working in the fertility industry, insufficient attention has been given to gathering empirical data and crafting responsible regulation of women's work that generates not a material "something" but a "someone." There is a resounding need for open debate, international collaboration and tighter regulation of this global marketplace.

The global commercial fertility industry has created a market for virtually every aspect of human reproduction: sperm ($275 a vial), eggs (up to $50,000 apiece), nine months' use of a womb ($20,000), and the creation of an embryo ($12,000 per cycle). The fertility-industrial complex is a stunning array of businesses. It includes the manufacturing of fertility hormones, harvesting of renewable natural resources (sperm and egg collection), international trade (foreign adoptions), expert services (IVF and other high-tech medicines), long-term storage (embryo banks), and even rental property (surrogate mothers).[7] As the industry continues to grow and affect more and more people's lives all over the world, awareness and understanding of its far-reaching implications for social justice should accompany the development of the technologies (IVF, Preimplantation Genetic Diagnosis-PGD, etc.) and practices of this global phenomenon. As all of the arrangements noted above amount to purchasing and securing the production of a child, a concern for the infants born within the surrogacy industry should also be on our radar screen.

Who looks after the interests of the children who are born through the gestational surrogacy industry? Do they have a right to know information about their biological parent(s) in cases where the egg and/or sperm donor is/are not the commissioning parents? What if they seek information about the surrogate mother or egg supplier? What rights do children have to information regarding any siblings they may have who are the offspring of the donor parents? In a more frightful scenario, who will look after the interests of the infants rejected or abandoned because of multiple births, or because of a serious disability deemed unacceptable by commissioning individuals, or when infants are unclaimed or abandoned or in the event of the death of a commissioning parent-to-be? Equally dire, what if a commissioning individual's home country will not accept the newborn infant or upon its birth the infant is placed in an orphanage in the country of physical birth? All of the above questions bear immediately on the well-being of children born within the fertility industry.

Lastly, addressing the issue of averting potential and real harm at a time when the fertility industry is bustling will be a challenge for Americans because advertising has cleverly linked increasing personal individual choices connected with parenting with letting the marketplace rule without regulation. As for women's interests, how do we defend abortion rights but also disentangle that issue from advocacy toward regulating risky technologies? While we have outlawed the market for selling organs in the United States, where are our heads about allowing young women's wombs, their ova with the all so very valuable totipotent embryonic stem cells, even their DNA and genetic sequences, to be thrown into a marketplace, patented and sold as a manufacturing "part" or "process"? Can we be under the illusion that our reproductive freedom will increase by allowing elites to patent or privatize the genetic sequences and processes embedded in women's reproductive biology? A really tough question is what are the moral and ethical consequences of transforming a normal biological function of a woman's body into a commercial production process with its attendant input costs, organized under a legally binding contract? Exactly how will that play out without our exerting a mitigating influence on the development of biotechnologies that reach deep within our bodies?

Eggs for Sale

BY JUDY NORSIGIAN

Judy Norsigian *is Executive Director and a founder of Our Bodies Ourselves, also known as the Boston Women's Health Book Collective. This article originally appeared in* GeneWatch, *volume 22, number 5, September-October 2009.*

Despite ongoing concerns about the safety of multiple egg extraction techniques, many young women see misleading ads in college newspapers and on the Internet that portray the risks of egg "donation" as quite minor and inconsequential. Most of these ads seek women who would provide eggs for reproductive purposes and often offer a hefty payment that brings the transaction well out of the realm of egg "donation." Some carry rather blatant headlines (for example, "Genius Asian Egg Donor Needed—$35,000 Compensation").[1]

Generally, regulations governing payments to women who provide eggs for research purposes (somatic cell nuclear transfer, or SCNT) have limited the compensation substantially, but that may change in the near future, especially since New York State set a precedent this past year in allowing much larger payments (up to $10,000) to women providing eggs for research purposes.

Elsewhere, we have noted the unresolved safety issues that need greater attention.[2] Dr. Jennifer Schneider, whose daughter developed colon cancer and died after several egg extraction procedures conducted during a two- to three-year period, has also written eloquently on the need for more long-term safety data.[3] *Time* magazine noted in its March 31, 2009 issue, "As egg donations mount, so do health concerns." Although a good deal is known about the potentially life-threatening complication of ovarian hyperstimulation syndrome—a rare response to drugs that hyperstimulate the ovaries to produce multiple follicles—there is no good long-term data on the risks of ovulation suppression from drugs like leuprolide acetate (Lupron). These drugs are typically administered first, in order to allow for more controlled hyperstimulation with the stimulatory drugs.

A recent editorial in *Nature* magazine unfortunately endorses the decision of New York State to pay women to undergo multiple egg extraction solely for research purposes, even after reporting in August 2006 that "Health effects of egg donation may take decades to emerge"[4] and that "More states should take New York's lead, and allow researchers to pay for egg donation. The potential for coercion, although real, is manageable. And the technique's move to the clinic would certainly be faster, and arguably more ethical, if donors were paid."[5]

The editorial glibly asserts that the potential for coercion is "manageable," while there is little evidence to support this claim. Moreover, it fails to mention that the lack of adequate long-term safety data continues to make it impossible to secure true informed consent from women undergoing these procedures.

In addition, some ethicists now argue that payments to egg donors for SCNT should parallel payments recognized as acceptable for IVF clinics by the American Society for Reproductive Medicine. Ethicist Nancy Dubler, a member of the Empire State Stem Cell Board, explained her view of the board's role this way: "I think that we are an ethics committee, and I actually think that, if good science demands these oocytes, that we have the obligation to provide them, and I'd like to see language like that. . . . I think that this will be a larger national discussion, and this might be an important statement to get out there."[6]

Prominent ethicist Art Caplan is among those criticizing the New York State's stem cell program for departing from the international consensus against paying women to provide eggs for cloning-based stem cell research: "The image of women having their eggs harvested in a market is one that the industry is going to find difficult to destigmatize . . . That notion of being treated as an object to derive those kinds of materials is not one that will sit well."[7]

Fortunately, the National Institutes of Health has enhanced access to funding for responsible embryonic stem cell research, while at the same time excluding federal funding for stem cell lines that are created with cloning techniques. This reflects an appropriately cautious stance given what we still need to learn about the potential risks of multiple egg extraction. Also, given that many aspects of this research have thus far failed to produce the hoped-for results, there is even more reason for discouraging other states and institutions from adopting the inappropriate payment incentives now being offered by New York State.

In addition, should scientists be able to demonstrate that cell reprogramming methods may be adequate for producing disease-specific stem cell lines, it may not ultimately be necessary to ask women to undergo the risks of egg extraction solely for research purposes.

Meanwhile, in the absence of well-designed, long-term clinical trials, more anecdotal experiences are appearing on the Internet and serve to offset the misleading ads. One woman's lengthy story ends with the following:

> When I decided to sell my eggs, I never thought I'd get cancer. . . . I know that breast cancers are hormone-sensitive and can be affected by hormone treatments. During my cancer treatment, two doctors mentioned that anecdotally they see more cancer in women who have had fertility treatments. I'll never know for sure if the egg donation caused my breast cancer, but now I know that it is likely to be a contributing factor. I think often about how much I love my husband, and it breaks my heart that my desperation for a couple thousand bucks has caused him such pain. A bad decision made seven years ago may cost me my life.[8]

A thoughtful Princeton student writing an article titled, "Truth in egg-donor advertising" for the *Daily Princetonian* noted the following key points:

> Proponents of a free-market egg-donation system may claim that cases of abuse are rare, but the simple reality is that no one knows the true scale of abuse. The Centers for Disease Control is only required to ask fertility clinics how many successful births result from donor eggs. They don't ask important questions like the amount of times a donor has previously donated, reported side-effects or long-term medical issues.

> With practically no government regulation and increasingly astronomical compensation prices, the blossoming egg donation system is rife for potential abuses. Donors have financial incentives to

> cover up their medical history, and growing evidence suggests that repeated egg donation may result in decreased fertility for the donor and damaged eggs for the recipient. Without a system to monitor and control egg donations, both donors and recipients could be abused.[9]

In an effort to improve the quality of informed consent for potential egg donors, the California legislature passed AB 1317 in 2009. This bill involves no cost to taxpayers and no new burdens on fertility clinics and brokers already operating in an ethical fashion. It would improve both the public oversight of embryo acquisition for stem cell research as well as the quality of informed consent for donors providing eggs for IVF purposes.

Larger Questions

The growing number of reproductive technologies offered by an expanding "fertility industry" has posed other challenges dealing with commercial, economic, and ethical aspects of these technologies. For example, abuses related to reproductive "tourism" are appearing more frequently in the news. In 2009, two Israeli doctors were arrested in Romania following allegations that their Sabyc fertility clinic in Bucharest made payments to human egg donors, a practice that is illegal there.

Internationally, Sama Resource Group for Women and Health in India organized a major consultation in January 2010 to address these challenges and to explore strategies for responding to the growing phenomenon of poorer women being sought as "suppliers" of cheap labor (as "gestational" mothers "renting" their wombs) or as providers of eggs used in IVF procedures. India, in particular, has become a "surrogacy" outsourcing capital because of the lower costs involved, the lack of regulation in provision of ARTs, and the highly regulated situation in many European countries. Some estimate that the surrogacy business alone is worth $445 million in India.[11]

Sama is also involved in action research focusing on social, medical, economic and ethical implications of the ARTs. They are exploring such issues as poverty, religion, caste, gender and state apathy. They note that medical care providers in India often develop their own eligibility criteria for the couples according to which, quite often, single or homosexual women/men fall outside the purview of these services. Their analysis of advertisements seeking surrogates and egg donors identifies stereotypes that reveal an obsession with particular kinds of physical features and an emphasis on eugenics. Because infertility has never been addressed as a public health issue in India, there has been little attention given to the underlying, sometimes preventable, causes of infertility.

Finally, it is also important to remember that SCNT lays the technical groundwork for human reproductive cloning. Until the United States joins the dozens of other countries that have already prohibited reproductive cloning, it is even more critical that we restrict SCNT research in this country.

IVF and Multiple Embryo Transfers: Everything in Moderation

Interview with Judy Stern

Judy Stern, Ph.D., *is director of the Dartmouth-Hitchcock Medical Center's Embryology and Andrology Laboratory. This interview originally appeared in* GeneWatch, *June-July 2011.*

GeneWatch: What kinds of procedures does your lab handle? Who are your patients?

Judy Stern: We do IVF and ICSI (intracytoplasmic sperm injection). We also do embryo freezing. We used to do GIFT (gamete intrafallopian transfer) years ago, but we don't do that anymore. We haven't done a lot of PGD (preimplantation genetic diagnosis), but we offer it. We basically do the gamut. We offer embryo donation and surrogacy as well, and other things that are out there, but a lot of the procedures like that we do less frequently than just standard bread-and-butter IVF.

GeneWatch: And when you say bread-and-butter IVF, you mean a couple comes in and.... ?

Stern: They use their own gametes, so we take the eggs from the woman and the sperm from the man and mix them together and they go back to the same couple. We do some donor eggs, sperm and embryos, but for the majority of what we do, the intended parents are also the gamete donors and the gestators.

GeneWatch: It seems like a lot of the talk about the success of IVF and a lot of other assisted reproductive technologies focuses on success rate in terms of live births. How do you measure the success of the procedure that your lab carries out?

Stern: We look at live birth rates; we submit our numbers to the CDC (Centers for Disease Control and Prevention) and to SART (the Society for Assisted Reproductive Technology) the way most other programs do. We're always evaluating live birth rates per cycle, per retrieval, and per transfer. Those numbers tell us different things. Live birth rate per cycle gives us an overall success of the procedure. Live birth rate per retrieval gives us more information about what is happening at the laboratory level. And live birth rate per transfer tells us about the quality of our embryos.

GeneWatch: Do you have any way of tracking what happens down the road after people leave the lab?

Stern: We don't specifically. We know whether or not there are live births, but we don't track long-term. I'm presently involved in a study, the Infertility Family Research Registry, an NIH-funded project in which we are trying to recruit volunteers nationwide who've been fertility patients to look at the long-term health of women and children from these procedures; but we don't do that specifically with the patients we have here. From the clinic perspective, that is a massive undertaking. We have a lot of trouble just keeping track of the people who have frozen embryos here, truthfully. And to track people—who very often don't want to be tracked, by the way—is a whole other endeavor.

GeneWatch: About how many embryos do you suppose you have at the lab at any given time?

Stern: Hang on, I'll look it up . . . seven hundred and fifty two embryos.

GeneWatch: You've mentioned that the goal of IVF is the delivery of a single healthy child. Can you explain where that goal comes from, and have you run into much opposition pushing for it?

Stern: We want a single healthy child because multiple pregnancies are much more risky than single pregnancies. So what we're trying to do is have singleton pregnancies be the result of our procedures—not even twins, and certainly not high order multiples. We've gotten better about higher multiples; twins, less so. We've had a fairly high twin rate, which on the one hand means that we're making very nice embryos; when we transfer two, we often end up with two. But we don't want that. We've really been pushing lately to go down to single embryo transfer, but it's very difficult, particularly in New Hampshire, where we don't have mandated coverage. You have patients who come in and are paying out of pocket. They have one chance to do a cycle. They're sitting there with two beautiful embryos. We can't tell them that both of them are going to be good and maybe one will be and the other one won't be . . . so which one do you transfer? There's a lot of push back from patients who very often are saying that they want to have both of them transferred. We're very careful, and we stick within the ASRM (American Society for Reproductive Medicine) guidelines. We never transfer more than two, for example, in women under thirty-five. But when it goes from two to one, it becomes very difficult sometimes to convince patients to do that. We're working very hard on that at the moment. We're offering it to everybody and we're pushing it especially for people who have good-quality embryos.

GeneWatch: If a patient asks to have six embryos transferred, can you say no?

Stern: Yes, we can, and we do.

GeneWatch: At what point can you say no?

Stern: We tell them that right up front. We tell them that we follow the national guidelines, so six is actually out of range of the guidelines at any age. We actually give them a table with the national guidelines and say, "These give you the max that we will do in whatever age category, and if the cycle has gone well and the embryos look good, we're going to recommend transferring the lower end of whatever that range is for your age."

GeneWatch: So for somebody who is older, the range is higher?

Stern: The range is higher, yes. We might transfer three or four in somebody 40 years old, which we would never do with someone under 35, but we stick within those guidelines. I mean, we are always willing to transfer fewer if somebody wants to transfer fewer, and there are occasionally patients in the older age ranges who say no, that they wouldn't want multiples no matter what. That's great, but that's not the way the conversation usually goes. It's usually the reverse.

GeneWatch: What do you do to convince people? How do you convince someone to do a single embryo transfer if she is, say, 30 years old?

Stern: I think that discussing the risks of multiple pregnancies and making that real to people is what is really important. People very often think when they go through this, when they're trying to have a child, that it looks as though having two at the same time would be great. Then they've gone through it once, they have their family and they're done. What people don't always realize is how risky those pregnancies can be. I think getting patients to understand that is really critical.

GeneWatch: You also do a lot of work with SART. What's your role there currently? You're the research chair, is that right?

Stern: I was the research chair for five years, and I'm still on the research committee. I've done a lot of work with that and what we did, and are still busy doing, are a lot of studies on outcomes of ART, looking at different categories of patients. We did a whole set on obesity; we've looked at patients' outcomes related to number of embryos transferred; we've looked at trying to maximize pregnancy rates while minimizing multiple rates in older patients by looking at the characteristics of patients and what we're transferring and that sort of thing.

We're also involved with several grants right now. We're trying to link information from the SART core database to existing vital record information. For example, in Massachusetts, we're working with a group at Boston University and the Massachusetts Department of Public Health on this. We have two NIH grants with them to try to look at long-term health of children from ARTs and to compare that to subfertile women and fertile women within Massachusetts. What we're really trying to get a handle on is long-term health of kids after these procedures, which unfortunately we know less about than we would like to know.

GeneWatch: Are there any procedures that seem to be particularly risky or that you would have some reservations about going through with?

Stern: Well, there's always concern about newer procedures when they come in. I mean, when we first started doing ICSI (intracytoplasmic sperm injection), there were concerns about whether or not it was an increased risk of a genetic or epigenetic abnormality, and we're still concerned that there might be a very small increased risk using this. I mean, when you think about that procedure, you're bypassing all of the physiology of fertilization and you're taking a sperm and putting it into an egg with a needle. You would imagine that there might be risks. So that's something that was a concern early on. There have now been thousands and thousands of births from ICSI, and while there are still papers that indicate that there may be a risk, we are pretty sure that it's fairly low. But, you know, that's still a concern.

As newer procedures are brought in, there's always some concern about this kind of thing. Egg freezing now is one that is raising these sorts of concerns—is this safe? And we're not sure. It's still being done under experimental protocol because we don't know how safe that is. Egg freezing is different than embryo freezing, which we've been doing since the late 1980s. Egg freezing is very new. It's just freezing the unfertilized egg, but that's where there is some concern, even though it's being marketed at a number of clinics for women who want to delay childbearing. They should be aware that there are some potential risks there and that we don't know the outcome of these procedures.

In vitro maturation is another one of those. It involves taking immature eggs from the ovaries and maturing them in vitro. Currently we give women lots of hormones to make multiple mature eggs in their follicles and we retrieve the eggs at the mature stage and use them for IVF; with in vitro maturation we don't do the hormone manipulation, or we do a minimal hormone manipulation and get the eggs out at an immature state and grow them up at the laboratory. This means that they're spending a lot more time in the laboratory, and it also means that the maturation process may be modified in some way from what it is in the body, *in vivo.* We don't know what the risk of that is yet, even though that's now being used a little bit more than it used to be.

So every time we bring in a new procedure, particularly new laboratory procedures, there's concern about the risk. It's kind of a Catch 22 in some ways, in that we can't really learn about the risk in people—assuming we have reasonable animal studies on all of these, which we usually do—until we really do it; and we can't do it, or we don't want to do a lot of it, until we know what the risk is. So we end up in a situation where if we really think it's valuable, we have to move forward cautiously to begin with. One of the things that we should include in that caution is human subject approval and the full consent of patients who realize how new the procedures are. But we won't learn about it unless we do it, you know . . . and we can't do it unless we overcome some of that caution.

GeneWatch: Another part of what SART does is create guidelines for assisted reproductive practices.

Stern: Yes, most recently in 2009. Every couple of years, they come out with slightly lower numbers.

GeneWatch: You mean a slightly lower number of embryos?

Stern: Yes, telling us to transfer fewer. Which is good. Keep in mind as I've been doing this for twenty-five years, when we first started doing this, the success rate was about 5 percent per cycle nationally. And over the past twenty-five years, we've gotten much, much better at keeping embryos alive and in good condition and at the point where they will implant. As we get better and better and better, the number that we've wanted to transfer has gone down.

GeneWatch: But these are still just guidelines, right? They aren't actually enforceable?

Stern: They're not laws. In some countries, the number transferred is legislated, but not in this country.

GeneWatch: So, do you think that there's good reason for it to be regulated here?

Stern: Well, I have concerns about anything that goes to a legislature in this country, given the state of our political system. I think that a much more effective way of dealing with this would be for insurance companies to recognize that covering IVF with a limitation on the number of embryos transferred makes far more financial sense to them than what they are doing now, which is insisting that people do multiple cycles of intrauterine insemination (IUI), where we have no control at all over multiple births, before they do IVF. And insurance companies may pay for some of the IVF procedure or not pay for any of it, but they end up paying for all of the neonatal intensive care unit (NICU) costs. They pay all of the NICU costs for multiple pregnancies, yet they won't, for the most part, come out with policies that say that "We'll pay for five or six cycles of IVF if you only transfer one embryo at a time." That makes a heck of a lot more sense to me than laws or any other way of doing it, and I think that it would make a huge difference in the multiple rate in this country because, as I said, one of the driving forces behind patients' desire to transfer more embryos is that they're paying out of pocket. So they pay for IVF out of pocket and transfer multiple embryos, then they end up with a million dollars in NICU costs that are paid by the same insurance company that refuses to pay for the IVF. That just doesn't make any sense to me.

The Fast and the Furious

BY LORI HAYMON

Lori Haymon, J.D., *works with the American Society of Human Genetics (ASHG), is a former intern with the Council for Responsible Genetics, and author of the CRG report, "Non-Invasive Prenatal Genetic Diagnosis (NIPD)." This article originally appeared in* GeneWatch, *June-July 2011.*

Non-invasive prenatal diagnosis, or NIPD, is scientifically groundbreaking.[1] It completely eliminates the risk of miscarriage, which has been associated with prenatal genetic diagnosis for the past forty years.[2] Based on the discovery of cell-free fetal DNA, researchers are now able to diagnose disorders and disease from just 10 ml of blood, and as early as six to ten weeks after gestation.[3] With the development of faster and more comprehensive DNA analysis, NIPD is projected to become the earliest, most comprehensive, and least expensive means of prenatal genetic diagnosis.[4] So why are so many scholars concerned by the prospect of clinical NIPD?

The ethical and social implications of NIPD are as extensive as its promised applications. Scholars and advocates from various fields argue that NIPD will not only exacerbate current ethical issues in prenatal diagnostics and screening, but it will also create entirely new issues. NIPD, some contend, will erode informed consent, blur the line between medically necessary and non-medical fetal testing, obviate the disability rights movement, undermine disability treatment efforts, and reshape consideration of reproductive freedom. Some scholars have gone so far as to call NIPD a sham and a cover-up for modern-day eugenics.[5]

Increasing fetal sex preference in the U.S.

Fetal sex determination is generally *not* for medical reasons.[6] Instead, even in the U.S., the more common reason for pre-conception and prenatal sex determination is preference.[7] The application of NIPD to fetal sex determinations means faster decisions, and potentially rash and irreversible decisions.

Most studies in the U.S. conclude that the majority of Americans prefer either the same number of children of either sex or have no preference at all.[8] Still, a "fundamental dynamic between technology and culture" exists, which is able to "coax cultures one way or the other by making it easier" to do what was difficult before.[9] To this effect, one study on sex preference in the U.S. has already shown that technology can affect these *currently* held values.[10]

Confronted by a hypothetical pill that would simplify and ensure fetal sex, respondents in this study changed their views on using preconception sex-selection technologies.[11] Once the hypothetical pill was introduced, respondents willing to use sex selection technology increased by 10 percent, and the number of respondents who answered "undecided" rose to above 20 percent.[12] While a majority of the respondents did not change their responses, those who did reveal an affect that is very likely to be seen with the widespread use of NIPD.

Certainly the cumbersome nature of prenatal genetic screening and diagnostics has acted as a "checkpoint," providing a reason to consider whether the sex of the child, or any other trait for that matter, was really worth the additional procedures, associated risks, and expense. NIPD is likely to eliminate that checkpoint. Moreover, many expectant parents may find themselves changing their views on what trait preferences they have and what technology they are willing to use from the comfort of their own homes.

Providing information of questionable benefit to patients who may not be adequately "informed"

Informed consent and informed decision-making are both cornerstones of medical care and medical research.[13] A few of the commonly agreed upon aspects of informed choice include information about the test itself, including the limitation and significance of the results; written consent; and reflection time.[14]

One study, performed in the UK, found that health professionals are more likely to follow the informed consent practices associated with prenatal screening, as opposed to those associated with invasive prenatal diagnosis.[15] In a multiple vignette study, researchers found that while 96 percent of clinician respondents indicated that written consent should precede invasive prenatal diagnosis, only 68 percent indicated the same for NIPD testing. Similarly, reflection time was important to 94 percent of the clinicians responding to the invasive prenatal diagnosis vignette, but to only 76 percent of those responding to the NIPD vignette.

Whether NIPD will provide medically relevant information for most conditions is also debatable. As David Litwack, AAAS Science and Technology Policy fellow, stated with regard to direct-to-consumer tests, "[i]n most cases, the results of genetic testing cannot be used for practical decisions about health care."[16] In part, this is because patients have the same options before as after genetic tests are performed. Yes, the medical information available has changed. But the medical actions that can be pursued have not. Patients are not being provided with additional or alternative medical courses of action—just *predictive* information.

As such, the ability of physicians and other medical professionals to make this information functional and to facilitate informed decision making becomes imperative. According to Theresa M. Marteau and Elizabeth Dormandy, "[t] hirty years after the routine introduction of prenatal diagnostic tests we remain unaware of how women are counseled, the information and support they receive, and how this affects the quality and type of decisions they make."[17] So whether patients will even benefit from the "information explosion"[18] that NIPD promises is uncertain.

Reinforcing the current genetic testing trends of faster, rather than accurate results

Moreover, despite confident promotion by researchers and providers, NIPD is not fully diagnostic. It is limited to identifying only paternally derived, or *de novo* genetic mutations.[19] Nor is NIPD less expensive than current invasive testing[20]—in part because isolating cell-free fetal DNA from the maternal serum is a "significant technical challenge."[21] Fetal cell-free DNA, which is the basis of NIPD, is outnumbered 20 to 1 by cell-free fetal DNA belonging to the pregnant woman.[22] This also affects the verification and confirmation of NIPD.

Thus far, NIPD results have only been verified through the use of amniocentesis and CVS.[23] But verification by the very technology that NIPD is supposed to replace would directly negate its benefit and is highly unlikely to transition to the clinical setting. If NIPD is to be clinically implemented, some procedure for confirmation is necessary, however, because NIPD has yet to reach diagnostic-level accuracy.[24]

In future developments of NIPD, researchers and developers are likely to find solutions to all of the listed limitations. What is troubling is that NIPD may be implemented clinically long before they do.

NIPD may become the standard of care for prenatal diagnosis, whether it reaches diagnostic-level accuracy or not.[25] For one thing, novel diagnostic tests can be introduced into the clinical setting without outside regulatory requirements, as long as they are developed and validated by the original providers.[26] Some applications of NIPD may bypass clinical validation completely by being offered as direct-to-consumer products. However, the DTC market has been anything but accurate in its promotion of valid testing results. Just last year, investigations by the United States Government Accountability Office (GAO) found that "identical DNA samples yield contradictory results," leading to their conclusion that DTC test results are misleading and of little or no practical value.[27]

Like any other genetic test, NIPD will provide "predictive," not "certain" tests results. But NIPD is likely to be even more misleading, if for no other reason than because of the impression of accuracy that submitting blood (as opposed to saliva) carries. Moreover, prospective parents may be more vulnerable and more likely to "seek a conclusive determination as to whether they are going to have a healthy baby."[28] The prospect of pregnant women making irreversible decisions based on inconclusive NIPD information is unacceptable.

Reshaping reproductive freedom

Developers and supporters of NIPD often describe the procedure as a corollary right flowing from the right to an abortion, a means of providing "prospective parents [with] preconception or prenatal information about the genetic characteristics of offspring, so that they may decide in a particular case whether or not to reproduce."[29] Indeed, clinicians are legally obligated to provide pregnant women with all sufficient information necessary to make informed reproductive decisions.

NIPD fits well into the current goals for advancing a woman's right to reproductive choice. But it is also likely to reshape current conceptions of reproductive freedom, raising new dilemmas for both pro-life and pro-choice communities.

"Pro-choice" may be defined as the recognition and defense of a woman's right to self-determination regarding sex, sexuality, reproduction, and motherhood and the promotion of equal access to abortive services, reproductive services, and sex education. The National Abortion and Reproductive Rights Action League (NARAL) web page advertises its belief "in reducing the need for abortion" by way of "improving access to birth control and teaching young people comprehensive sex education." Clinical implementation of NIPD raises new questions. Will sex education now include genetic inheritance and disability if women will be routinely confronted with decisions based on prenatal genetic information? How different is the issue of choosing to have *this* pregnancy, and choosing to have *this particular* pregnancy?[30] To what extent does trait-selection abortion coincide with being pro-choice? And is a "healthy pregnancy" destined to become a list of wanted versus unwanted traits?

For the pro-life community, NIPD brings to the forefront the question of whether one *should* bear children with life-altering or life-threatening genetic conditions. It may also shift the pro-life focus away from the usual adoption alternative. NIPD will most often be used by expectant mothers of higher income, whose motivation for considering abortion may not be "whether a woman can raise this child" but rather *if* this is "the child she wants to raise." Hence, the usual adoption alternative (*i.e.,* bringing the pregnancy to term and putting the child up for adoption) will no longer suffice. NIPD may also "normalize" abortions based on prenatal diagnostics, and, according to some, trivialize any significance associated with terminating a pregnancy. That "[e] very pregnancy becomes a 'tentative pregnancy' pending the results of prenatal screening"[31] seems to be the eventual result of NIPD.

Wherever one stands on the pro-choice, pro-life debate, NIPD will change the medical context in which women make reproductive choices. As one commentator notes, "[e]ach accurate detection presents the possibility of aborting a very sick fetus, which, if born, could cost its parents (and its parents' insurer) large amounts of money. Thus, the medical necessity standard may be a moot point when it comes to coverage of [NIPD]."[32] That NIPD provides a benefit to third-party payers is also troubling.[33] In light of these observations, the question must be asked whether NIPD will provide a woman with information that she can use to make her decisions, or information that makes the decision for her. As George Annas states, "[w]hen non-invasive prenatal genetic testing is available and reasonably priced, there will be tremendous pressure from many sources to use them."[34]

Exacerbating the tensions between disability rights and prenatal diagnosis

Although the general consensus is that "the purpose of prenatal testing is to enhance reproductive choice for women and families—not to decrease the number of children with disabilities who are born," it is difficult to ignore the "tension between the goals of enhancing reproductive choice and preventing the births of children who would have disabilities."[35]

The long history of invidious discrimination against persons with disabilities rightly instructs fears about continuing developments in prenatal genetic diagnosis. Like discrimination generally, prenatal diagnostic testing focuses solely on a specific trait, allowing that trait to stand in for the entire worth of an individual.[36] Disability advocates argue, among other things, that prenatal diagnosis techniques *reinforce* the medical model that disability itself is the problem to be solved.[37]

NIPD is certainly part of a medical model that focuses on "solving the problem" of disabling genetic conditions. And even in the situation where test results are used merely to prepare for having a child with a disability, testing may still "send the message that there's no need to find out about the rest"—whether or not the trait is present is all that matters.[38]

Many, if not most, of us would have trouble with the idea that "someone like you will never be born again." With NIPD, the option of early termination not only means that someone with Down syndrome, or Tay-Sachs, or sickle-cell anemia, may never be born again, it may also imply that someone with such a condition should never have been born. Harriet McBryde Johnson expressed the same in her critique of the writings of Peter Singer (a well-known advocate of selective abortion of disabled pregnancies):

"He insists he doesn't want to kill me. He simply thinks it would have been better, all things considered, to have given my parents the option of killing the baby I once was, and to let other parents kill similar babies as they come along and thereby avoid the suffering that comes with lives like mine and satisfy the reasonable preferences of parents for a different kind of child. It has nothing to do with me. I should not feel threatened."[39]

Thus, while discrimination against persons living with disabilities is no longer tolerated, the termination of diseased/disabled pregnancies is not only promoted, it threatens to become common practice.[40] How will we resolve this contradiction in contemporary goals?

And what about "less than life-altering" genetic conditions, such as cleft lip, or hereditary deafness, or ectrodactyly (the deletion of digits on the hands and/or feet)? The account of Bree Walker Lampley, whose choice to have children with her genetic condition, ectrodactyly, and the backlash she encountered from those who saw her choice as "irresponsible" and "cruel," is informing. It reflects just how far we are from a consensus on the definition of serious medical conditions.[41]

One study of 1,481 certified genetic professionals[42] showed substantial overlap among what genetic/congenital conditions were considered lethal," "serious but not lethal," and "not serious."[43] Sixty-four percent of the conditions listed as lethal by some respondents were considered "serious but not lethal" by others. Fifty-one conditions appeared in all three categories. These conditions included Down syndrome, cystic fibrosis, and Huntington's disease, as well as ectrodactyly and hereditary deafness.[44]

How the widespread use of NIPD will affect our perspective on disability must be addressed. Whether NIPD should be used to select against a pregnancy with the potential to have such a seemingly minor genetic condition as a missing finger or toe must also be considered.

Prenatal genetics and eugenics: a slippery slope

NIPD is likely to increase the frequency of selective abortions. As one commentator notes, " . . . from a dollars-and-cents perspective, [prenatal screening] is an unnecessary expense. Developers of prenatal testing, however, have justified these unnecessary expenses by demonstrating through cost-effectiveness studies that such costs can be offset to the private insurer or the public healthcare system, provided that enough children prenatally identified with Down syndrome are terminated."[45]

The orientation of current developments in the prenatal field has led many to argue that technology like NIPD may lead to institutionalized eugenics.[46] Other scholars note that procreative liberty provides for the right *not* to use reproductive genetics, as well as the right to use reproductive genetics,[47] and that we are not likely to enter a situation of forced genetic selection.

The "old eugenic" practices of involuntary sterilization and confinement[48] are not likely to result from the widespread use of NIPD. Still, there is a crucial difference between "preventing children from being born with Down syndrome" and "preventing children with Down syndrome from being born."

According to many scholars, the "new eugenics" movement focuses on the offspring directly. Rather than involuntary sterilization, new eugenic techniques are based on individual choice. Choices made based on *scientifically* accurate screening, diagnosing, and selective termination. The purpose of new eugenic techniques is to give "parents power over their children that parents cannot exercise once the children are born,"[49] or to provide parents with the "discretion to select—or not—the characteristics of [one's] offspring."[50] Even in the most favorable light, these goals are very difficult to separate from "improving the human population by controlled reproduction and decreasing the occurrence of undesirable characteristics and conditions,"[51] the very definition of eugenics.

The ways in which NIPD and other forms of genetic diagnostics threaten our preservation of a non-eugenic society cannot be taken lightly. There appears to be no end to the possible uses or clinical applications of NIPD. In fact, the only plateau foreseeable for prenatal genetic diagnosis now will be policy-based. For the public and policy makers, it is important to recognize that each step taken toward more controlled forms of reproduction must be implemented with caution, if implemented at all.

35

The Southern Baptist Ethics of Arts

INTERVIEW WITH BARRETT DUKE

Barrett Duke, Ph.D., *is vice president for Public Policy and Research and director of the Research Institute of the Ethics & Religious Liberty Commission. The Ethics & Religious Liberty Commission (ERLC) is the Southern Baptist Convention's agency for social and moral concerns. This interview originally appeared in* GeneWatch, *June-July 2011.*

GeneWatch: Where do the Southern Baptist positions come from? Who decides where you fall on the issues and what you'll be pushing for in Washington?

Barrett Duke: There are a couple of guides for us. One is the resolution process that Southern Baptists engage in each year in their annual meeting. We have a committee that considers resolutions on a whole range of social, moral and theological issues. Oftentimes they will propose, and the annual meeting will adopt, a resolution that addresses a moral issue in some kind of way that has policy implications. My organization, the Ethics & Religious Liberty Commission, has the responsibility to basically make sure that the values that resolution calls for are promoted in Washington, D.C., and around the world as well. Since the Convention often only meets once a year and there are a lot of issues that come up during the rest of the year, our commission will often need to look at what the issue is, talk to professionals in that particular area, develop a position that we believe reflects the teachings of the Bible and the commitments of Southern Baptists, and begin to work toward it.

GeneWatch: Would I be right in thinking that in the Southern Baptist positions on reproductive technologies, the central theme is protecting the sanctity of life?

Duke: That is the overriding theme for most Southern Baptists, and certainly the Ethics and Religious Liberty Commission: to protect life at all stages.

GeneWatch: Beginning at fertilization?

Duke: Yes, beginning at fertilization, and even including the process toward fertilization. We believe that there are certain ethical boundaries even prior to fertilization that couples and societies ought to respect as well.

GeneWatch: Does that apply then to egg donation, sperm donation, surrogacy, or other assisted reproductive technologies?

Duke: On most of these issues there isn't a so-called "official" Southern Baptist position; but most Southern Baptists are in agreement that while we certainly sympathize with couples who have difficulty having children, and we want to support them, we still believe there are certain boundaries that need to be maintained. The issue of surrogacy is a serious one for us. It raises serious questions about who the parents are, who has legal rights to the children, the parents' relationships with the children and how they are raised. We have serious concerns about the whole concept of surrogacy, and with egg and sperm donation as well.

GeneWatch: So in this case, it's less about sanctity of life than about sanctity of marriage?

Duke: No, I would say the issues are still primarily about the sanctity of life, but they certainly have implications for marriage as well, especially with the issue of surrogacy. But when you're talking about sperm donation, are you talking about sperm donated from the husband or from another male? So there are also implications here for what exactly is the nature of marriage itself.

GeneWatch: And the Southern Baptist Convention encourages adoption as an alternative for infertile couples.

Duke: We do encourage adoption, not only for infertile couples but for all couples who would consider the needs of thousands of children who are unable to experience the important relationships with a mother and a father. So we encourage adoption for all couples, but we certainly encourage adoption for infertile couples. And that can be adoption not only of children who have already been born, but also adoption of embryos in fertility clinics, where they have been frozen and are waiting to be given the chance to fully develop—what's referred to as a "snowflake adoption."

GeneWatch: The clinics are there, the embryos are already there, but how do you feel about the purposes of the fertility clinics themselves?

Duke: Well, we certainly have concerns about how fertility clinics operate. Again, on this question there is no official Southern Baptist position. We do know that there are Southern Baptist couples who have resorted to fertility clinics and in vitro fertilization, and what we encourage is that couples who are considering that option fertilize no more eggs than they are prepared to have implanted, so they make sure that they are at least not deliberately planning to basically conceive children that are never going to be given a chance to fully develop.

GeneWatch: I imagine that would be an issue in pre-implantation diagnosis or sex selection, where a couple might choose a fertilized embryo that has their desired traits. Of course, genetic manipulation can go well beyond that. Is there an official Southern Baptist position on human genetic manipulation?

Duke: The Convention is on record in opposition to cloning, and our members are clearly in opposition to sex selection and to any genetic selection criteria. The Convention has significant concerns with any effort to determine which humans have the right to life and which ones do not.

GeneWatch: Are there certain passages of Scripture that you find yourself bringing up to illustrate these positions?

Duke: I think our arguments generally begin with the teaching in Genesis that human beings are created in the image of God. It doesn't tell us that some human beings are created in the image of God, or that some human beings have more of His image than others; the Biblical teaching is that every human being is created in the image of God and thus deserves equal respect and, we believe, equal protection under the law.

GeneWatch: Issues like human genetic manipulation and assisted reproductive technologies often seem to create strange bedfellows. Have you found yourself working with unexpected allies on Capitol Hill?

Duke: We do. In this work, you find groups that you have an issue in common with, and you work with them on that issue. Maybe you don't agree on hardly anything else, but you do find agreement on that issue. When you can find common cause, you should work together. Right now we're working on gene patenting, which has us working with some groups considerably to the left of us on many social issues, but we're all in agreement that it's unethical to allow anyone to own naturally occurring genetic material.

GeneWatch: Do you also encounter any unexpected opposition from folks you might otherwise be in agreement with?

Duke: You know, on the gene patenting issue, I can't say that we've encountered opposition from any surprising groups; but there are occasions where, when we join in a coalition, we don't get everybody who is a natural ally with us. Often they have concerns about possible implications of a policy that they just can't support, so they end up opposing it. So yes, on occasion we end up in opposition with groups who tend to be natural allies on other issues.

GeneWatch: Some of our other contributors have addressed the reluctance of some pro-choice advocates to involve themselves too deeply in issues surrounding assisted reproductive technology, for fear of muddling their message on women's reproductive rights. How do the Southern Baptist positions on assisted reproductive technologies fit into the pro-life point of view?

Duke: I think that the Southern Baptist Convention is fairly mainstream on its commitment to the protection of life. There are some groups that find it more difficult to engage in incremental strategies toward protecting life—they feel that they have to protect all of life if they are going to protect any life. So it makes it more difficult to engage in some strategies that would at least limit, for example, the number of abortions. We might back some policy that moves us closer to ending all elective abortion, and if it doesn't accomplish the entire goal, we might still feel that it is significant progress. Some groups don't feel like they can make that step. We understand that, and we sympathize with that concern. So yes, there are times when the particular issue that we're involved in doesn't get agreement from everybody; but the long-term goals are the same between Southern Baptists and most other pro-life groups as well, and that is to end the current regime of elective abortion as it exists in this country today.

GeneWatch: How high do reproductive technologies rank on your priority list?

Duke: The sanctity of life is our top priority issue; everything from conception to natural death, anything that touches on the sanctity of human life is a top-tier issue for us. But oftentimes there just isn't any movement at the moment in any of those areas, and I would say that right now, there just isn't any movement on reproductive technologies. There is movement on some abortion questions, and we're dealing with this gene patenting question, but I'm not aware of any significant movement of trying to reign in the current Wild West practices in the United States regarding reproductive technologies. If there were—if we found some interested parties who were trying to get movement on that—we would certainly partner with them and try to make some progress on it. It is a sanctity of life issue, so it is a top-tier issue for us, and we are ready to begin working on it as soon as we feel that there is an opportunity.

Willful Ignorance

BY ABBY LIPPMAN

Abby Lippman, Ph.D., *is a Professor in the Department of Epidemiology, Biostatistics, and Occupational Health at McGill University and a Board Member of Fédération du Québec pour le Planning des Naissances (FQPN) and Alliance for Humane Biotechnology.* *This article originally appeared in* GeneWatch, *volume 14, number 2, June-July 2011.*

It was almost 40 years ago that I first began to question and explore what were then the "new" reproductive technologies. Reflecting back on those decades reveals how much—and how little—has changed. Many of the technologies themselves are now the realities only imagined then. However, the questions and concerns about them are basically still the same, and still just as hotly debated. Are we (again) on the verge of "designer babies?" Will women's bodies be commercialized to provide children for those contracting surrogacy services? Will women's eggs be high-priced commodities for profit-making international brokers? Why do we still know so little about the environmental and occupational determinants of infertility? Perhaps the most notable change in the discussion has been a subtle shift in emphasis from concerns about their eugenic implications to the economic ramifications of the use of the technologies, as what Debora Spar has called the "baby business" proliferates and as "choice" has been reduced to a consumer option.

Assisted human reproduction comprises a range of technologies that were initially developed to make it possible for some women considered infertile because of biological problems to have children. Over time, many of these technologies have become primarily ways to circumvent social factors that prevent a woman or a man from being the biological parent of a child without outside assistance. And yet, despite being viewed as merely a "choice" for women, there continues to be grossly inadequate regulation and surveillance—or full public discussion—of their use, so that information necessary for informed decision making is woefully inadequate.

Canadian feminists began calling for the regulation of the range of reproductive (and genetic) technologies in the 1980s. This led to the formation of a Royal Commission to examine the "New Reproductive Technologies" and to the long-delayed release of its final (two-volume) report, "Proceed with Care," and its almost 300 recommendations in November 1993. Another decade passed before the federal government approved (in 2004) the Assisted Human Reproduction (AHR) Act and the creation of an agency to oversee activities in this area.

The jurisdiction of the federal government to act in what was labeled a matter of "health" (a provincial matter in Canada) was rapidly contested by Québec and no real progress—not even with regard to the practices the AHR Act proscribed as criminal offenses—was made to regulate the expansion and increasing use of the NRTs while the matter was (for too many years) before the courts. Thus, when the Supreme Court of Canada supported most of Québec's claims and thereby rejected the AHR Act in December 2010, it became a matter for the provinces to draft legislation to protect the health and interests of women and society.

In Quebec, this has turned out to be fairly shallow protection at best. Yes, this province now covers in vitro fertilization (IVF) with public funds as part of the universal medical care coverage offered to almost all residents. However, none of the regulatory and surveillance measures long demanded are in place. And while this financial coverage gives the appearance of promoting equity, since IVF technologies may now be available to any woman accepted as a candidate and not only to those with the financial resources to pay for the procedures involved, this "equity" may actually be more apparent than real.

The political decision to fund IVF and not midwifery programs, proper school-based sex education, or to expand publicly-funded daycare demonstrates the allures of—and lobbying for—technological fixes over societal and structural change that can truly reduce inequities. In this regard, funds allocated for these high-tech procedures could have been better used to remove some of the major determinants of "infertility" and to develop policies that will address those situations that now lead to women delaying their childbearing.

Thus, what has happened in Québec reflects what too often has happened with reprogenetic technologies everywhere: willful ignorance of the processes, policies, and other "upstream" societal, environmental, and structural determinants that lead to demands for interventions to circumvent infertility. And, while we lack the proper downstream regulation of assisted repro technologies necessary to ensure the health and safety of users and the children to be born, this is not all that's needed. Essential as they are, having laws—and even registers that track users and their offspring—must not deflect attention from what women need for social justice and for their complete sexual and reproductive health, measures that will also prevent infertility: safe places to grow up, live, work, and play; a system in which social and economic rights are secured; and conditions in which bodies and body parts are not commercialized.

It remains a priority, unaddressed even after 40 years of arguing for it, that we frame assisted reproduction within an overall reproductive health policy—one that would, for example, attend to ways to reduce infertility and not just ways to manage it once it was diagnosed. High-tech interventions should be last resorts at the margins, not the center of things. And to refocus, we need to reduce the space given to the scientific, commercial, and medical interests that tend to dominate the discussions and consider what the technologies and the technological approach means for individual women and women collectively.

Reproductive technologies are not a treatment of infertility but merely a way of circumventing some of the problems that keep women from having the children they may want. Unfortunately, there is still overemphasis on expanding access to these technologies rather than on seeking the causes of the underlying problems and ways to prevent them. It's not just my impatience at how slowly things change that makes me hope we will have the discussions we sorely need to set in place the social systems that will truly promote and protect the reproductive and sexual health of *all* women, those who do and those who do not want children. We don't have another four decades to act.

Prenatal Diagnosis and Selective Abortion

BY ADRIENNE ASCH

Adrienne Asch, Ph.D., M.S., *is the Edward and Robin Milstein Professor of Bioethics at Yeshiva University and professor of epidemiology and population health and family and social medicine at Albert Einstein College of Medicine.* *A longer version of this article originally ran in the* American Journal of Public Health, *November 1999 (Vol. 89. No. 11), p. 1649-1657. This article originally appeared in* GeneWatch, *volume 14, number 2, March 2001.*

> "Although sex selection might ameliorate the situation of some individuals, it lowers the status of women in general and only perpetuates the situation that gave rise to it. . . . If we believe that sexual equality is necessary for a just society, then we should oppose sex selection." —D. C. Wertz and J. C. Fletcher, *Feminist Perspectives in Medical Ethics*

> "The very motivation for seeking an 'origin' of homosexuality reveals homophobia. Moreover, such research may lead to prenatal tests that claim to predict for homosexuality. For homosexual people who live in countries with no legal protections these dangers are particularly serious." —Udo Schüklenk, *Hastings Center Report,* 1997

The tenor of the preceding statements may spark relatively little comment in the world of health policy, the medical profession, or the readers of this book, because many recognize the dangers of using the technology of prenatal testing followed by selective abortion for the characteristic of fetal sex. Similarly, the medical and psychiatric professions, and the world of public health, have aided the civil rights struggle of gays and lesbians by insisting that homosexuality is not a disease. Consequently, many readers would concur with those who question the motives behind searching for the causes of homosexuality, which might lead scientists to develop a prenatal test for that characteristic. Many in our society, however, have no such misgivings about prenatal testing for characteristics regarded as genetic or chromosomal diseases, abnormalities, or disabilities, as shown in the following quote from *Mapping Our Genes*:[1]

> "Human mating that proceeds without the use of genetic data about the risks of transmitting diseases will produce greater mortality and medical costs than if carriers of potentially deleterious genes are alerted to their carrier status and *encouraged* to mate with non-carriers or to issue other reproductive strategies [emphasis added]."

If public health frowns on efforts to select for or against girls or boys or to select for a particular sexual orientation but promotes people's efforts to avoid having children who would have disabilities, it is because medicine and public health view disability as extremely different from and worse than these other forms of human variation. At first blush this view may strike one as self-evident. To challenge it might even appear to be questioning our professional mission. Characteristics like chronic illness and disabilities (discussed together throughout this article) do not resemble traits like sex, sexual orientation, or race, because the latter are not in themselves perceived as inimical to a rewarding life. Disability is thought to be just that—to be incompatible with life satisfaction. When public health considers matters of sex, sexual orientation or race, it examines how factors in social and economic life pose obstacles to health and to health care, and it champions actions to improve the well-being of those disadvantaged by the discrimination that accompanies minority status. By contrast, public health struggles to eradicate disease and disability or to treat, ameliorate, or cure these when they occur, because, for medicine and public health, disease and disability is the problem to solve. So it appears natural to use prenatal testing and abortion as one more means of minimizing the incidence of disability.

Rationales for prenatal testing

The medical professions justify prenatal diagnosis and selective abortion on the grounds that the *costs* of childhood disability—to the child, to the family, and to society—are too high. Some proponents of the Human Genome Project from the fields of science and bioethics argue that in a world of limited resources, we can reduce disability-related expenditures if all diagnoses of fetal impairment are followed by abortion.[2]

On both empirical and moral grounds, endorsing prenatal diagnosis for societal reasons is dangerous. Only a small fraction of total disability can now be detected prenatally. Even if future technology enables the detection of predisposition to diabetes, forms of depression, Alzheimer's disease, heart disease, arthritis, or back problems—all more prevalent in the population than many of the currently detectable disability conditions—we will never manage to prevent most disability in the population. Rates of disability increase markedly with age, and the gains in lifespan guarantee that most people will deal with disability in themselves or someone close to them. Laws and services to support people with disabilities will still be necessary, unless society chooses to eliminate disabled people in addition to preventing the births of those who would be disabled. Thus, when we consider the total social costs of addressing disability there is small cost-saving in money or in human resources to be achieved by even the most vigorous determination to test every pregnant woman and abort every fetus found to exhibit disabling traits.

My moral opposition to prenatal testing and selective abortion flows from the conviction that life with disability is worthwhile and the belief that a just society must appreciate and nurture the lives of all people, whatever the endowments they receive in the natural lottery. I hold these beliefs because—as I show throughout this article—there is abundant evidence that people with disabilities can thrive even in this less than welcoming society. Moreover, people with disabilities do not merely take from others, they contribute as well—to families, to friends, to the economy. They contribute neither in spite of nor because of their disabilities, but because along with their disabilities come other characteristics of personality, talent, and humanity that render people with disabilities full members of the human and moral community.

Implications for people with disabilities

For children whose disabling conditions do not cause early degeneration, intractable pain, and early death, life offers a host of interactions with the physical and social world that people can be involved in to their and others' satisfaction. Autobiographical writings and family narratives testify eloquently to the rich lives and the even richer futures that are possible for people with disabilities today.

Nonetheless, I do not deny that disability can entail physical pain, psychic anguish, and social isolation even if much of the psychological and social pain can be attributed to human cruelty rather than to biological conditions. In order to imagine bringing a child with a disability into the world when abortion is possible, a prospective parent must be able to imagine saying to a child, "I wanted you so much and believed enough in who you could be that I felt you could have a life you would appreciate even with the difficulties your disability causes." If parents and siblings, family members and friends can genuinely love and enjoy the child for who he or she is and not lament what he or she is not; if child care centers, schools, and youth groups routinely include disabled children; and if television programs, children's books, and toys take children with disabilities into account by including them naturally in programs and products, the child may not live with the anguish and isolation that have marred lives of disabled children for generations.

Implications for professional practice

If prenatal testing indicates a disabling condition in the fetus, the following disability-specific information should be given to the prospective parents: information about services to benefit children with specific disabilities in a particular area, and which of these a child and family are likely to need immediately after birth, including contact information for a parent-group representative; and information about members of disability rights groups and independent living centers. In addition, the parents should be offered a visit with both a child and family, and an adult living with the diagnosed disability.

Although some prospective parents will reject some or all of this information and these contacts, responsible practice that is concerned with genuine informed decision-making and true reproductive choice must include access to this information. In addition, the provision of this information should be timed so that prospective parents can assimilate general ideas about life with disability before testing and obtain particular disability-relevant information if they discover that their fetus carries a disabling trait. These ideas may appear unrealistic or unfeasible, but a growing number of diverse voices support similar versions of these reforms to encourage wise decision making. Statements by Little People of America, the National Down Syndrome Congress, the National Institutes of Health workshop, and the Hastings Center Project on Prenatal Testing for Genetic Disability all urge versions of these changes in the process of helping people make childbearing decisions.

Ideally, such discussions will include mention of the fact that every child invariably differs from parental dreams, and that successful parenting requires a mix of shaping and influencing children and ruefully appreciating the ways they pick and choose from what parents offer, sometimes rejecting attitudes, activities, or values dear to the parents. If prospective parents cannot envision appreciating the child who will depart in particular, known ways from the parents' fantasy, are they truly ready to raise would-be athletes when they hate sports, classical violinists when they delight in the Grateful Dead? Testing and abortion guarantee little about the child and the life parents create and nurture, and all parents and children will be harmed by inflated notions of what parenting in an age of genetic knowledge can bring in terms of fulfilled expectations.

Public health professionals must do more than they have been doing to change the climate in which prenatal tests are offered. Think about what people would say if prenatal clinics contained pamphlets telling poor women or African American women that they should consider refraining from childbearing because their children could be similarly poor and could endure discrimination or because they could be less healthy and more likely to find themselves imprisoned than members of the middle class or than whites. Public health is committed to ending such inequities, not to endorsing them, tolerating them, or asking prospective parents to live with them. Yet the current promotion of prenatal testing condones just such an approach to life with disability.

Practitioners and policymakers can increase women's and couples' reproductive choice through testing and counseling, and they can expend energy and resources on changing the society in which families consider raising disabled children. If families that include children with disabilities now spend more money and ingenuity on after school care for those children because they are denied entrance into existing programs attended by their peers and siblings, public health can join with others to ensure that existing programs include *all* children. The principle of education for all, which is reforming public education for disabled children, must spread to incorporate those same children into the network of services and supports that parents count on for other children. Such programs, like other institutions, must change to fit the people who exist in the world, not claim that some people should not exist because society is not prepared for them. We can fight to reform insurance practices that deny reimbursement for diabetes test strips; special diets for people with disabilities; household modifications that give disabled children freedom to explore their environment; and modification of equipment, games, and toys that enable disabled children to participate in activities comparable to those of their peers. Public health can fight to end the "catch-22" conditions that remove subsidies for life-sustaining personal assistance services once disabled people enter the workforce, a policy that acts as a powerful disincentive to productivity and needlessly perpetuates poverty and dependence.

In order to make testing and selecting for or against disability consonant with improving life for those who will inevitably be born with or acquire disabilities, our clinical and policy personnel must communicate that it is as acceptable to live with a disability as it is to live without one and that society will support and appreciate every person with the inevitable variety of traits they bring with them. We can assure prospective parents that they and their future child will be welcomed whether or not the child has a disability. If that professional message is conveyed, more prospective parents may envision that their lives can be rewarding, whatever the characteristics are of the child they are raising. When our professions can envision such communication and the reality of full actualization and appreciation of people with disabilities, then prenatal technology can help people to make decisions without implying that *only one decision is right.* If the child with a disability is not a problem for the world, and the world is not a problem for the child, perhaps we can diminish our desire for prenatal testing and selective abortion and can comfortably welcome and support children of all characteristics.

PART IV

Artist: Sam Anderson

A recent article in *The New York Times* entitled, "Genetic Basis for Crime: A New Look," raised a backlash of criticism for its unsubstantiated implication that criminal behavior is in our genes. Just four months earlier, the U.S. Court of Appeals for the Second Circuit reassigned the case of defendant Gary Cossey from Northern District Judge Gary Sharpe after finding the judge committed plain error by ordering a longer sentence because he said Mr. Cossey had a genetic propensity to re-offend.

Such incidents are hardly an aberration. We are regularly inundated with claims linking genes to everything from promiscuity to philanthropy, from social networking and gang membership to a trove of studies connecting genetics to various attributes of human "political behavior"—from the fervor of one's partisanship to whether or not one turns out to vote—to a genetic basis. Such claims come from a long history of thinking that proposes that people can be fundamentally understood through their genetic makeup.

Two common themes emerge from many of these claims: a reliance on studies involving twins, and a failure to properly take into account environmental factors. Often these studies rely heavily on the "shared environment" assumption: that there is no significant difference between the way identical twins and non-identical twins grow up; that a pair of non-identical twins share the same environment growing up, just as a pair of identical twins do.

Whatever limits there are on the reliability of the methods, one of the big questions one has to ask, after reading studies trying to reduce complex human behaviors into a very narrow causal explanation, is: What is really being accomplished here? Why are we studying this? With more people in fields from psychology to political science turning to genetics, the question is more relevant than ever.

From criminal activity to intelligence to the pillars of behavioral genetics research, the chapters that follow explore the very basis for extravagant claims about genetic causation of human behavior.

Criminal Genes

BY EVAN BALABAN AND RICHARD LEWONTIN

Evan Balaban *is a member of the board of the Council for Responsible Genetics and associate professor of psychology at McGill University.* **Richard Lewontin** *is Professor of Zoology in the Museum of Comparative Zoology, Emeritus at Harvard University. He is the author of numerous works on evolutionary theory and genetic determinism, including* The Genetic Basis of Evolutionary Change and Biology as Ideology, The Dialectical Biologist *(with Richard Levins) and* Not in Our Genes *(with Steven Rose and Leon Kamin). This article originally appeared in* GeneWatch, *volume 20, number 2, March-April 2007.*

The idea that crime is "in the blood" is an old one. It was characteristic of social theory in the nineteenth century, especially reflected in the most popular literature of the time, where authors passionately explored its major facets. To what extent were criminal tendencies inherent in a person's makeup, and, once there, to what extent could they be changed? How did criminal tendencies get into people in the first place: from chance events during pregnancy or early childhood that left an imprint, from a systematically lawless environment, or from like begetting like?

Charles Dickens' Artful Dodger in *Oliver Twist* and the entire Macquart family described by Emile Zola in his sequence of twenty novels owed their criminal tendencies to hereditary defects. As Zola said in his preface to the first of the series, "heredity has its laws, just like gravity," although we would regard Zola's notion of those laws as rather quaint.[1] In the late nineteenth and early twentieth centuries, the rise of psychological theories of personality development, social reform movements, and the emergence of behaviorism ushered in an age where the psycho-social environment became the chief causative agent in the development of anti-social and criminal behavior. The reaction to biological theories of the National Socialists in Germany seemed to make hereditary notions of social behavior widely unacceptable. Over the last half-century, the immense expansion of knowledge and technique in genetics and the extraordinary publicity given to genetic research accompanied by the repeated claims that genes "determine" the organism have refocused attention on heredity as a major determining factor causing individuals to commit crimes.

One might suppose that genetic determination of criminal acts would present a serious challenge to the legal system, for how can society hold a person responsible for an act that was the inevitable consequence of his or her biological nature? Many legal scholars, however, are not convinced that biological information purporting to predict criminal behavior would present any new problems for legal systems. This is because insanity defenses and the doctrine of diminished responsibility are well-established parts of modern legal systems and behavior-genetic or other biological information can be accommodated under these rubrics.[2] The legal issue is whether a person is regarded as sane and therefore responsible for his or her actions, irrespective of the underlying causes of his or her insanity. The issue of causes is regarded as relevant only to the course of action to be taken in sentencing, but not for the guilt or innocence of the accused. So, if the possession of a given genotype only increased the probability of performing a criminal act but did not absolutely determine the action, there would be no serious challenge to legal doctrine.

The status of genetic arguments about "free will" is therefore not a cause for special concern in the immediate future; however, there is a faulty premise that may lead to more serious social problems. There is an uncritical acceptance by many non-scientific scholars of the notion that there has been or will soon be some kind of breakthrough in the ability to predict human behavior stemming from a better understanding of genetics.[3 4] Problems may occur when this faulty premise leads to the "informal application" of tentative behavior-genetic correlations, especially in the context of DNA databases. Individual rights and the rights of disadvantaged populations will be put at risk at early phases of criminal investigation, when such databases are used to select potential suspects in criminal cases.

Behavior genetics: the new hype

Most practicing human behavior geneticists have abandoned former research designs based on twin studies and embrace newer methodologies based on direct correlations between DNA sequence variation and variation in performance on behavioral measurements. The first stage of such an investigation is to survey a very large number of known human genes that exhibit many variations to find statistical correlations between variant gene alleles and the incidence of some behavioral syndrome, such as schizophrenia or abnormal aggressiveness. Among the genes for which statistically significant correlations are found, those known to code for gene products that would reasonably seem to be related to the central nervous system are identified as "candidate genes" for further investigation. There are two serious problems with this approach.

The first is a statistical fallacy. A correlation between two variables is considered "statistically significant" if a correlation as large or larger than the one observed would rarely occur by pure chance in a sample of observations, without any real causal connection. In conventional practice, "rarely" means 5 percent of the time or less. But if one surveys, say, 10,000 genetic polymorphisms for a correlation with a behavioral syndrome, then by chance up to 500 of those polymorphisms could turn out to be significantly correlated to the behavior at the 5 percent level of significance even if none of them has any real causal relation to the syndrome. How is one to decide which, if any, of these genetic variations is causal? One possibility is to run a second, independent, study on a new sample and see how many of the original 500 significant correlations will show up again. But by pure chance 25 (5 percent of 500) of the original correlations will appear in both samples. How many of these are real? And how many real connections have been missed because a relevant gene polymorphism only increases the probability of the behavior by a small amount too weak to be detected by the study design?

The next step is to look for candidate genes among those identified statistically, based on knowledge of which of these codes for central nervous system proteins are likely to be related to behavior. The second problem arises at this stage. Behavior is more complicated than brains dictating a series of behavioral instructions. Like other vertebrates, humans have a neuroendocrine system in which there is a constant feedback between nervous system states and the secretion of a variety of hormones to which neurons are sensitive. These hormones, in turn, have their rates of production modulated by more general physiological cues, such as blood sugar levels and chemical products of muscle fatigue. One of the most serious errors in understanding human behavior is to suppose that the brain is an isolated and autonomous causal organ. Reciprocal causal interactions between the brain and the rest of the body make the number of candidate genes for any behavioral syndrome very large, encompassing a large part of any organism's metabolic functions. This problem is exacerbated if the genes that have a real causal connection to the target syndrome do not completely determine the behavior but only make it somewhat more likely, say, by a factor of 25 percent.

An example of the candidate gene approach that has become a poster child for the new human behavioral genetics as applied to criminal behavior is a study that involved the comparison of individuals with DNA sequence differences in a chromosomal region just in front of the gene for an enzyme present in many parts of the brain called monoamine oxidase A, which is involved in the metabolism of several different brain neurotransmitters.[5] A cohort of about 540 "Caucasian" boys growing up in New Zealand was monitored every few years from the ages of three to twenty-one years for a multidisciplinary health and development study. Evidence of maltreatment during the first eleven years of life was gathered from these periodic observations, from parental reports, and from retrospective reports by study members and their families. Court records of violent convictions in adulthood were obtained (11 percent of the subjects had criminal records). A "disposition toward violence" was ascertained by using a personality questionnaire (containing items like, "When I get angry I am ready to hit someone," and "I admit that I sometimes enjoy hurting someone physically"), to which subjects provided a number on a scale indicating how much they agreed or disagreed with the statement. Finally, subjects provided the investigators with the name of someone who knew them "well," and the investigators mailed a questionnaire to these persons.

Molecular genetic variation was split into two categories: variants of this region of the monoamine oxidase A (MAOA) gene that previous studies had suggested to have "high enzyme activity," and ones that previous studies had suggested to have "low enzyme activity" (based on studies in cancerous or non-neural cell lines that do not entirely agree with one other). The investigators found that when their subjects were classified according to the presumed activity of their alleles, and the extent of their maltreatment as children ("no maltreatment," "probable maltreatment," "severe maltreatment"), the convictions for violent offenses, proportion of conduct disorders, extent of disposition toward violence, and antisocial personality symptoms all varied positively with maltreatment, but there were larger increases in these behaviors with maltreatment for possessors of "low activity" alleles than for possessors of "high activity" alleles. According to the authors, "These findings provide initial evidence that a functional polymorphism in the MAOA gene moderates the impact of early childhood maltreatment on the development of antisocial behavior in males."

Since this study also clearly shows that childhood maltreatment doesn't do the holders of any allelic MAOA gene form any good, it is unclear what the proponents of this type of research would have us do based on allelic differences, especially since subsequent research disagrees on how big the effect of allelic variation is and the extent of psychopathologies in possessors of the "high-activity" allele.[6] While the original study statistically excluded the result of "no difference" between the two allelic variation categories, four other subsequent studies either do not (2 studies) or are very close to not doing so (2 studies).

Modern molecular behavior genetic work remains as tentative and inconclusive as its historical, non-molecular counterparts do. Yet its proponents have become much more sophisticated about discussing the complexity of the link between genes, brains, and behavior, even if a vocal sector of this community remains dedicated to advancing the proposition that individual behavioral variation can be meaningfully predicted from genetic sequence variation, and that we will not commit any egregious violations of individual rights by starting to apply such information. We believe that such points of view reflect a fundamental conflict of interest, are scientifically flawed, ideologically motivated, and, if generally accepted, could potentially lead to significant setbacks in the rights of individuals in democratic societies, disproportionately affecting those individuals in socially and economically disadvantaged groups.[7 8]

What is criminal behavior?

Another problem with the studies of the genetics of "criminal behavior" lies in their extremely narrow view of what constitutes crime. In practice, genetic studies have identified crime with crimes of violence against individuals, a bias that is inevitable when there is an attempt to connect genes and behavior. What would be the criteria for a candidate gene whose coded product led to an increased probability of, say, embezzlement, securities fraud or automobile theft? While these are undoubtedly a consequence of mental states, what sort of biochemical or developmental variation can we identify that is conducive to a mental state that contravenes the Eighth Commandment (*i.e.,* thou shall not steal)? Crimes of violence against individuals account for only about 10 percent of all reported criminal offences. About 10 percent of arrests are for violation of drug laws, but there is no common agreement about the mental states that lead to immersion in a culture of drugs, nor the genetic variations that could be the proximate causes of drug-dependency. Even for crimes of violence against individuals it would be naive to suppose that these overwhelmingly arise from an "aggressive personality" as opposed to calculated or unforeseen acts of violence that arise in the commission of other offenses. The gap between mental states and the coding of gene products does not require simply more work on biochemical and molecular genetics and neurobiology. It requires establishing a fundamental causal connection between brain states and states of mind that modern neuroscience is unable to forge convincingly.

The overwhelming danger of underformalized investigative procedures

While the discussion so far has concentrated on crime and genetics, there is an analogous problem developing in the brain sciences and their applications to criminal behavior—the "my criminal brain made me do it" argument.[9] The increasing attempts to use brain imaging data to bolster biological arguments of culpability in courts, and to drive public policy decisions on how to treat criminals, is a problem which is receiving good critical coverage in the scientific press, but not in the popular press.[10 11] Two groups of researchers have recently upped the ante by claiming to provide a scientific framework for criminal brain arguments, hypothesizing that there is an identifiable collection of connected brain regions that constitutes a discrete organ for "moral sense" and that criminals have defects in just these regions of the brain.[12 13] Nikolas Rose has previously provided a cogent review and analysis of these newer forms of "biologicization" of criminal behavior and the future challenges these will present in terms of discrimination and screening.[14] He also argued that new ideas of biologically based criminal conduct raise the prospect of not entirely consensual "treatments," increased use of preventive detention and other "pre-emptive" actions on the part of authorities.

Ian Hacking a Canadian philosopher of science born in 1936 echoed many of Rose's sentiments, making the point that a shift in language from "crime" to "criminal behavior" signals a shift in the association of these human phenomena from the perpetrators of crimes, their victims and the criminal justice system to criminologists, psychologists, sociologists and biologists.[15] He pointed out how this process of public reclassification could influence the behavior and attitudes of the classified and their classifiers but did not think that biological classifications used by intellectual elites had much feedback on the behavior of the average person. Troy Duster, on the other hand, reinforced Rose's concerns about the effects of biological ideas primarily on their new owners—the criminologists, psychologists and sociologists who teach and consult with law enforcement officials.[16] Through them, tentative, theoretical ideas may easily shape the largely hidden procedures police forces and government agencies use to identify "likely" suspects of crimes via burgeoning information databases that may disproportionately scrutinize disadvantaged groups within larger societies. No matter if many of the results of these procedures are later thrown out—it is here that introduced biases lead to unequal treatment and rights violations, which, if allowed to become accepted procedure, can erode fundamental rights and freedoms for all members of society.

We share and amplify Duster's concerns about the "dangerous intersection of allele frequencies in special populations and police profiling via phenotype" and the many ways that DNA database information can be abused that would selectively disfavor those segments of the population most likely to run afoul of the criminal justice system, which in many countries is highly correlated with race and ethnicity.[17] To make matters worse, much of the future research correlating genetic variation with behavioral attributes is likely to be corporate, profit-driven and secret. Many of the achievement tests that academic institutions and professional schools now rely on, effectively controlling young people's access to opportunities in the U.S., are corporate products whose composition and scoring procedures are not open to scientific and public scrutiny. We worry that the potential exists for corporations or entrepreneurial academics to offer similar services in the areas of law enforcement involved in solving crimes. How much access would the public have to the contents of such tests, or to indiscriminate data-mining from DNA databases that could be used by law officials or government agencies? Take the example of the lie detector, which is based on flawed scientific premises and is inadmissible evidence in court but which continues to be used as a tool of intimidation. It was perhaps most effective in intimidating suspects during the period when its unreliability was not widely known. If the idea of brains predisposed to criminality or criminal genes is widely accepted, criminologists may consider trying to formalize the strategy of using meaningless biologically-based tests to leverage confessions. Suppose "biological suitability" services were offered to the government or employers to screen people for their "natural potential" for withstanding the stresses of particular lines of work—what "proof" would a private corporation have to offer that any of their methods actually worked?

The greatest danger of behavior-genetic applications to crime is not in the courtroom but in the back rooms during the initial phases of law enforcement investigations where bizarre pseudoscientific theories, fads, and unproven technologies can become mixed with bias and prejudice free from explicit scrutiny. We believe this is the arena where the interaction between crime and behavior genetics is poised to have its most pernicious effects, and where counteracting these effects poses the greatest challenge.

Recommendations

We recommend that organizations with social justice concerns develop new initiatives to, first, combat the idea, which is already well-entrenched, in a variety of academic, public policy and media circles, that molecular behavioral genetics will inevitably reveal new information about the control of human behavior that, once accepted, will of necessity change the way societies deal with crime. Second, organizations and individuals concerned with social justice issues must explore the variety of ways in which biological information can be abused in the initial phases of crime investigations and develop effective strategies for regulating such abuses. The first goal can be accomplished in part by aiding the formation and public effectiveness of groups of scientific professionals like the Neuroethics Society, a multidisciplinary group set up to address inappropriate use of brain scans and other applications of neuroscience. The second goal, however, will require much more concerted efforts to identify particularly pressing issues and place them on the agendas of civil liberty and public policy organizations.

Deconstructing Violence

BY COREY MORRIS, AIMEE SHEN, KHADIJA PIERCE, AND JON BECKWITH

Corey A. Morris, Ph.D., *is co-founder and director of Primary Care Progress;* **Aimee Shen, Ph.D.,** *is an assistant professor in the Department of Microbiology at the University of Vermont.* **Khadija Robin Pierce, J.D.,** *is on the faculty of Applied Sciences, Department of Biotechnology at Delft University of Technology;* **Jon Beckwith** *is American Cancer Society Professor in the Department of Microbiology and Molecular Genetics, Harvard Medical School. This article originally appeared in* GeneWatch, *volume 20, number 2, March-April 2007.*

". . . a man pushed to acts where his own free will stood for nothing." –Jacques Lantier, in Emile Zola's La Bête Humaine, 1889

Using the latest genetic technologies, scientists have identified certain gene variants in human populations that they say predispose individuals to antisocial behavior. Most prominent among these genes is monoamine oxidase A (MAOA). Here, we explore the available research on MAOA, scientists' claims about their findings, and the media's reaction to these studies. A review of the research reveals that an MAOA-antisocial behavioral link is far from conclusive and has garnered media attention disproportionate to its significance. These MAOA studies illustrate how relatively minor and unconfirmed genetic findings, when poorly communicated, can be misinterpreted, amplified by the media, and inappropriately incorporated into discussions of legal and social policy.

The notion that criminal behavior is heritable is a long-standing societal belief predating any knowledge of genetics. One can see the seeds of this popularized thinking as far back as 1889 in *La Bête Humaine The Human Beast*, Emile Zola's story of a family that passed on abnormal behavioral defects from generation to generation.[1] Zola derived much of his thought on the subject of human antisocial behavior from Cesare Lombroso's influential proposals that human physiognomy could predict criminal behavior. With the rediscovery of Mendel's laws of inheritance in 1900 and the subsequent dramatic successes of genetics came claims that criminal behavior, and indeed much of human behavior, could be explained by genes. This view so permeated popular thinking that laws passed in most of the United States promoted sterilization of those considered hereditarily criminal as well as individuals with low I.Q.s. Eugenics propaganda and legislation reached its zenith in the 1920s; however, the later horrors of Nazi Germany's eugenics programs caused many to turn to more environmental explanations of human social behavior.

Significant public focus on a genetic basis of criminality was not revived until the 1960s when Patricia Jacobs and her coworkers published an article in *Nature* entitled, "Aggressive Behavior, Mental Subnormality and the XYY Male," presenting the results of chromosomal screening of inmates in an institution for the criminally insane.[2] Finding a frequency of XYY males in this institution of about 3 percent and assuming that this frequency was much higher than in the general population (later found to be about 0.1 percent), the authors suggested a correlation between possession of an extra Y chromosome and aggressive behavior. Despite the very preliminary nature of the results and numerous criticisms of the conclusions by other scientists, the study received extensive public attention.[3] This quickly-dubbed "criminal chromosome" study provoked widespread discussion about the genetic basis of criminal behavior, leading to the use of the "XYY defense" in murder trials and screening for XYY males by prison officials in some states. Since its publication, the original study has been criticized for its small sample number and for problems of ascertainment bias. (Ascertainment bias is the systematic bias introduced when nonrandom criteria are used to select individuals and/or implicated genes in which genetic variation is assayed.)

By examining patients in a special kind of institution as opposed to the population at large, the authors could not draw any valid conclusions of a significant correlation. In fact, in 1975 the first of many studies to follow was published, showing that when researchers screened the general population for males with the extra Y chromosome, they did not observe the claimed associated aggressiveness. The only replicable features of XYY males in these studies were greater than average height and certain deficiencies in language and motor skills.

In one of the few studies to report an increased likelihood of incarceration of XYY males, researchers in the United Kingdom reported in 1999 that criminal behavior was mediated by the lower intelligence of XYY subjects rather than a direct consequence of the extra Y chromosome. Of course, the correlation of lower intelligence and criminal behavior could also reflect an interaction of poorer school performance, etc., with class, social attitudes and social values. In addition, the crimes committed by the subjects of this study were crimes against property, not the physically aggressive crimes suggested in the title of the original paper. In fact, Patricia Jacobs, in her 1982 speech upon receiving an award from the American Society of Human Genetics, publicly regretted the title of her 1965 paper that had provoked the media interest. Despite the subsequent moderating studies and Jacobs' apology, the myth of the criminal chromosome died a slow death. As late as 1993, the movie *Alien 3* featured a remote planet populated by "extra Y chromos" who had been exiled from Earth and who were, according to their leader, "thieves, murderers, rapists and child molesters . . . all scum."

Nevertheless, crime is becoming genetic again, but with a new twist. In contrast to earlier research on antisocial behavior, some of the current studies, including those on MAOA, have begun to examine the interactions between people's genes and their environment, specifically, the particulars of their upbringing. Does this more expansive view, which attempts to account for the environment as a factor influencing human behavior, truly represent a change from the past in this area of behavioral genetic research?

Do altered MAOA genes contribute to antisocial behavior?

Monoamine oxidases (MAOs) are a class of enzymes that catalyze the deamination of certain amine-containing molecules. One member of this class, MAOA, plays a significant role in the metabolism of the neurotransmitters serotonin, dopamine, and norepinephrine. (Neurotransmitters are chemicals that are used to relay signals between cells of the nervous system.) Because they function in breaking down key neurotransmitters in the brain, the role of MAOs in modulating behavior has been extensively studied. Research on the role of MAO in violent behavior began in the mid 1970s with work in rats.[4] By the early 1980s, researchers were studying human thrombocyte (blood platelet) MAO levels and looking for correlations between low MAO activity and social introversion, aggressive behavior, antisocial behavior, and criminal behavior.

In 1993, scientists reported on a Dutch family in which many of the males exhibited a specific MAOA mutation.[5 6] (The MAOA gene is on the human X chromosome; thus, males have only one copy of the gene (males have only one X chromosome), while females have two copies (females have two X chromosomes). Therefore, most of the MAOA studies conducted in humans, and described here, examine the behavior of males because they are able to inherit only one copy of the gene, while studies in females are complicated by the fact that they can inherit two different copies of the gene.) These males exhibited mild mental retardation and occasional aggressive or violent behavior. Although this correlation was intriguing, the applicability of this finding to the general population was questionable. The mutation was exceedingly rare—it completely eliminated MAOA enzyme activity and, in the years since, has not been reported to have been found in any other men tested. Given that the complete loss of MAOA activity resulted in mild mental retardation, some suggested that the violent behavior might be attributable to the individual's frustration experienced upon being unable to effectively communicate.[7] Indeed, loss of MAOA activity causes severe behavioral abnormalities in mice including increased aggressiveness, tremulousness and fearfulness, as well as defects in brain development.[8] Thus, the variety of structural abnormalities caused by the complete loss of MAOA activity may be responsible for the observed behavioral abnormalities and may not be the direct effect of monoamine metabolism in neuronal function per se.[9]

In 2002, the journal *Science* published research that suggested a broader connection between MAOA levels in human populations and antisocial behavior (referred to hereafter as "Caspi").[10] In contrast to the Dutch report on the rare MAOA mutation that eliminated the enzyme activity completely, this study was based on the existence of common human variants (polymorphisms) of the MAOA gene that affect the levels of the enzyme made in cells. Surprisingly, several studies had shown that the two major human versions of the MAOA gene—a high activity and a low activity polymorphism—represent approximately 65 percent and 35 percent, respectively, of males tested in diverse populations. It should be noted that the effect of the polymorphism on levels of activity was measured not in brain cells but in fibroblast cells in the skin, which raises concerns about the conclusions of this and subsequent studies. Do the polymorphisms have the same effect on MAOA levels in brain cells? In addition, while the average difference in levels of MAOA between the two classes was approximately 7-fold in one study, the differences were quite variable (2- to 10-fold in another study) and the levels between the two classes overlapped.

Using knowledge of the MAOA polymorphisms, Caspi and coworkers assessed the correlation of MAOA activity with antisocial behavior of 442 males followed in a longitudinal study conducted in Dunedin, New Zealand. The researchers used various criteria to assess the behavior of those who exhibited the low and high activity versions of the MAOA gene. They found those individuals with the low activity MAOA allele and who had been subjected to severe maltreatment as children were more likely to exhibit antisocial behavior than those with the high activity version. In contrast, those children with the low activity version but who had not been subject to abuse did not show any greater degree of antisocial behavior than those who had the higher activity MAOA gene. Moreover, those children with the higher activity MAOA gene who experienced child abuse were also more likely to exhibit antisocial behavior, albeit not to the same extent as the low activity group. The authors concluded that the high activity form of the gene might be "protective," preventing a continuing cycle of violence for maltreated males: the potential for antisocial behavior in males with the lower expressed version of MAOA only appeared when the subjects had experienced significant child abuse. The authors concluded that the correlation between MAOA polymorphism, maltreatment as a child, and subsequent antisocial behavior represented what is known as a gene-environment interaction (G x E).

Importantly, the Caspi research team noted that replication of their findings was needed before any solid conclusions could be drawn. Subsequent replication attempts have yielded mixed results; some studies have reached similar conclusions,[11] [12] [13] another reported a "non significant trend" toward the Caspi conclusions,[14] and yet others completely failed to replicate the findings, including the gene-environment interaction.[15] [16] [17] Moreover, in addition to those studies that failed to replicate the Caspi findings, at least two studies found a contradictory inverse relationship—that is, an association between the high activity MAOA genotype and aggression in males.[18] [19] Furthermore, even the 2004 Foley *et al.* study, which is frequently referenced as a replication of Caspi, can be interpreted differently upon closer reading; the authors acknowledge that on controlling for early adversity and the interaction of adversity and MAOA genotype, low activity MAOA was associated with lower (not higher) risk of conduct disorder.[11 20]

Studies of MAOA function in non-human primates have also been inconsistent with the original Caspi report. For example, Rhesus macaque monkeys with the low MAOA activity allele, raised in the absence of parental input, were even less aggressive than mother-raised macaques with either the low or high MAOA activity alleles.[21] Curiously, despite this contradictory observation, studies have cited the Newman *et al.* study as supporting the findings of the Caspi *et al.* (2002) study.[12 21]

Recently as of 2006, several authors of the original Caspi study published a meta-analysis,[22] which included new data, the original Caspi data[10] as well as three other reports.[11 12 14] However, the criteria used to choose studies for this meta-analysis resulted in inclusion of only the original Caspi study, two studies that replicated the findings, and only one study that had a partial failure to replicate. Thus, given the included data, the meta-analysis replication of the Caspi data is relatively unsurprising. In addition, one should consider that published results are frequently biased toward positive findings, skewing such meta-analyses. Indeed, at a Ciba Foundation Symposium in 1995, one scientist reported finding the inverse correlation between the low activity allele and delinquent behavior but did not seek publication because "it was contrary to what he had predicted." Han Brunner, the lead author on the original MAOA Dutch study, responded: "It's extremely important that these sorts of negative findings are published to avoid the meta-analysis later coming up with the wrong answer."[23] Nevertheless, the contradictory study was never published.

Even if the Caspi results were eventually confirmed, such findings are not de facto translatable to all populations. For example, while the earlier studies had focused on "white" males[10 11], a recent study was the first to directly compare "white" and "non-white" populations.[13] For "non-white" individuals, the authors found no significant interaction between MAOA allele, maltreatment as a child, and subsequent violent and antisocial behavior. Interestingly, the authors chose to reanalyze the data using only "blacks." Again, they found no significant interaction. In addition to the mixed results from attempted replications of the Caspi *et al.* study, the failure to identify significant association in "non-white" individuals further challenges the generalizability of any results obtained in select "white" populations.

This same study also found that for "white" subjects, a statistically significant interaction was only found between MAOA genotype and maltreatment with juvenile violence and was not significant for subsequent adult behavior.[13] This result suggests that knowing the genotype and childhood environment has no predictive power for the propensity for violence of the adult, arguing that life history modulates adult behavior irrespective of MAOA genotype. Indeed, as the recent meta-analysis study notes, "both scientists and the public are becoming increasingly aware that, like many developmental processes, the nature of gene effects on behavior, too, is often contingent upon experience."[22]

Despite the many discrepancies and failures to replicate, two results were clearly consistent among all of the studies: first, childhood maltreatment is the strongest predictor of violent or antisocial behavior, and, second, variation in the MAOA gene is not predictive of antisocial behaviors later in life. From the research completed to date, there are no well-established relationships between MAOA genotype and violent or antisocial behaviors. Furthermore, maltreated children who are at the highest genetic risk predicted by MAOA genotype appear to comprise a relatively small fraction of maltreated children overall. Even if the gene-environment interaction suggested by Caspi and others is eventually demonstrated to be correct, the relationship is merely probabilistic. That is, one would not be able to predict which children with the low expressing version of the MAOA gene will exhibit antisocial behaviors.

We know from psychological and sociological research, which long predates behavioral genetic studies on the subject of antisocial behavior, that, in general, individuals exposed to childhood maltreatment are more likely to have behavioral problems as adults (*i.e.,* continue "the cycle of violence"). As even some of the behavioral geneticists studying the effects of MAOA polymorphisms note, "eradicating child maltreatment is clearly the preferred way to combat risk for psychiatric problems . . . social support can protect even genetically vulnerable children from the negative sequelae of maltreatment."[22]

Even with this data, the authors of the original Caspi report and many of the replication studies allude to the possibility that understanding MAOA gene variation may permit the development of "improved pharmacological treatments" for children with the low-expressing version of MAOA.[10] Some may point to studies in MAOA-deficient mice that show that administration of serotonin synthesis inhibitors can counteract the development of behavioral abnormalities resulting from the loss of MAOA activity.[8] However, given that these findings are limited to MAOA-deficient animals, and loss of MAOA activity in humans is extremely rare,[5] it is unclear how such pharmacological interventions would be useful. Taken to a logical conclusion, some may suggest that pharmacotherapy should be used to treat susceptible children living in abusive homes so that they can withstand the effects of the abuse.

Numerous questions about the utility of these reports are raised by the uncertainty of the science conducted so far. These uncertainties are confounded by the lack of predictive power of the proposed correlations, the stunning variability in MAOA levels among males with the same "low-expressing" polymorphism, the absence of understanding of the mechanism for the effects reported, the contradictory studies that attempted to replicate the findings, and the much deeper knowledge of the overall role of child abuse in the development of antisocial behavior. At the very least, we are far from any certainty about the significance of this research field and similarly far from conclusions relating to any intervention, let alone rational pharmacological treatment.

A number of authors of the behavior genetic studies have suggested that knowledge of the MAOA allele variant could permit the development of programs to reduce antisocial behavior by, for example, "leading to more focused interventions"[13] or "screening" for children who are highly likely to develop severe conduct problems."[24] Such proposals raise serious ethical and social questions. These proposals disregard the fact that the findings of replication studies are mixed and that children with the "protective" high MAOA activity polymorphism may also exhibit violent behavior as a result of maltreatment. Thus, these polymorphisms are unlikely to have any real predictive power, calling into question the utility of targeting only for "intervention" or "screening" those maltreated children with the "low MAOA activity" polymorphism for experiential (or still non-existent pharmacological) therapies. Such a strategy seems to be an unfair and dangerous approach allowing avoidance of the real problem—child abuse. Maltreatment has consistently been demonstrated to predispose children to violent behavior,[25] and experiential therapeutic interventions have been shown to reverse the effects of maltreatment on behavior.[26] Thus, it is irresponsible to focus treatment on a subset of maltreated children based on a possible genetic susceptibility to experience-induced violent behavior. Directing societal efforts to reduce child abuse and to provide experiential therapies for both victims and their families is the most effective, most ethical, and most socially rational method for counteracting the negative effects caused by maltreatment.

MAOA and society

Since the late 1980s, the monoamine oxidase (MAO) enzymes have taken a star role in several psychological dramas played out in the press. In addition to numerous articles on MAO's role in depression and depression-related disorders, *The New York Times* alone has profiled studies reporting MAO level hypotheses in risk-taking behavior, thrill-seeking behavior, alcoholism, drug addiction, cigarette addiction, and violent behavior.[27 28 29 30 31 32]

The news reports of the early 1990s on the MAOA-deficient Dutch family were no exception. The authors of the scientific studies, in their conclusions, hinted at a possible generalization of their finding in this one family to the larger problem of aggression in society. This hint was not missed. A *Science* reporter suggested "it might be possible to identify people who are prone to violent acts by screening for MAOA gene mutations...."[33] The provocative conclusions in the papers and the *Science* news article aroused tremendous media interest. A *Newsweek* article entitled, "The genetics of bad behavior," included a photograph of a violent confrontation between Palestinians and Israelis, implying a genetic basis for world strife.[34] A television news report used films of U.S. street gang violence as a back-drop for a report on the MAOA study.[35] An issue of *U.S. News and World Report* that carried a story reporting on the Dutch study featured an infant dressed in prison clothing to indicate the supposed deterministic relationship between heredity and criminal behavior.[36]

Not surprisingly, these media reports on genetics and criminality have had their impact on the legal system. A *New York Times* article recalled that "researchers were besieged by calls from lawyers, who wanted their clients tested for the genetic defect to use as a possible defense." Indeed, in at least two court cases defendants have attempted to claim genetics as a mitigating defense for violent crimes. In *Turpin v. Mobley,* the court rejected the defendant's claim in the sentencing phase that his behavioral/personality problems had a genetic basis. In this 1998 case, the defense attorney initially contacted Dr. Xandra Breakefield, one of the authors of the Dutch study; however, the lawyer subsequently decided not to have the defendant tested for the MAOA gene defect because he doubted that his client actually had the same genetic trait.[37] In November 2006, a defendant charged with murder unsuccessfully attempted to use the MAOA gene as a defense.[38] Despite the ultimate finding in this case, the language of the court is disturbing for its unquestioning acceptance of a genetic basis for violence: "Genetic testing did not show that the appellee had the MAOA gene, *which is the gene related to violent behavior*, but revealed that the appellee had a genetic vulnerability to becoming depressed and dysfunctional, especially in stressful, crisis-type situations" emphasis added. The Dutch study illustrates the way in which a relatively minor genetic finding can rapidly be transmitted to the public, be misinterpreted, and then be incorporated into discussions of legal and social policy.

The reporting by the media on the 2002 Caspi article in many ways demonstrates a more careful approach to the subject than the reporting of the 1993 Dutch study. The later study, by its very nature an attempt to show that genes and environment interact to generate behavioral problems, had a less genetically deterministic flavor. Generally speaking, the press articles qualified the study findings at some point and emphasized the importance of environmental factors, with at least two of the articles stating that the discovery is not of a "gene for violence."[36 40] The *Chicago Sun-Times* also stressed the environmental element and expressly referred to the need to replicate the study. Such attention to key details of the science is encouraging.

Other media outlets recognized the social and ethical confusions of the study. The *New Scientist* categorically rejected the Caspi study's suggestion of an eventual drug intervention, stating that these people are "victims of child abuse, not bad genes."[41] ABC *Health Minutes* concluded soberly that the solution is not to move toward genetic testing but to prevent child abuse.[42] The *Guardian* (UK) predicted a day when the military, firefighters and police would screen recruits and otherwise echoed a possible pharmacologic intervention for higher risk people yet at the same time cautioned against the medicalization of social problems.[39] The *Hindu* suggested that the Caspi findings highlighted uncertainty about how much social practices might influence individual behavior and "why racial and caste discrimination and child abuse are not just uncivilized but even dangerous" and ended with a declaration that "genetics is telling us how to behave."[43]

Nevertheless, the Caspi article's suggestion of pharmacological approaches to the problem based on genetic knowledge focused attention on the genetic side of the interaction between genes and environment and some in the media used unqualified declarative headlines like, "Study Finds Genetic Link to Violence," "Study Links Past Abuse, Gene to Violent Acts,"[44] and even a mathematical equation, "Gene + Abuse = Trouble."[36] *Popular Mechanics,* which provides "informative articles on automotive technology," carried the story under the headline, "Criminals Share a Common Genetic Flaw" and employed various analogies using car keys to illustrate the relationship of the MAOA gene to the outcome of violent behavior.[41]

The tendency to explain human behavior as largely genetic in nature, even in the absence of supporting evidence, is often willingly propagated by scientists themselves. The *San Francisco Chronicle* quoted a Stanford psychiatrist, "people have to get used to the idea that there are probably genetic influences on many kinds of behavior."[44] The *Boston Globe* quoted one scientist not associated with the study as saying that "If the results can be replicated, their public policy applications could be 'explosive.'"[45] With such hyperbole, it is little wonder that judges would convene, as they have, to begin to prepare themselves for the onslaught of legal issues that could arise from a genetic link to violence, however premature that assertion.

The era of strict genetic determinism of human behavior and aptitudes is rapidly disappearing. While the Human Genome Project advocates initially argued that genes for human behaviors would be rapidly identified, the opposite has been the case. This has led many researchers to emphasize the complexity of such behaviors and to suggest that multiple genes and environmental factors, difficult or impossible to separate, are involved in these human traits.[46] Researchers are increasingly considering these multiple factors and presenting more nuanced explanations when proposing genetic influences on behavior. Nevertheless, many scientists still lack sufficient information outside of their narrow fields of study and are not cautious or scholarly in presenting results, prematurely drawing conclusions about complicated and preliminary results that are difficult to replicate, and suggesting treatments that raise serious ethical and social issues. These sorts of presentations may direct the media to present more dramatic reports on the studies that often feature old-style deterministic headlines. In the case of antisocial or criminal behavior, the claims of genetic correlates are introduced relatively rapidly into the legal system in an unwarranted fashion. Over a century of history of scientific attempts to find these correlations illustrates the sometimes severe impact of this field on society. The social consequences of this area of research both in terms of the legal system and how society, in general, deals with antisocial behavior call for a much more cautious and informed presentation by both scientists and the media.

The Folly of Geneticizing Criminal Behavior

BY TROY DUSTER

Troy Duster, Ph.D., *is a Chancellor's Professor of Sociology at UC Berkeley and professor of Sociology and Director of the Institute for the History of the Production of Knowledge at New York University. This article originally appeared in* GeneWatch, *volume 24, number 3-4, June-July 2011.*

On June 20, 2011, *The New York Times* published an article entitled, "Genetic Basis for Crime: A New Look," in anticipation of the start (that day) of the annual National Institute of Justice Conference in Arlington, Virginia. The article noted: "On the opening day criminologists from around the country can attend a panel on creating databases for information about DNA and 'new genetic markers' that forensic scientists are discovering."[1] The reporter noted how much the climate has changed in the last two decades, when social scientists studying crime generally eschewed any attempts to bring genetics into their explanatory models.

The general frame for the *Times'* piece is that careful, thoughtful and reasonable social scientists are finally acknowledging that genes play a role in criminal behavior. By innuendo, only those social scientists that are ideologically bound to ignore the role of genetics hold to an outmoded view. This is very reminiscent of an article that appeared in the *Chronicle of Higher Education* two years earlier, in which a very similar frame was used even more dramatically, with the jaw-dropping opening lines: "Genetic research finally makes its way into the thinking of sociologists. If sociologists ignore genes, will other academics—and the wider world—ignore sociology?"[2]

For both articles, the reporters interviewed me. The earlier piece cast me in the role of the backward looking Luddite trying to push back against the inevitable tide of progress in science. Indeed, this article stated that I go ". . . so far as to suggest that any sociologist who embraced genetic approaches was a traitor to the discipline."

Actually, I said nothing of the kind. Rather, in the article that the reporter references, I assiduously described the scaffolding of knowledge production around criminal statistics of both violent and property crimes, indicating how those who begin theorizing about such crimes with already collected crime statistics are likely to make a fundamental mistake in scientific inquiry.[3] The first principle of data collection in science is the assurance of the accuracy and integrity of baseline data. If there is a systematic skew or flaw in the database, theories built upon that database are unreliable. This is as true for sociologists as it is for geneticists. Before I turn to the matter of bringing in genetics to help explain criminal behavior, some historical context is necessary for how sociologists have made this error in theorizing about explanations of crime.

In the middle of the twentieth century, there were two major competing sociological approaches to the study of deviance, law, and the criminal justice system in the United States. The first was centered at Columbia and could be generally described as dominated by Robert Merton and his students using the functionalist paradigm. Drawing from both Durkheim and Parsons, Merton and his colleagues approached the study of deviance by exploring the putative functional relationship between "the deviant world" and "the normal world." Their basic assumptions pivoted on an empirical strategy relying on statistical aggregates of deviant and criminal behavior as reflected in local police records and the Uniform Crime Reports of the FBI. While there was an occasional foray into field site investigations, the overwhelming tendency of this approach assumed that the data from official statistics sufficiently reflected the empirical world and that theorizing from these databases warranted little in the way of caution or concern.

The University of Chicago was the alternative and competitive approach, with a long tradition of field site investigations of particular forms of "deviance," from "the gang" to "the hobo" to the prostitute, gambler, and other vice sets and settings. Their practitioners literally went to those places in the society that any competent actor would presume to be the site of deviance.

While there were significant differences between Chicago and Columbia in their respective renderings of explanations of crime and deviance, in very important ways both adopted a "taken-for-granted" world.

Onto this stage would step a third set of key players who would shift the focus to challenging the epistemological foundations of the whole playing field, and not just of theory and research on deviance. Egon Bittner and Aaron Cicourel would ride around in police cars and see just how and when police used discretion in their arrest procedures.[4 5] David Sudnow would go to the Public Defender's office and record the way in which the Public Defender and the Prosecuting Attorney worked in concert to secure guilty pleas from "certain suspects" but let others bargain "harder" when their legal representation took a private turn.[6] Erving Goffman would venture into mental hospital wards and break the path for the next generation that would study intake decisions for mental institutions.[7]

In some ways, the methodology of this new breed of work seemed parallel to the earlier field research traditions. However, in fundamental ways the domain assumptions were strikingly different. Chicago was mainly trying to find out and explain what "the deviants were really like" in their natural setting. The new approach was to raise another whole order of question: "What are the social processes that explain why some get classified and others don't, even though both are engaged in the same or similar behavior?" When Cicourel and Bittner were riding around in police cars, observing and recording how official statistics are compiled, they were asserting that the point at which rate construction occurs is the preferred site for an investigation of those otherwise "taken for granted" statistics that had been the structural underpinning database for theorizing about deviance and crime. It would not be much of a leap to conclude that, if the site of rate construction was the preferred focal point for inquiry and theorizing about these data, then surely it would have bearing and impact on all manners of rate construction, from epidemiological work in public health to coroners' collective accounts of the causes of death.

The importance of data collection at the site of knowledge production

There are powerful organizational motives for police departments to demonstrate effectiveness in "solving crimes." It is a considerable embarrassment for a police department to have a long list of crimes on their books for which no arrest has been made. No police chief wishes to face a city council with this problem. Thus, there are organizational imperatives for police departments to clean up the books by a procedure known as "cleared by arrest."

Few matters count as much as this one when it comes to reporting to the public about what the police are doing.[8] To understand how arrest rates are influenced by "clearing," it is vital to empirically ground this procedure by close observation.

Here is the pattern: Someone, "P," is arrested and charged with committing a crime, let's say burglary. There are a number of other burglaries in this police precinct. The arresting officers see a pattern to these burglaries, and decide that the suspect is likely to have committed a number of those on their unsolved burglary list. Thus, it sometimes happens that when "P" is arrested for just one of those burglaries, the police can "clear by arrest" 15 to 20 crimes with that single arrest. This can show up in the statistics as a "repeat offender," even though there may never be any follow-up empirical research to verify or corroborate that the "rap sheet" accurately represents the burglaries now attributed to "P".

But only if one is "riding around in police cars" or doing the equivalent close-up observation of police work can this be corroborated as a pattern.[9] And yet, if social theorists take the FBI Uniform Crime Reports as a reflection of the crime rate, with no observations as to how those rates were constructed, they will make the predictable "policy error" of assuming that there are only a very small number of persons who commit a large number of crimes. The resulting error in theorizing would be to then look for the "kind of person" who repeatedly engages in this behavior (as if it were not "cleared by arrest" that generated the long rap sheet). So the "bad apple" theory, that only a very few people commit the bulk of the crimes, may indeed be a function of the bureaucratic imperatives of police work. The "bad apple" solves a problem for police reporting of crimes "cleared by arrest." But if sociologists, criminologists, and then behavioral geneticists begin their inquiry at this point, what they will have missed is the actual empirical behavior—in this case, the police requirement to clear-by-arrest.

Bringing in genetics—compounding the error at the base of data entry

It is at this very point of knowledge production that we see the methodological problem of bringing in geneticists to help explain criminality. By definition, geneticists are seeking to find genetic markers, the usually very complex matter of finding the genotype to explain the phenotype. But this presumes that the Uniform Crime Reports represent the actual, concentrated behavior of single individuals with those genetic markers and is not an artifact of bureaucratic resolution to an organizational imperative to "clear by arrest." And it is *this* that I said sociologists should be focusing their research on, namely, the site of knowledgc production in crime prevention, control and explanation! Indeed, in that article, I explicitly cited the work of anthropologist of science Duana Fullwiley, who goes to the laboratories where geneticists deploy concepts such as racial admixture to examine how they are using the idea of race and ethnicity in their search for genetic markers.[10] A recommendation for the preferred focus on an empirical site where scientists can more accurately assess rate construction is a far cry from calling colleagues "traitors" for pursuing collaborative research. Instead, that framing of the problem is a red herring diverting attention from the problem of "bringing genetics in to help explain crime."

I conclude with a dramatic example to illustrate the basic problem. We have reasonable grounds to conclude that although whites consume more marijuana per capita than do African Americans, it is African Americans who are routinely arrested for marijuana possession at least four to nine times the rate as are whites, depending upon the region of the country.[11]

If sociologists begin with these current arrest data for marijuana possession and then join with geneticists to help explain the sharply distorted over-representation of African Americans in the criminal database, they will have made the most fundamental of errors in scientific theorizing: never realizing the source of their error is based upon inattention to how crime rates are constructed. Even worse, they will have erroneously concluded that they have "advanced scientific inquiry" by partnering with a neighboring discipline in a thoughtful, ecumenical embrace to better explain the genetic basis of drug crimes.

Genetics without Ideology

BY KENNETH WEISS

Kenneth M. Weiss, Ph.D., *is Evan Pugh Professor of Anthropology and Genetics at Penn State University.* *This article originally appeared in* GeneWatch, *volume 24, number 3-4, June-July 2011.*

Ever since Darwin, there has been great controversy over the degree to which genes determine who we are. The reason is that if genes determine individuals' or groups' nature, value judgments can be assigned to those genes—and the people who carry them—and social policy can be implemented accordingly. Much good has come from the pursuit of genes that cause disease, but racism and eugenics based on uncritical genetic determinism have also visited incalculable harm upon people.

There's no legitimate doubt about the importance of genes in the making of organisms, or of the role of genetic variation among them. The theory of evolution led us to understand that life has diverged from a common origin, by way of inherited variation that has been screened by luck and its bearers' ability to reproduce. One hundred and fifty years of research have revealed much about the nature and importance of genes, and today the view of life that Darwin spawned holds genes to be primary instruments of the development of organisms.

Organisms develop from single cells. There is more in a cell than DNA, but only DNA carries the specific information that controls biological function. In a mechanistic sense, DNA is ultimately responsible for the structures of individuals. Differences in traits among individuals are affected by differences in DNA, and, by extension, populations that are isolated from each other gradually accumulate genome-wide differences. These differences affect both functional and non-functional sites in the genome, some of which are harmful.

It may thus seem unexceptionable to look for harmful genes and hope to intervene. The idea begins with detection, treatment and prevention of readily identifiable genetic diseases. That is now routine for some diseases. Hospitals screen newborn infants for a number of genetic conditions, and many couples submit to DNA tests to determine the possibility that they might pass a condition on to their children. The hope is that harmful alleles may even be removed from the human gene pool. Today such measures are largely voluntary, though screening newborns for some genetic disorders, notably PKU, has been mandatory for many years in the U.S. and widely accepted because there is effective treatment.

Things become problematic when "public good" is used to advocate mandatory social policy. The belief that all traits are meaningfully genetic opens the extension of policy concerns beyond clear-cut genetic disease. Historically, that has stressed behaviors that experts and policymakers have chosen to define as undesirable, with special attention given to traits like homosexuality, insanity, drug and alcohol addiction, immorality, and aggression. Historically, that's when the wheels came off.

Post World War II revulsion over abuses made in the name of eugenics and its racist nadir in Nazism led the nature-nurture cycle to swing to an environment-centered view of human traits, accompanied by declarations of human rights to prevent such kinds of discrimination. Physicians were even advised to avoid specializing in genetics because genetic diseases were rare and nothing could be done about them. It was felt that the common diseases and the diseases of older age were caused by environmental exposure and lifestyles; they were not genetic and could be addressed by therapeutic or preventive measures. Similarly, psychologists and social scientists took behavioral traits off the genetic table.

But there has been a resurgence of genetic determinism. Modern genetic technologies have made it increasingly possible to document the structure of DNA and to characterize gene function. Along with this came increased ability to search for genetic effects underlying disease. In the process, awareness of raw abuses of the past has faded into a rather stunning complacency in the belief (once again) that our actions are entirely value-neutral and benevolent.

The fact is that most clearly measurable traits show substantial heritability. Offspring resemble their parents, even when environmental and statistical errors are accounted for. It is typical for roughly 30 to 50 percent of variation in a trait to be statistically accounted for in terms of genetic variation.

Despite well-known problems with heritability as a measure, this clear-cut familial correlation feeds the determination to identify the genetic basis of any disease. There is considerable controversy about how efficacious such gene 'mapping' approaches are, but nobody wants to stand in the way of prevention and treatment of disease.

If disease traits that everyone agrees should be addressed clearly show genetic influences, then there is no scientific reason to think that other traits, including even culturally sensitive behavioral traits or group differences, should not also have some genetic basis as well. To deny that genetics is a part of the causal spectrum of inconvenient traits is to be an ostrich to the facts.

This may be difficult to accept in light of the brutal history of abuses in the name of Darwin, Mendel, the inherent value assigned to persons on account of their genes, and the patently facile way in which tiny indications of difference are extrapolated across vast evolutionary time periods. Reconciling the legitimate scientific facts with the legitimate social fears is not easy.

Meanwhile, researchers are now excitedly proclaiming they have found the genes 'for' all sorts of behavioral traits, including such things as athletic or musical ability. Unfortunately, much of the stress is on the same kinds of "undesirable" traits that led to the abuses of the eugenic era. This is a return to the acceptance of inherency ushered in with the Darwinian view of life. As in the earlier reasoning, no one wants people to suffer needlessly from genes predisposing them to disease, and no one wants society to suffer needlessly from genes predisposing an individual to harmful social behavior. And as before, a belief in genetic inherency labels people from birth. It is entirely predictable that genetic screening that identifies people with a genotype supposedly predisposing them to crime will lead to various forms of surveillance (or mandatory preventive pharmaceutical "therapy") before they commit their predestined offense. It is clear that this is only statistical risk, usually weak, and most of the individuals targeted as having the "criminal gene" will never commit an offense.

It is not easy to decide how to strike a proper balance between genetic determinism and civil society. At least, we should take a closer look at the facts. The issues are subtle and perhaps that is why they are not easily absorbed, even by professionals. But they are important.

People and groups will always differ, on anything one can measure

No two people or populations will be exactly identical on any measure. There will always be some difference. Whether it is statistically significant depends on the sample, the precision of measurement, and subjectively chosen probability criteria (like the typical 5 percent level). No one can pretend that such differences cannot, in principle, affect any trait. This is why if one chooses to sample from different continents one can make "races" seem so real, even though we know that the evolutionary processes are more continuous in nature.

However, one has to be careful about what such differences are. Samples are usually compared by mean values, like average IQ scores or blood pressure levels. But usually there is a large overlap in the distributions where the distributions differ by a typical half-standard deviation. The difference in the mean may be statistically "significant," but that is not the same as importance, or even the main story. Similar statistical issues apply to comparisons of individuals defined as affected versus unaffected by disease or behavioral traits. Whether the mean differences are important relative to the overlap is a societal judgment. Such judgments have been, and are still, widely used to support racist views.

Why is genetic causation so elusive?

Mapping studies have been done on almost any trait one can imagine and in every major research organism, from bacteria to human. They have found very similar results that are entirely consistent with the genetic effects expected from our understanding of the evolutionary process and the nature of the way genes work.

Generally, there are one or a few strong effects at individually identifiable locations in the genome. The effects of these sites achieve at least suggestive statistical significance according to standard (though subjective) criteria and have replicable association with the presence of the trait. However, such sites usually account for only a small fraction of the estimated heritability of the trait. Accompanying them, ever-larger studies have shown that a host of sites—often estimated in the hundreds or more—have individually minute effects. Even in aggregate, these minor sites, or "polygenes," still only account for a small fraction of the heritability. They rarely are replicable among studies or populations.

How predictive are genes?

Such findings reinforce the idea that genetic variation contributes overall to the trait in question but do not directly address the question of the predictive power of individual genotypes, which is crucial to an understanding of genetic causation. The same mapping results unambiguously show that the non-genetic environmental effects account for the bulk of variation.

Furthermore, heritability implies genetic involvement but says little or nothing about genotype-specific prediction. Individual multilocus genotypes are unique, and therefore we usually have little way to predict their specific effects because statistical association depends on replication. More fundamentally (or even fatally) serious is that environmental exposures contribute the bulk of effects but are both poorly understood and unknowably changeable, meaning that the causal effects of all but the few, usually rare, strong genetic variants are contextually relative and quantitatively unpredictable.

The same is even truer of behavioral traits. For behavioral traits, much depends even on the purely cultural nature of their definition. As an obvious example, it is those in charge who define what is "antisocial" or criminal behavior. Unscrupulous hedge fund managers walk the streets freely, while a large fraction of the US criminal population is incarcerated for drug-related offenses. A search for criminality genes would compare these "normal" and "criminal" groups to find genetic differences between them—genes "for" criminality. But if drugs were legalized, what had been offenders' genotypes would overnight become "normal." This cultural component is so obvious it is a wonder anybody accedes to the kinds of studies searching for genes "for" behavior increasingly being funded in our gene-centered age.

One must acknowledge that there is, at present, none of the racist or vitriolic judgmental fervor that drove the classic eugenics movement. But the genetic ideology and target traits are similar, and that's what raises caution about where the enthusiastic promises of finding the genes conferring specific value-laden behaviors may go.

What, if anything, can be done to recognize the effects of genes and yet protect against exuberant application of genetic determinism? Could one identify a line beyond which research is banned because its findings might be too likely to be misused? That may seem like unacceptable or unenforceable censorship, but, in fact, society routinely makes such decisions. For example, investigators are not allowed to torture people as part of a scientific experiment (even if, for some reason, they could be persuaded to consent).

We already make informal discriminatory decisions based on traits that have some genetic basis. Intelligence, however you want to measure it, surely is in that category. Bright students are assigned faster academic tracks, are admitted to the best universities, or are given special mentoring and encouragement. It seems acceptable to base such discrimination on test scores, perhaps because that at least reflects achieved ability. But many would object if we used genetic tests instead, even if, for example, it might identify those in whom even earlier investment was warranted.

The search for simple answers affects society and science alike. The complexity and uncertainties of nature encourage both progress and problems, but the lesson is clear: Genes are involved in everything, but not everything is "genetic."

The Crumbling Pillars of Behavioral Genetics

BY JAY JOSEPH

Jay Joseph, Psy.D., *is a licensed psychologist practicing in the San Francisco Bay Area. His published work in professional journals, books, and book chapters has focused on a critical appraisal of genetic theories in psychiatry and psychology. This article originally appeared in* GeneWatch, *volume 24, number 6, October-November 2011.*

Schizophrenia researcher Timothy Crow wrote in 2008 that molecular genetic researchers investigating psychotic disorders like schizophrenia had previously thought that "success was inevitable—one would 'drain the pond dry' and there would be the genes!" But as Crow concluded, "The pond is empty."[1] Four years later the psychiatric disorder and psychological trait "gene ponds" appear to have been completely drained, and there are few if any genes to be found. Twenty years ago, however, leading behavioral geneticists had high expectations that molecular genetic research would soon "revolutionize" the behavioral sciences.

During that heady period of the early 1990s, leading behavioral genetic researchers like Robert Plomin attempted to shift the field's focus toward gene-finding efforts. After all, they reasoned, "quantitative genetic" studies of families, twins, and adoptees had established beyond question that variation in such "normally distributed" psychological traits as personality and cognitive ability (IQ), as well as psychiatric disorders and abnormal behavior, had an important genetic component. The decade of the 1990s did in fact witness an explosion of molecular genetic research attempting to pinpoint the genes believed to underlie these traits and disorders. This was followed by the publication of the initial working draft of the human genome sequence in 2001, which many researchers believed would lead to rapid gene discoveries in psychiatry and psychology. According to a pair of prominent researchers, writing in 2003, "Completion of the human genome project has provided an unprecedented opportunity to identify the effect of gene variants on complex phenotypes, such as psychiatric disorders."[2] As we approach 2012, however, behavioral genetics and the allied and overlapping field of psychiatric genetics are attempting to come to grips with the stunning failure to discover genes. These fields appear to be approaching a crisis stage, if they are not there already.

Critics, on the other hand, have argued all along that both twin studies and family studies are unable to disentangle the potential roles of genes and environment. They have pointed out for decades that the validity of equal environment assumption (EEA) of the twin method is not supported by the evidence, and that the much more similar environments experienced by reared-together monozygotic (MZ) versus reared-together dizygotic (DZ) twin pairs confound the results of the twin method. Therefore, both family studies and twin studies prove nothing about genetics, and their results can be *completely* explained by non-genetic factors.[3] Most behavioral geneticists agree with this assessment as it relates to family studies but continue to maintain that twin studies provide conclusive evidence that genes play an important role.[4] Critics have also pointed to the massive methodological problems and untenable assumptions found in psychological and psychiatric adoption studies, as well as the major problems and environmental confounds in studies of purportedly reared-apart twins.[5]

Behavioral geneticist Eric Turkheimer described the competing positions of behavioral geneticists and their critics in 2000: Gene discoveries to come would signify behavioral geneticists' "vindication," whereas "critics of behavior genetics expect the opposite, pointing to the repeated failures to replicate associations between genes and behavior as evidence of the shaky theoretical underpinnings of which they have so long complained."[6] Turkheimer, however, recognized in 2011, "to the great surprise of almost everyone, the molecular genetic project has foundered on the . . . shoals of developmental complexity...."[7]

Behavioral genetics and the related fields have recently adopted the "missing heritability" position to explain the ongoing failure to uncover genes.[8] Proponents of this position argue that genes ("heritability") are "missing" because researchers must find better ways to uncover them, as opposed to some critics' contention that the failure to discover genes indicates that these genes do not exist.[9] By the summer of 2011 it had reached the point where 96 leading psychiatric genetic researchers, in an open letter, asked potential funding sources not to "give up" on genome-wide association (GWA) studies.[10]

In light of the ongoing failures of molecular genetic research, it is worthwhile to look back at the way that behavioral geneticists have written about the search for genes, including numerous claims and predictions published in textbooks and leading scientific journals. Here I focus mainly on the writings of the world's leading and most influential behavior geneticist, Robert Plomin of King's College of London, Institute of Psychiatry, who is the lead author of a frequently cited multi-edition textbook on the subject: *Behavioral Genetics*.[11]

Three decades of claims and predictions

As far back as 1978, DeFries and Plomin claimed that "Evidence has accumulated to indicate that inheritance of bipolar depression involves X-linkage in some instances."[12] Although these and other claims were not replicated, psychiatric molecular genetic research took off in the 1980s, a decade that witnessed many more highly publicized, yet subsequently unsubstantiated, gene-finding claims. Nevertheless, another group of prominent behavioral genetic researchers wrote in a 1988 *Annual Review of Psychology* contribution, "We are witnessing major breakthroughs in identifying gene coding for some mental disorders."[13]

In the 1990 second edition of *Behavioral Genetics*, Plomin and colleagues wrote, "During the past decade, advances in molecular genetics have led to the dawn of a new era for behavioral genetic research."[14] They argued that "these techniques are already beginning to revolutionize behavioral genetic research in some areas, especially psychopathology." However, these "revolutionary advances" were not actual replicated gene findings. Also in 1990, Plomin predicted in *Science* that "the use of molecular biology techniques will revolutionize behavioral genetics."[15]

Plomin and his colleagues published a 1994 molecular genetic study in which they found DNA markers associated with IQ.[16] However, this study, as well as all subsequent molecular genetic IQ studies, was not replicated.[17] In another *Science* publication, Plomin and colleagues reported genetic linkages and associations for reading disability, sexual orientation, alcoholism, drug use, violence, paranoid schizophrenia, and hyperactivity.[18]

In the third edition of *Behavioral Genetics*, published in 1997, the authors repeated their position that psychology is "at the dawn of a new era" on the basis of "molecular genetic techniques."[19] For these authors, "nothing can be more important than identifying specific genes involved" in psychological traits and psychiatric disorders. In the same year, Rutter and Plomin wrote that although gene discoveries had not yet been made in psychiatry, "it is obvious that these are likely to be forthcoming very soon as findings with respect to schizophrenia . . . affective disorder . . . and dyslexia . . . all show."[20]

Plomin and Rutter published a 1998 article in *Child Development* in which they informed developmental psychologists that "Genes associated with behavioral dimensions and disorders are beginning to be identified."[21] They added, "as associations between genes and complex behavioral traits are found, they are beginning to revolutionize research." The authors were attempting to prepare psychologists for gene discoveries-in-the-making which they believed would soon revolutionize their field. In another 1998 publication, Plomin and colleagues wrote that a pair of 1996 studies claiming an association between genes and the personality trait of "novelty seeking" constituted a "watershed" event for the field.[22]

At the dawn of the new millennium, Plomin and Crabbe predicted in 2000 that "within a few years, psychology will be awash with genes associated with behavioral disorders as well as genes associated with variation in the normal range." They also predicted that in the future, clinical psychologists would routinely collect patients' DNA "to aid in diagnosis and to plan treatment programs."[23] Elsewhere in 2000, Plomin wrote that genes "are being found for personality . . . reading disability . . . and *g* general intelligence . . . in addition to the main area of research in psychopathology."[24] In 2001, at the time of the publication of the first draft of the sequence of the human genome, McGuffin, Riley, and Plomin published an article in *Science* entitled, "Toward Behavioral Genomics,"[25] repeating the 1994 claim that gene linkages and associations had been discovered for such traits as aggression, schizophrenia, attention-deficit hyperactivity disorder (ADHD), male homosexuality, and dyslexia. In the same year, Plomin and colleagues published the fourth edition of *Behavioral Genetics*.[26] Here they claimed that "ADHD is one of the first behavioral areas in which specific genes have been identified," and they continued on their theme that "one of the most exciting directions for genetic research in psychology involves harnessing the power of molecular genetics to identify specific genes responsible for the widespread influence of genetics on behavior."

Having entered the "post-genomic era," in 2003 Plomin and McGuffin claimed "progress . . . toward finding genes. . . . although progress has nonetheless been slower than some had originally anticipated."[27] They wrote that the identification of genes for schizophrenia "remains elusive," and the "story for major depression and bipolar depression is similar to schizophrenia." Nevertheless, they continued to believe that the future of molecular genetic research in psychiatry "looks bright because complex traits like psychopathology will be the major beneficiaries of postgenomic developments,"[28] although they wrote a year later that researchers would need "very large samples" to uncover genes.[29]

In the period from 2003 to 2004, Plomin began to write more about gene discoveries as something that had not yet occurred, and less about discoveries that had been made or were in the process of being made. He wrote about "future . . . molecular genetic studies of DNA that will eventually identify specific DNA variants responsible for the widespread influence of genes in psychological development."[30] Elsewhere, he recognized that "no solid" gene associations for IQ "have yet emerged" and that "the road ahead will be much more difficult than generally assumed. . . ."[31]

Plomin's frustration became more apparent the following year, when he publicly asked, in relation to gene-finding attempts, "When are we going to be there?" Plomin answered, "Being an optimist, my response is 'soon,'" and recognized that his readers might be "skeptical, because they have heard this before."[32] Although Plomin claimed as always that the field was moving toward gene discoveries, he believed that behavioral genetics remained only "on the cusp of a new post-genomic era. . . ." He and his colleagues had decided not to produce a new edition of *Behavioral Genetics*, he wrote, "until we had some solid DNA results to present."[33] Although the next (fifth) edition did report some purported gene associations, the fact that none was replicated meant that they were not so "solid" after all. Currently, some continue to believe, and lament, that we are still on this "cusp."

In a 2010 publication, Haworth and Plomin appeared to give up hope that GWA studies would uncover genes anytime soon, writing that "it seems highly unlikely that most of the genes responsible for the heritability for any complex trait will be identified in the foreseeable future." They added, "we hope that our prediction about GWA research is wrong. . . ." In the process, they fell back on the *ad hoc* "missing heritability" theory to explain GWA failures.[34] Indeed, they recognized that genome-wide association studies "are struggling to identify a few of the many genes responsible for the ubiquitous heritability of common disorders" and psychological traits. In the face of the unexpected and disappointing failures of GWA studies and previous molecular genetic research methods, Haworth and Plomin argued that the field should return its focus to quantitative genetic studies of families, twins, and adoptees, which have a "bright future." Thus, they called for a retreat to previous kinship studies in light of the failures of molecular genetic research, never considering the possibility that the critics were right all along that the massive flaws and untenable theoretical assumptions of these methods explain these failures.

Plomin could not name any replicated gene findings in a 2011 publication and continued to explain these negative results on the basis of "missing heritability."[35] According to Plomin, "The big question now in molecular genetics is how to identify the 'missing' heritability; the big question for non-shared environment is how to identify the 'missing' non-shared environment." As critics have argued, both are "missing" because behavioral geneticists have mistakenly interpreted twin studies as providing unequivocal evidence in favor of genetics. Plomin and his colleagues continue to place total faith in twin research and continue to ignore the implications of other evidence, which includes Plomin's own carefully performed 1998 longitudinal adoption study that found a non-significant .01 personality test score correlation between birthparents and their 245 adopted-away biological offspring. According to Plomin and his colleagues, this birthparent-biological offspring correlation is "the most powerful adoption design for estimating genetic influence," which "directly indexes genetic influence."[36]

Conclusion

Science writer John Horgan published a critical appraisal of behavioral genetics in a 1993 edition of *Scientific American*.[37] Horgan noted that although there were many gene-finding claims for such traits as crime, bipolar disorder, schizophrenia, alcoholism, intelligence, and homosexuality, none of these claims had been replicated. He presented the results under the heading, "Behavioral Genetics: A Lack-of Progress Report." We can now update Horgan's "progress report" and issue the field of behavioral genetics its apparent final report card:

The evidence suggests that genes for the major psychiatric disorders, as well as for IQ and personality, do not exist. As Turkheimer concluded in 2011, in light of the failures of molecular genetic research, it is time to develop a "new paradigm."[38]

Simply put, the gene-finding claims and predictions by Plomin and other leading behavioral geneticists turned out to be wrong. The best explanation for why this occurred is not that "heritability is missing" but that previous and current claims that psychiatric and psychological twin studies prove something about genetics are also wrong.

We cannot expect the proponents of behavioral genetics to recognize that the historical positions of their field are mistaken, that their prized research methods and "landmark" studies are massively flawed and environmentally confounded and that family, social, cultural, economic, and political environments—and not genetics—are the main causes of psychiatric disorders and variation in human psychological traits. Because most leaders of the field will not allow themselves to see this, it is left to others to show that the pillars of behavioral genetics are crumbling before our very eyes.

We are indeed at the "dawn of a new era," but it will be an era very different than the one that Plomin and his colleagues envisioned.

Behavioral Genetics Research

Interview with Jonathan Beckwith

Jonathan Beckwith, Ph.D., *is a professor of Microbiology and Molecular Genetics at Harvard Medical School. This interview originally appeared in* GeneWatch, *volume 24, number 6, October-November 2011.*

GeneWatch: Do you see a trend in the amount of research oriented toward finding a genetic basis for human behaviors?

Jonathan Beckwith: At this particular moment, yes. It seems to go in waves. What surprises me is that the wave seems to be peaking again right now, at a time when, at least in terms of using all the new genetic technologies, there is an increasing question about the utility of what's being done.

GeneWatch: Do you have any idea why that is?

Beckwith: I think a component of it is that people outside the field of contemporary research in genetics are getting into it. I think certain groups of psychologists have always been interested in genetics, but now it's also people in other fields, like political science.

GeneWatch: It seems like, at least up until recently, twin studies have been critical for a lot of behavioral genetics. Do you see that as a cause for concern?

Beckwith: Yes, I've always thought it's an area of concern. The history of twin studies has gone through these ups and downs; they've been considered very important at points, then people have found problems with them, and then people in the field have come back and claimed to have overcome the problems. Twin studies are generally done by people who are not actually in the field of genetics but are usually in psychology departments; and I think today, geneticists who are looking for genes associated with behaviors have relied on the finding of high heritabilities for certain traits from the twin studies, even though I don't think they have looked at them carefully enough to realize all the problems of twin studies.

I think that comes out very strongly in a current favorite topic among geneticists, the "missing heritability." The point is that people are looking for genes that are associated with certain behavioral conditions, or to mark chromosomal regions that are associated with those conditions, encouraged by the claimed high heritability obtained in twin studies. But when they find genes or regions of chromosomes that they think contribute to behaviors, their contribution to the heritability of the traits is much less than that predicted by the twin studies. So there's this supposed puzzle: what's happened to the missing heritability?

GeneWatch: Of course, supposing it's missing is supposing it was there in the first place, based on the twin studies.

Beckwith: Exactly.

GeneWatch: Are there any genes in particular that get a lot of attention in behavioral genetics?

Beckwith: Certain genes are favored for explaining all sorts of things, like the Monoamine oxidase A (MAO-A) gene and the dopamine receptor gene. There's a whole list of behaviors for MAO-A, but the ones that get particular attention are aggression, and it's come to be called "the warrior gene."

GeneWatch: Is that similar to the "criminal gene?"

Beckwith: Oh, yes—it goes back a long way. It started in the 1990s. The very first paper that got a lot of attention and was discussed in the media as a "criminal gene" study was of one family, where a number of the men had a mutation in a gene that completely knocked out its function. The study found an association with "antisocial behavior." That was picked up by the media to be called a "criminal gene," and even the senior author on the paper publicly stated, at least at a scientific meeting, that it was ridiculous to call it that.

Then, in the early 2000s, there was a group that presented a more subtle analysis, basically on some very common polymorphisms of the MAO-A gene, two of which are present in a significant part of the population. They found, by various measures, increased "antisocial activity," but only when the subjects had been subjected to child abuse. The problem is that a number of people have tried to replicate this, and a good portion of them have not been able to; it's not clear why.

It is argued that the reason the MAO-A and dopamine receptor genes have been focused on is that they are well-studied genes involved in brain function. Because this is one of the few genes that there were very early studies on, a lot of people who are studying various behavioral traits look at those candidate genes to begin with. However, since there are so many people looking at those particular genes to associate them with a behavior, inevitably some people are going to find a correlation just by statistical chance. That may explain why many of these findings cannot be replicated.

GeneWatch: Do you think that if someone is going into the study set on finding heritability that it can ever be a sort of self-fulfilling prophecy? If by going in to a study looking for something, could that make you more likely to find it (or think you've found it)?

Beckwith: I think it may be, and if enough people are looking for the same thing, eventually some people are going to find it—but then it's not replicated, which goes into the long history of this field. I particularly think that's a problem because of people coming in from other fields who haven't been through the experiences of the last 20 or 30 years of research in genetics and who are unaware of all the criteria that are necessary to really define things. That is, some of the work is just simply of very poor quality, not correcting for all the things that geneticists learned to correct for after a lot of false alarms.

Back in the 1980s and 1990s, there were a whole set of reports about single genes for homosexuality, schizophrenia, manic depression, risk-taking, happiness, etc. Those reports turned out to be wrong or have not been replicated. At that point, geneticists were very unsophisticated about statistics. They were using inappropriate statistical approaches and made other incorrect assumptions that required continuing revisions of the criteria needed in order to draw conclusions. When people used more appropriate approaches, if they found any genes or chromosomal regions correlated with such traits, they found, at best, large numbers (sometimes hundreds) of genes (rather than a single gene). Even then, all of these genetic variants were still only contributing a small percentage of the variation in expression of that particular trait.

GeneWatch: Why do you think there has been an influx of people from fields outside of genetics conducting studies to find a genetic basis for certain behaviors?

Beckwith: Maybe they start with the belief that genetics is very important in these behaviors; but also, like a lot of other fields, people have moved into genetics because there's so much money available. There are people from the social and political sciences who are getting grants from the National Institutes of Health and National Science Foundation to do science that ignores the lessons learned from previous failed attempts to find genes for complex human traits.

GeneWatch: Do you have any sense of why the NIH would be putting money into finding a genetic basis for things, even if it doesn't seem, in a lot of cases, any more successful than finding environmental influences?

Beckwith: Well, that's what NIH does, mainly. And of course there's just been a very big push for genetics recently. Now with the National Human Genome Research Institute, it's a major segment of what the NIH is doing, and a lot of promises have been made about what's going to come out of it. But I don't particularly know why they and the National Science Foundation are funding studies that deal with genes for moral judgment or voting habits, for example.

GeneWatch: Right—it's hard to see the utility of figuring out whether or not there's a genetic basis for whether or not people vote. Do you have a sense of what the political scientists and other people studying those things—"genopolitics"—are hoping to get out of it?

Beckwith: I'm not sure . . . certain people in genetics see this field taking over, more and more, in the way in which E. O. Wilson, when he came out with his book *Sociobiology* in the 1970s, made the argument that sociobiology was going to subsume many other disciplines.

GeneWatch: After reading about certain claims linking a behavior to a genetic basis, the question I always find myself asking is: If something as complex as whether or not we show up to vote can be explained largely by our genes, what behaviors can't?

Beckwith: First of all, I just have to say, that particular paper is one of the worst, I think. It's a surprisingly bad paper—at least to somebody who has been in the field or watching it carefully for the last twenty years or more.

What you asked reminds me of a conference I went to many years ago about behavior and genetics that was mainly based on twin studies in those days. The meeting was attended by both critics and practitioners in the field. One of the people got up, a quite well-known geneticist who had been doing twin studies for ages, and he talked about the studies they were conducting on various behavioral issues, using identical twins and so on. And he said, "We realized we needed a control in this study of something that wasn't genetic. So we decided to ask about people's religious attitudes as a control," since that wouldn't be genetic. Instead, he said: "We found that was genetic too!"

That was the most bizarre thing. It's kind of an elemental scientific problem—you can't just change your control into an actual subject that you've proven something with—but he wasn't joking. He's published on that, in fact.

GeneWatch: Do you think this is sometimes a symptom of people coming into the field of behavioral genetics from outside of genetics—this idea of not just problematic study design, but going out and presenting those results without waiting to see if they are replicable?

Beckwith: It's not really the same thing. I would say that, more than other fields, people come in with preconceived notions. And I don't think they were thinking about science, in terms of what it means to be doing science, the way that people in other fields were.

For example, one of my favorite beefs about the identical twin studies is a major assumption behind these studies, the equal environment assumption. That was unchallenged—it was a hidden assumption, I don't think they even felt they had to state it—that is, that if you looked at fraternal twins and identical twins, since they were born at the same time and grew up in their families and the world at the same time, that the two types of twins essentially share the same environment to the same degree. From someone outside the field, it's an obvious question whether that's true or not, but that was an unchallenged assumption until maybe 30 years ago, although twin studies have been going on since the early part of the twentieth century. It was only then that they decided they had to do something to test those assumptions—which I don't think they did very well—and then they published a bunch of studies saying, "Well, we've proven the equal environment assumption is correct, therefore it's not an issue."

GeneWatch: How has the equal environment assumption been tested?

Beckwith: What they did when they tried to test the assumption, once it had been challenged, was pretty simple, I thought. If it was a genes and intelligence study, they would go and count the number of books in the house of the twins, for example. Supposedly, one is asking the question: What are the influences that determine intelligence? If you don't know what they are, it's not clear exactly what you should be looking for, and just looking at a few things—the parents and their education, or the number of books in the house—isn't necessarily telling you much, because you don't know if you're looking at the right things.

GeneWatch: Another question that seems to sometimes be missing after a gene for something has apparently been identified is: What is the mechanism? In the political participation study, it's not surprising that although they claimed to identify a genetic basis, they didn't have an explanation for why those genes would influence whether or not someone votes. Is that a reason to raise an eyebrow?

Beckwith: I think because of the false alarms, people have to be very careful. Before jumping into dramatic conclusions about what it means, they have to know more about the function of the gene.

I'll give you an example. A researcher, actually someone associated with our department, was doing studies on mice. I think they knocked out a gene in the mice, and they found that the mice no longer nurtured their progeny. It actually was a big story, at least in the local papers, that "the gene for nurturing" had been found. But it turned out that this gene was a major regulator of a huge number of different processes, and without it the mouse was just not a healthy mouse in general—so it's hardly a "gene for nurturing."

That kind of criticism can certainly come up when people are looking at a gene and see an effect on a behavior and don't even know what the gene does, how many processes in the body are affected and where in the body the gene is expressed. Such correlations are simply not very meaningful without much deeper study, and reporting of them is extraordinarily premature.

GeneWatch: I guess when you start hearing things like "the gene for nurturing," the researchers might not have had that in mind or might not have expressed it that way—is it often the case that other people, at least in the media, are sort of taking the idea and running with it?

Beckwith: I would say that most of the time, when the media presents these findings in a dramatic way, there are at least hints from the scientists themselves that it should be taken that way. That's not always the case, though—sometimes scientists who produce the work become quite dismayed at its interpretation by the media.

The MAO-A "criminal gene" study I mentioned before is an example. I spoke to one of the researchers afterwards, and the publicity was so horrendous to her . . . for example, there was a *Newsweek* article that discussed the finding of this gene, called it a "criminal gene," and had a picture of Arabs and Israelis fighting. It got really blown out of proportion, and she was really very upset. She actually said she would never work on that subject matter again.

And the senior person on that study did make a statement about how it shouldn't be called the "criminal gene" or the "aggression gene." For these rueful scientists, some of this might have been avoided if the education of the scientist could prepare us to be more sensitive to the social implications of our work. On the other hand, I think there are many other instances where it's very clear what the intent of the authors is, if not in their publications, then in their press conferences or public statements.

In Our DNA?

BY STUART NEWMAN

Stuart A. Newman, Ph.D., *is Professor of Cell Biology and Anatomy at New York Medical College. He was a founding member of the Council for Responsible Genetics.* *This article originally appeared in* GeneWatch, *volume 24, number 6, October-November 2011.*

Human behaviors are either innate or environmentally induced. They can be novel or habitual actions, which, at a minimum, involve interactions between the nervous system and other organs. A gene specifies the sequence of subunits of a single RNA or protein molecule. How is it possible for specific behaviors to be associated with particular genes? Can anything significant be learned from such associations?

One persistent view, increasingly discredited in recent years, is that genes collectively provide a blueprint or software program for the generation of all organismal characters—anatomical structures, but also behaviors—and variant genes are lines of code of alternative programs. So, for example, two recent reports in the journal *Nature Genetics* described evidence that variations within any of 11 DNA regions in the human genome have a strong association with schizophrenia, bipolar disorder, or both.[1] One of the studies' authors was quoted as stating, "Our findings are a significant advance in our knowledge of the underlying causes of psychosis—especially in relation to the development and function of the brain."[2]

The expectation that such genetic associations will provide insight into how the brain generates normal or abnormal behaviors is directly connected to belief in the notion of a genetic program. No one would assert that learning the chemical composition of a piece of tile from a medieval mosaic represents a significant advance in understanding the meaning of the art work. But when it comes to living organisms, which operate individually and socially on multiple spatial and temporal scales and have assumed their present forms and behavior patterns over billions of years, genetically reductionist explanations can unfortunately still be advanced without evoking derision.

The acceptance of this bizarre way of thinking, which is even more prevalent in the scientific and medical professions than in lay society, derives from a particular theory of evolution that has prevailed over the past century. Referred to as the Modern Synthesis, it links Darwin's idea of natural selection as the generator of all inherited traits to the notion that genes and their variants are the only significant determinants passed from one generation to the next. All standard features of an organism, for example, the five fingers of the human hand or the propensity of humans to live in social groups, are consequently considered *adaptations*. By this argument, pathological behaviors that are endemic to many societies, like racism and rape, are likely to be adaptations as well (or sometimes negative but understandable misappropriation of adaptive behaviors).[3 4] And now that a "genetic basis" is claimed to have been discerned for psychosis, it is all but certain that some evolutionary psychologist is laboring to uncover its benefits on the prehistoric savannah.

Fortunately, a new concept of evolution is now taking hold. In various subdisciplines like "evolutionary developmental biology"[5] and "ecological developmental biology"[6] there is increasing receptivity to the idea of a loose, nonprogrammatic relation between the phenotype (particularly the behavioral phenotype) and the genotype. This includes an openness to the notion that the external environment may influence the development of inherently plastic living systems,[7] not in arbitrary ways, but in ways constrained by the forming systems' inherent modes of action.[8]

The "Baldwin effect," the name given to a concept put forward by the psychologist J. Mark Baldwin in the post-Darwinian period of intellectual ferment before the Modern Synthesis consolidated its grip, in a paper titled "A New Factor in Evolution,"[9] has gained renewed prominence. This is a "phenotype first" scenario in which a character or trait change occurring in an organism as a result of its interaction with its environment becomes assimilated (often by Darwinian selection) into its developmental repertoire. The descendent organisms are then born with the novel phenotype rather than having to acquire it through each generation.

In this emerging theoretical framework not every persistent phenotype is an adaptation; organisms with novel anatomies or behaviors need not sink or swim in pre-existing niches but can construct new ways of life compatible with their new biology;[10] and genes often play catch-up, consolidating behavioral or other phenotypic changes after the fact.[11] This has permitted conceptual accommodation of older, puzzling findings and has energized previously proscribed research programs.

Genetics, since it deals with an intrinsic set of determinants of all living systems, is far from being sidelined in the new approach, but genes must now take their place alongside other key factors. A set of studies in Siberia, beginning in the 1950s, on farmed foxes, for example, showed that docile, human-friendly behavior could be propagated from parent to offspring, along with a suite of morphological attributes (shortened snouts, floppy ears, patchy coat color) seen in other, unrelated, independently domesticated animals, by a selective breeding protocol that acted not on gene frequency but on the level of stress hormones in the gestational environment.[12] A more recent study demonstrated the passing on of grooming behavior from mother to daughter mice by the effects on the offspring's biology of the behavior itself.[13] In both studies an effect on DNA was one of the steps in trait transmission, but the DNA was not irreversibly changed, nor was it the first or unique event in the behavior modification.

Sexual dimorphism is one of the best known examples of the above-mentioned plasticity, and of the fact that an animal embryo can potentially follow alternative routes of development. In humans, this decision, with both anatomical and behavioral consequences, is specified by genes. Males differ from females by having an entire chromosome, the Y, with a set of genes which are absent in the female genome; with two X chromosomes a biological female takes form. Only one of these male-specific genes, *SRY*, is needed for development of a male body and male gender identity, however.[14] If, as in XX male (de la Chapelle) syndrome, *SRY* winds up on the X chromosomes and no Y is present, the resulting anatomical and behavioral characteristics are still typically within the standard male range.

From the above it would seem that *SRY* is the gene for maleness. Indeed, its production sets the male developmental pathway in motion in most mammals, including marsupials. But some rodents, such as spiny rats, just use extra copies of the *CBX2* gene, which is also present in females, to perform the same function.[15] So maleness can be generated without a specific gene for it, perhaps just needing *more* of some gene. But even this notion is refuted if we cast our zoological comparisons a little wider. In some birds and fish, and in many reptiles, particularly turtles, sex determination is controlled by the egg incubation temperature. That is, with respect to prospective sexual phenotype the genotype of the embryo is a matter of indifference. If it develops at a low temperature it will become a male, and if at a high temperature, a female. (Or vice versa, for some turtle species; or, in the case of other turtle species, female if high or low and male in-between.)[16]

So for behaviors, reproductive and otherwise, and even for characters as deeply dimorphic as sexual anatomy, specific genes make a difference, or other genes make the same difference, or the difference is made by the social or physical environment acting on the organism during development—including, of course, the expression of many of its other genes. This reality consigns to the realm of tea-leaf reading any notion that we can infer the underlying cause of a behavior solely from its correlation to some gene variants.

45

Wising Up on the Heritability of Intelligence

BY KEN RICHARDSON

Ken Richardson *was formerly Senior Lecturer at the Centre for Human Development and Learning, The Open University, UK (now retired). He is the author of* Understanding Psychology; Understanding Intelligence; Origins of Human Potential: Evolution, Development and Psychology; Models of Cognitive Development *and* The Making of Intelligence. *This article originally appeared in* GeneWatch, *volume 22, number 3-4, July-August 2009.*

Behavioral genetics has covered a wide range of topics and, in animals, a respected history of scientific work. In humans, though, its work has been more controversial, dominated by "fitness" for social roles and rank and, therefore, cognitive ability or "intelligence." Indeed, it was Darwin's cousin, Francis Galton, who argued that fitness in humans depended on "General Ability or Intelligence" and proposed "to show . . . that a man's natural abilities are derived by inheritance." To do this he set up the world's first mental test center in London in 1882. Using simple tests of mental speed, memory, sensory acuity, and so on, he wanted to show that scores would match social status or "reputation." Unfortunately for his theory, they didn't; but Galton's strategy—using test scores as surrogates for differences already "known"—laid the foundations of human behavior genetics. It only remained to find the "right" test.

This was indirectly provided by Francis Binet's work in Paris. He had been devising school-type tasks—general knowledge, comprehension of sentences, simple arithmetic, and so on—for screening children in school for special treatment, and produced his first test in 1904. Because scores depended on family background, these tests *did* correlate with social class. Galton's followers in Britain and America seized upon them as what they had been looking for—their test of "innate" intelligence—and quickly translated them for use in their anti-immigration and eugenics policies. But this still didn't actually prove that score differences are due to inheritance.

The problem was partly solved by R.A. Fisher in 1918, who introduced the statistical concept of "heritability," the proportion of trait variance attributable to genetic variance, from experimental breeding programs in agricultural species.[1] Its validity lay in knowing the genetic background and environmental experiences of the organisms, which we don't tend to have in humans. But Cyril Burt thought he had solved *that* problem when he estimated the heritability of intelligence from IQ correlations in pairs of twins. Since then, results of a number of twin studies of cognitive ability have suggested a sizeable heritability of between 0.5 and 0.8, meaning that 50 to 80 percent of the variance in cognitive ability is genetic in origin. So twin studies and correlations between IQ test scores, became the dominant paradigm of human behavior genetics.

Definition of phenotype and the IQ test

Since we cannot measure intelligence directly—not least because psychologists cannot really agree what it is—the validity of Binet-type IQ tests depends entirely on inferences from correlations. Since tests were constructed to predict school achievement, which determines entry to the job market, there is an inevitable correlation between test scores and occupational and social status. Behavior geneticists draw considerable conviction from these correlations, as if they represent proof of a causal mental power. However, scores have little if any association with job *performance* and, as Joan Freeman's studies have shown, are not reliable indicators of adult careers. High achievers in adulthood did not tend to shine above the average as children; and we don't find members of MENSA, the high IQ society, dominating the ranks of high achievers in society.

What this "intelligence" is, or what actually varies, therefore, is still not clear after more than a century of scientific inquiry. As prominent behavior geneticist Ian Deary puts it, "There is no such thing as a theory of human intelligence differences—not in the way that grown-up sciences like physics or chemistry have theories."[2] Or, as Carl Zimmer put it in *Scientific American*, "intelligence remains a profound mystery . . . It's amazing the extent to which we know very little."[3]

Instead, cognitive behavior-geneticists rely on a kind of mystique around test demands, as if they were equivalent to cognitively complex tasks. However, the vast majority of IQ test items are simple tests of memory and general knowledge with a high learned literacy/numeracy (and, therefore, social class) content: "What is the boiling point of water?"; "Who wrote Hamlet?"; "In what continent is Egypt?" and so on.

Much weight is placed on the "Raven" test (Raven's Standard Matrices), and other non-verbal tests, said to measure "abstract reasoning," detached from cultural learning. As for complexity, there seems little to distinguish test items from the complexity of reasoning required in everyday practical and social tasks carried out by nearly everyone. Analyses have found little evidence that "level of abstraction" (defined informally) distinguishes item difficulty. As Téglás and colleagues have shown, even 12-month-old infants are good at "integrating multiple sources of information, guided by abstract knowledge, to form rational expectations about novel situations never directly experienced."[4]

As for being "culture-free," what is overlooked is that, like languages, cognitive styles differ according to the kinds of activities most prominent in different cultures and social classes. In studies of formal logic and reasoning, it is a classic finding that different problems of equal complexity can be of widely different difficulty to different people. Western societies are deeply class-stratified along occupational lines, which create starkly different activities and habits of thought. As cognitive psychologist Lev Vygotsky argued, such activities "determine the entire flow and structure of mental functions."[5]

Accordingly, much research shows how "ways of thinking," and even brain networks, are shaped by cultural activities. What is clear about the Raven test is that the cognitive processes demanded are those most common in middle class cultural activities: reading from top-left to bottom right; following accounts, reading timetables, and so on. It is not testing an individual's rank on some fixed scalar power so much as their "distance" from forms of knowledge and thinking deemed to be the norm by test designers. This is indicated by the massive gains in average IQ scores (including Raven scores) over time as more people have moved from working class to expanding middle class occupations (the so-called "Flynn effect").

Methods—Twin studies and heritability estimation

Strong claims about IQ heritability suggest that behavior geneticists have firm measures of the genetic variance underlying the (not so clear) phenotype of intelligence. On the contrary, neither the genetic nor environmental values are actually known. Rather these are (again) inferred from correlations among relatives, on the basis of a host of unlikely assumptions.

Identical, or monozygotic (MZ), twins share all their genes, and tend to correlate around 0.7-0.8 on IQ. We can infer that this resemblance is due to their common genes, and the rest is due to differences in environmental experiences. But the correlation could be due to the environments that they also share, and the remainder due to errors of measurement (which are often forgotten). If the twins are reared apart, in completely different environments, then, in theory, the correlation would provide a direct estimate of heritability. In practice, it has been extremely difficult to find suitable samples of twins reared apart in completely uncorrelated environments, so estimates derived from them have been highly dubious.[6]

Consequently, most heritability estimates have come from comparisons between the resemblances (correlations) of MZ and dizygotic (DZ, nonidentical) twins. We might expect MZ pairs, who share all their genes, to be more similar then DZ pairs, who share only half their genes on average, so it can be inferred that differences in average resemblance or correlation between kinds of twins is related to differences in genetic similarity. Through formulae explained elsewhere, heritability is usually estimated from twice the MZ-DZ difference in correlations:

$$h^2 = 2\,(r_{mz} - r_{dz}).$$

More recently, some sort of statistical modelling (usually, structural equation modelling) has been used, which has some advantages. But such modelling also makes a number of assumptions that may or may not be valid.

Now let us look at some of those assumptions. The first is that human intelligence can be treated exactly like a simple quantitative trait such as height or weight, and that the relevant genes, although possibly numerous, exert effects additively (independently of each other). Otherwise—if there were interactions between genes or genes and environments—it would be impossible to determine what correlations to expect. It is also assumed that "environments" contribute to differences in the same additive way. As we shall see, this flies in the face of what we now know; yet there are few serious attempts to rule out such interactions in the twin IQ data.

What has drawn the most attention of critics, though, is the "equal environments assumption." The twin method requires that the environments of MZ pairs are no more similar than environments of DZ pairs, on average; otherwise, those variations could partly or entirely explain the correlation differences. As it happens, the assumption is flatly contradicted by numerous studies. In one review, David Evans and Nicholas Martin said, "There is overwhelming evidence that MZ twins are treated more similarly than their DZ counterparts."[7] Studies reveal that MZ twins are more likely to share playmates, bedrooms, and clothes, and to share experiences like identity confusion (91 percent vs. 10 percent); being brought up as a unit (72 percent vs. 19 percent); being inseparable as children (73 percent vs. 19 percent); and having an extremely strong level of closeness (65 percent vs. 19 percent). Parents also hold more similar expectations for their MZ than DZ twins. Behavior geneticists have a tendency to wave these differences aside as if they don't really matter, but they can easily explain part or all of the differences in IQ resemblance and grossly inflated heritability estimates.

There are other problems that might distort heritability estimates. A major—and problematic—prior assumption of these estimates is that there *will be* substantial (additive) genetic variance underlying all individual differences. In actuality, it is one of the laws of natural selection that, for traits important to survival, additive genetic variance tends to be reduced across generations, creating cohesive, interactive genotypes. Indeed, many experimental and observational studies have confirmed this.[8] Over a decade of genome-wide molecular studies meant to take us "beyond" heritability have failed to find genes for IQ. One explanation often proposed is that the gene effects are cumulatively present, but individually too small to be separately detected. So investigators have returned with greater resolve, recently, to "proving" IQ heritability. For example, the recent report by Davies and colleagues, brought out with press reports and much media coverage, claims to "establish" and "unequivocally confirm" (surely unedifying terms in any science) just that. As usual, the study involves a host of "ifs" and assumptions, including the dubious one that genetic effects can be treated as a random (independent) variable. But it also uses a device for extrapolating from identified to non-identified variances which David Golan and Saharon Rosset recently describe as "a questionable heuristic."[9] Moreover, subjects were in their upper sixties and seventies (and, therefore, non-representative in other respects); and heritability estimates varied from 0.17 to 0.99(!), depending on combinations of samples and tests.

Intelligent systems

What is definitely unclear is why this enterprise continues. It is now widely accepted that heritability estimates of intelligence are, in the human context, of little practical relevance. Even if accurately achieved (which so far seems unlikely), they do not predict the likely developmental endpoints for individuals or groups, or the consequences of interventions; they have told us little reliably about genes or environments; and they have not helped to provide a "grown-up" theory of intelligence. As the originator, Ronald Fisher himself, said, "(heritability) is one of those unfortunate short-cuts which have emerged from biometry for lack of a more thorough analysis of the data."

One reason is that, behind all the controversies, there are different world views of the nature of genes and environments, traits and their development. The "genes" and "environments" of the behavior geneticist are abstract, idealistic entities with little interaction, a linear determinism that defines limits on individual development and, therefore, social status and privilege. On the contrary, the recent "omics" revolution—the creation of a broad range of research areas, including genomics, proteomics, metabolomics, interferomics, and glycomics—suggests the very opposite of such independent, linear effects. It suggests how processes and systems utilize higher information structures geared to changing environmental contexts.

Various discoveries now show how intense cross-talk between multitudes of gene-regulatory pathways provide complex non-linear dynamics. These dynamics can create novel developmental pathways, often proposing new targets for selection. They integrate the transcription of genes contextually, often "rewiring" the gene network in response to changing environments. In addition there are the vast regulatory functions of alternative splicing, messenger RNA, vast numbers of non-coding RNAs, and so on, all depending on cooperative interactions. These explain why many different phenotypes can develop from the same genotypes, or the same phenotype from different genotypes; and why a population of individuals of identical genes developing in identical (or closely similar) environments can exhibit a normal range of behavioral phenotypes.

Even at this level, the "dumb" independent factors and simple quantitative traits of the behavior geneticist have disappeared into highly interactive intelligent systems. Metabolic networks evolved into nested hierarchies of still more intelligent systems: physiological systems; nervous systems and brains; cognitive systems; and, finally, the human socio-cognitive system. On "top" of this nested hierarchy the socio-cognitive system differentiates according to dominant cultural activities, making humans far more adaptable than any system of independent genes. IQ tests simply collapse this enormous diversity into a (pretend) scalar trait. What is allocated to the category "genetic variance" is, in reality, variation in the expression of nested dynamic systems. This is why a leading behavior geneticist of IQ, Eric Turkheimer, has had to admit, recently, that "The systematic causal effects of any of these inputs are lost in the developmental complexity of the network."[10] It seems ironic that the current unfolding of the real nature of intelligent systems is leading to the eclipse of the Galton paradigm.

Race, Genes, and Intelligence

BY PILAR OSSORIO

Pilar Ossorio, Ph.D., *is Associate Professor of Law and Bioethics at the University of Wisconsin at Madison and Program Faculty in the Graduate Program in Population Health at the UW. Dr. Ossorio is a fellow of the American Association for the Advancement of Science, a member of the editorial board of the* American Journal of Bioethics, *chair of an NHGRI advisory group on ethical issues in large-scale sequencing, and a member of UW's institutional review board for health sciences research. This article originally appeared in* GeneWatch, *volume 22, number 3-4, July-August 2009.*

Over the past two centuries biomedical science has, at times, provided justification for white privilege. Science has been used to support the proposition that differences in achievement reflect innate differences in ability among racial groups. Broadly speaking, the view that differences in academic achievement, IQ scores, employment status or wealth primarily reflect innate differences is called "biological determinism."[1] As the late Stephen J. Gould pointed out, at its core, biological determinism is "a theory of limits. It takes the current status of groups as a measure of where they should and must be (even while it allows some rare individuals to rise as a consequence of their fortunate biology)."[2]

Biological determinism lost most of its scientific credibility by the mid-twentieth century and lost much of its social and political power after World War II; however, it never entirely disappeared. Today, some people believe that persistent racial gaps in, for instance, school achievement, family income, and wealth must reflect innate differences in ability. One human trait that is postulated to play a role in many kinds of achievement is intelligence, and some commentators theorize that racial differences in average levels of intelligence explain achievement gaps.

At the same time, the new molecular genetics has captured the public imagination and has provided tools for conducting large-scale genetic comparisons between individuals and between human groups. Some people will look to modern genetics to provide scientific justifications for racial inequalities. Genetics is particularly appealing in this role because of its apparent precision, authority, and high-tech chic. Many people reason that if groups vary with respect to innate cognitive abilities, then the differences between groups must be attributable to differing racial patterns of genetic variation. To disentangle claims about race, genetics and intelligence, we must examine beliefs about race and intelligence and understand what role genes reasonably could or could not play with regard to the intersection of these two concepts.

In the contemporary world, beliefs about racial difference, and racial superiority or inferiority, may be articulated in the language of molecular genetics and genomics. Modern genetics has great authority, and beliefs about race that once relied on vague notions of innate difference can be made to sound more precise and credible by framing them as genetic explanations. Genes can be viewed as the substrate by which God or natural selection rendered some groups superior and others inferior. Educational achievement, wealth, and other measures of status often run in families, a fact that may increase the intuitive credibility of genetic explanations. However, societal institutions operate to entrench groups who wield power into self-perpetuating dynasties. From the Tudor monarchical dynasty in sixteenth-century England to the Bush and Kennedy family dynasties in the twentieth and twenty-first centuries in the United States, families with access to power pass their positions of privilege on to succeeding generations through processes that have little to do with hereditary transmission of genetic traits.

Just as there is no unitary definition of race, there is no agreed-upon or single definition of intelligence; one aphorism holds that intelligence is what intelligence tests measure. Psychometricians argue that intelligence tests measure reasoning skills, although the tests also measure knowledge. Some innovative scholars have developed theories of emotional intelligence and multiple intelligences—multiple types of cognitive function that are valuable and measurable, and that may manifest differently in different contexts.[3] The typical IQ test does not measure multiple intelligences; instead, the test produces a single intelligence quotient (IQ).

Some scholars argue that one's IQ indicates one's general cognitive ability, often referred to by the letter "g".[1 2] Many other scholars argue that the notion of a single, general quality that underlies performance on all cognitive tests is incoherent.[3] Stephen Jay Gould has provided a thorough explanation and critique of the concept of *g* in *The Mismeasure of Man*.[2] The measure "g" has been a useful concept for commentators who seek to create social hierarchies based on intelligence, because ". . . ranking requires a criterion for assigning all individuals to their proper status in a single series."[2]

Alfred Binet, the developer of the first intelligence test in the early twentieth century, rejected the notion that his test measured a person's inborn or fixed cognitive ability. He also declined to use his test to rank individuals according to cognitive ability. The purpose for which he devised the test, and the only purpose for which he thought it appropriate, was to measure the intellectual capacity of children who were performing poorly in school, to determine which children had cognitive deficits for which remedial instruction might be helpful. Later psychologists, particularly those in the United States, took up and modified Binet's test, and were willing to embrace the view that intelligence was an inborn and fixed attribute of a person. We can call this view the hereditarian theory of IQ.

Over the past decade, some contemporary proponents of the hereditarian theory have argued that 1) IQ is the most important determinant of academic success; 2) academic success is the most important determinant of high status and wealth-generating employment; and therefore 3) the economic elite have their positions and wealth as a matter of merit (intellectual contribution to society), and conversely, members of the economic underclass also deserve their position at the bottom of the social hierarchy.[4] These commentators argue that programs aimed at raising the academic achievement of disadvantaged students are misguided because those students are, on average, biologically incapable of significant academic success. Racial gaps in test scores, from IQ to the SAT, are interpreted by hereditarians as evidence of inherent and immutable racial or ethnic differences in underlying cognitive capacity.

Many claims of contemporary hereditarians have been critiqued and debunked in such books as *Measured Lies*, *Inequality by Design*, *Whitewashing Race*, and *Intelligence and How to Get It*. These books describe mistakes of fact, method and logic made by the hereditarians.

A significant problem in debates about hereditarian theories of IQ is that correlations are often treated as proof of causation. If one observes that people in lower socioeconomic brackets, on average, score lower on IQ tests than people in higher socioeconomic brackets, this does not mean that low IQ causes poverty. It could be that poverty causes low IQ, or that something else causes both outcomes. If IQ test scores correlate with race (however race is defined), this does not mean that some inborn racial essence causes particular IQ test scores. One reasonable alternative explanation is that race is correlated with other factors, such as quality of schools, exposure to lead, or malnutrition, and these other factors are causing the observed differences in test scores.

Many scholars question the entire enterprise of treating heritability statistics as though genes and environment are actually separable influences on IQ or any other trait. Genes always function within particular environments to shape the developing human organism. The developmental interaction among many genes and numerous environmental factors, is complex, varies over time, and is susceptible to chance events.

Researchers have, in fact, found evidence that some environmental factors are strongly associated with IQ and other measures of cognition. Malnutrition and exposure to environmental toxins, such as lead from paint, are strongly correlated with IQ. The quality of a person's school significantly impacts her IQ score—children who begin their education in poor-quality schools and then move to better ones show increases in their IQ scores.[5]

A study published in 2009 found that long-term stress is negatively associated with young adults' performance on cognitive tests.[6] This study measured levels of several physiological properties associated with stress, including blood pressure, cortisol, and epinephrine levels. The researchers collected data throughout their participants' childhood years, then administered tests of cognitive performance when the children turned 17 years old. Young adults whose bodies exhibited the highest levels of chronic stress had the least effective working memories and poorer cognitive performance. These data only show correlations between stress and IQ scores, they do not prove causation. But, they suggest an alternative theory that is at least as plausible as the theory that genetic differences are the primary cause of group differences in IQ scores.

Research also undermines the hereditarian claim that IQ is the primary determinant of achievement. Many environmental variables predict achievement as well or better than IQ, except for people whose IQ scores are at the abnormally low end of the scale. For instance, a person's social environment may be an important determinant of her achievement, yet variables that capture a person's social environment are often, literally, left out of the equation in work done by hereditarians. The social environment includes the expectations of one's peers, encouragement by one's parents and teachers, enrichment opportunities available in the neighborhood, etc. A decades' long study that included social environment variables found that a 15-point difference in IQ scores among high school boys only explained 6 percent of the variability in their earnings at age 35. The greater the number of social factors taken into account, the less important IQ became.[7] Social context variables were still significantly correlated with earnings by age 55.

In a related analysis, Fisher *et al.* demonstrated that if all adults in the country had the same score on an IQ test, the variation in household income would only decrease by about 10 percent. Contrary to hereditarian claims, these data suggest that differences in IQ do not explain much about professional achievement and wage inequality, including wage inequality between racial groups.[5] On the other hand, factors external to an individual can greatly influence her or his lifelong course of achievement.

Because race comprehensively structures people's lives in the United States, it is correlated with many environmental factors that can influence IQ and achievement. People of different races tend to live in different neighborhoods, so they may be exposed to different levels of lead, different quality schools, different diets and different levels or types of stress. They may be exposed to different attitudes about achievement. People of minority groups may routinely experience racism, a kind of stress that can have long-term physiological consequences. On average, people of different races receive health care at different institutions, and the care they receive is not of the same quality. In sum, racial groups differ with respect to so many environmental factors that it is very likely that environmental differences explain current racial gaps in mean IQ scores.

The environment can be modified in ways that genes cannot. When the environment is changed, the trait of interest (in this case intelligence) may also change even though genes also play a role in shaping that trait. In one study, African American children in Milwaukee who were thought to be at risk for cognitive disability, were randomized so that half received intensive day care and early, enriched education, while the other half received ordinary day care and schooling.[8] By age five, children who received the intensive intervention averaged 110 on a standard IQ test (above average), while children in the control group averaged 83 (well below average). The effects of early, intensive education were still apparent by adolescence, when the children from the intervention group scored, on average, 10 points higher on IQ tests than the children from the control group.

There is some evidence that differing environments have influenced the entire human population's IQ scores over time. People's average IQ scores have risen by about 3 IQ points per decade over the last century.[8] The average IQ score from 1917 would amount to about 73 on today's tests. This effect almost certainly is not due to changes in human genetics, because there has not been enough time for new intelligence-related mutations to arise and spread throughout human populations. The most likely explanation for the rise in IQ is that some relevant environmental factors have changed, causing people to develop in ways that are reflected in higher average IQ scores.

Another piece of evidence concerning widespread environmental influences on IQ is that the mean difference between black Americans' and white Americans' test scores has narrowed since the 1970s. Using data from several different IQ tests that were administered in a standard manner to black and non-Hispanic white people, Dickens and Flynn showed that blacks have narrowed the IQ gap by one-third to one-half of what it was in the 1970s.[9] If IQ were a fixed, intrinsic quality of races, then the IQ gap should be stable over time, but it is not.

The binary formulation of "genes vs. environment" is misleading. Cognitive abilities are complex and will likely be influenced by a myriad of environmental factors and genes. Given the complexity of brains and cognition, one ought not expect that a few genes will play a dominant role in shaping the normal range of human cognitive abilities; numerous genes will be involved. It is statistically implausible that variants of numerous genes relating to intelligence would be distributed among racial groups in a manner that systematically conferred cognitive advantage on one group or disadvantage on another. Furthermore, there is no evidence to support the claim that current racial differences in mean IQ scores are caused by racially distinctive patterns of genetic variation.

There is evidence that IQ scores are influenced by environmental factors that are pervasively and systematically patterned along racial lines in the U.S. Nonetheless, mean IQ differences among racial groups have been decreasing over the past few decades, perhaps in response to improved educational opportunities for some minority individuals. Taken together, the evidence suggests that differences in IQ scores are the result of social inequality rather than its cause.

Gene Association Studies

BY EVAN CHARNEY

Evan Charney, Ph.D., *is Associate Professor of the Practice of Public Policy and Political Science; Fellow, Duke Institute for Brain Sciences.*

Why are some people politically liberal while others are conservative? Why do some have a sunny disposition while others seem perpetually irritated?

According to many psychologists, as well as a growing number of social scientists—economists, political scientists, and sociologists—to answer these questions we must consider not only one's environment and life history but his or her genes as well. According to numerous studies, differences in almost all aspects of human behavior—everything from how frequently one votes[1] to time spent texting on a cellphone[2]—are due, in part, to genetic differences. Furthermore, studies have identified particular common variants of particular genes that predispose an individual to everything from criminal behavior to job satisfaction. Claims of this sort typically receive a good deal of attention in the media and exert a powerful influence upon the popular imagination. Who would not want to know, before choosing a career, if he or she had the gene that predisposes one to success as a Wall Street trader[3], or if their partner had the gene that predisposes one to infidelity[4]? And what if a judge, on sentencing a criminal, knew that he or she possessed the gene that predisposes criminal behavior[5]?

Claims of an association between a particular gene or genes and a particular behavior are more accurately claims of an association between polymorphisms and a particular behavior. Persons possess two copies of each gene (one inherited from the mother and one from the father), called alleles. Polymorphisms are slight structural variations in alleles that occur in less than 1 percent of the population. Hence, a claim of the sort, "a gene predicts voter turnout," is the claim that a specific polymorphism predicts voter turnout: Those who possess the polymorphism are more likely to vote than those who possess a different polymorphism of the same gene. But can a single gene, or 10 or 100 genes, really predict such things? The answer is almost certainly "no."

In conducting gene association studies, researchers typically make use of large, widely available data sets. The National Longitudinal Study of Adolescent Health (Add Health) is a representative data set and has been used in many gene association studies. This data set contains participant responses to thousands of questions about every aspect of their behavior, from age at first sexual intercourse to whether or not they voted in the last presidential election. It also contains a limited amount of genetic information. Due to technical and monetary constraints, the data set contains information for only 8 genes (out of an estimated 25,000 to 30,000 genes), and 20 or so polymorphisms for these genes (out of an estimated 15 million in the human genome). These genes were chosen because of a belief that they were associated in some way with adolescent problem behaviors such as depression, anti-social behavior, and smoking.

A typical study utilizing the Add Health data set is that of Fowler and Dawes[6], according to which two genes predicts voter turnout. One of the questions study participants were asked was "Did you vote in the last presidential election?" Fowler and Dawes found, through a statistical analysis of the data, that those who possessed a polymorphism of the MAOA gene were 5 percent more likely to answer this question "yes," while those who possessed a certain form of the 5HTT gene *and* reported that they attended church frequently were also more like to answer this question "yes." Hence, the claim that "Two genes predict voter turnout."

Association studies of this sort are problematic for two main reasons: Methodological (*i.e.,* relating to the research technique itself), and what I shall call "foundational" (*i.e.,* relating to the underlying assumptions as to how genes impact human behavior). Let us begin with the former.

Studies that associate specific genes with specific behaviors most often fail to be replicated, or are replicated by some researchers but not others. Replication refers to the attempt by different researchers studying different groups of people or different populations to come up with the same results as the initial study. In science, consistent replication is considered essential for demonstrating that results are valid and not the result of errors of one kind or another.

Because there are very few large data sets that contain both genetic information (no matter how limited) and behavioral information (in the form of self-reporting), the same data sets have been widely used. All available data sets of this sort contain genetic information for only a handful of genes, and, more often than not, they contain information for the *same* handful of genes. With hundreds or thousands of researchers scouring the same data sets for associations between the same genes and any and all possible behaviors, the result has been that one and the same genes have been associated with, or are said to predict, hundreds of different traits. For example, the same polymorphism that was found to predict voting has also been found to predict alcoholism, Alzheimer's disease, anger/aggression, anorexia, brain activation by colorectal distension, breast cancer, chronic fatigue syndrome, cleft lip, epilepsy, attitudes toward individualism and collectivism, insomnia, irritable bowel syndrome, job satisfaction, number of sexual partners, obesity, periodontal disease, premature ejaculation, schizophrenia, and utilitarian moral judgments. (This list is by no means complete. For a list—also not complete—of associations—and failures to replicate those associations—for four genes commonly found in large data sets, see http://tinyurl.com/962vfdd.) How could one and the same gene simultaneously predict so many disparate behaviors? And what is the likelihood that the very same handful of genes for which information is available in large data sets like Add Health will serendipitously turn out to be the key to all of human behavior?

If investigators search widely enough, they will almost certainly discover associations that exist purely by chance. For an association study to be statistically sound, a researcher is not supposed to look for an association between a single polymorphism and thousands of different traits (this is called "data-mining"). Yet, when a thousand researchers search for connections between the same polymorphism and a thousand different traits, the result is no different than if a single researcher looked for correlations between the same polymorphism and one thousand traits. Thus, it is no surprise that most studies of this sort either fail to be replicated or attempts at replication yield a mixture of positive and negative results.

As significant as these methodological problems are, the foundational problems are more significant because they cannot be solved by any refinements in statistical analysis. They call into question the cogency of the entire endeavor to search for genes that predispose for complex human behaviors in the first place.

There is a growing consensus that, to the extent that differences in genes do influence human behavior, thousands of genes are involved in interacting with each other, the environment, and the epigenome (the complex biochemical system that turns genes on and off) in complex ways. First, consider a non-behavior trait that is known to be highly heritable (*i.e.,* differences in the trait in a given population at a given time can be ascribed to genetic rather than environmental differences). Height is estimated to be ~80 percent heritable. Current estimates are that differences in anywhere from ~1500 to over 7,000 genes are involved in explaining 45 percent of this genetic variance[7].

Or consider differences in aggression in fruit flies (we know much more about the genetics of behavior in flies than we do humans because we can experiment with flies in a way impossible with humans). Scientists bred a strain of hyper-aggressive fruit fly. Using advanced DNA expression analysis they found differences in the transcription levels of 4,038 genes in hyper-aggressive flies versus controls. These same genes were involved in a host of basic physiological processes[8 9 10]. These studies reveal not only that there is no such thing as an "aggression gene" in fruit flies, but that no single variant of one gene or 10 genes or a 100 genes could predict, or be a risk factor for, aggression. Yet we are told that a single gene can predict voting behavior in humans.

Some researchers who now accept that one or two genes will not predict complex behaviors have embarked upon a search for thousands of genes that in combination predispose for complex behaviors. Utilizing genome-wide association studies (GWAS), which sequence the entire genome of typically thousands of persons, they search for thousands of genes of small effect that those who for example, are liberals, tend to share more often than non-liberals. But this search is misguided. The lesson of cutting edge research in molecular genetics is not that, for example, thousands of genes in combination predispose to being a liberal, but that genes do not predispose to complex behaviors. Once again, fruit flies provide some insight. For all of the difference in levels of genes being transcribed in hyper-aggressive v. normal fruit flies, heritability of aggression—the amount of variation in aggression in fruit flies attributable to genetic as opposed to environmental variation—was found to be only 0.1. This means that 90 percent of the variation in fruit fly aggression was due to environmental differences *even though the researchers assumed that they raised the flies in identical environments*.

The lesson to be drawn from this is not that it's mostly environment ("nurture") as opposed to genes ("nature"). Rather, it is that the dichotomy between genes and environment, at least as traditionally conceived, needs to be reconsidered. Gene association studies, whether we are hypothesizing that a single gene predisposes to a particular behavior, or that 1,000 genes predispose to that behavior, assume that we can equate a particular polymorphism with a certain kind of gene "behavior." For example, certain alleles MAOA are deemed less transcriptionally efficient than others, with the more transcriptionally efficient alleles designated as "high" and the less transcriptionally efficient as "low." The assumption is that "high" MAOA will produce more of a given protein and more or less of the neurotransmitter associated with that protein and more or less of a given behavior associated with that neurotransmitter.

Genes, however, do not "produce" anything. It is, rather, the cell that produces proteins, mobilizing numerous responses to internal and external environmental stimuli that will enable it to produce more of a given protein that is coded for in one of its genes. On its own, DNA is incapable of producing anything or expressing anything: All the biochemical machinery necessary for transcribing DNA is external to it. Gene association studies depend on the assumption that the high and low alleles of all who possess them are "turned on," that is, they are capable of being transcribed in the manner associated with those alleles. Before any gene can be transcribed, however, it must be "turned on," that is, it must be accessible to "transcription factors," special proteins that enable DNA transcription. Whether or not a gene can be transcribed is determined by a complex biochemical regulatory system called the epigenome. Hence, we cannot simply associate a polymorphism with more or less of a given protein, because the gene itself might be "turned off" by the epigenome and incapable of being transcribed, or it might be partially turned off, or it might be completely "turned on." What is more, the epigenome can be highly environmentally responsive, changing the extent to which particular genes can be transcribed in response to different environments. Studies have shown that, for example, early environmental experiences can permanently change the extent to which genes can be transcribed via changes in the epigenome with lifelong behavioral consequences.

In sum, gene association studies, whether we are discussing one gene or a thousand, assume a conception of the relationship between genotype and phenotype (all observable traits) that is both static and deterministic. The assumption is that complex behaviors "reside" in the structures of genes, whether 1 or 1,000, and that the structure of genes determines or predisposes to, complex behaviors. But the structure of genes cannot be equated with the effect they will have on complex traits. What really matters, the creation of the proteins for which genes are transcribed, involves not simply the genes themselves but everything that affects the extent to which these genes will be transcribed by the transcriptional machinery of cells. This includes the epigenome and potential aspects of the life environment from conception on. The structure of neither 1 nor 1,000 can provide such information.

PART V

Forensic DNA: Why Would the FBI Want Your Genes?

Artist: Sam Anderson

DNA testing was initially introduced into the criminal justice system of the United States as a method of developing supplemental evidence to be used in convicting violent felony offenders. Later it was introduced by defense attorneys to free incarcerated felons who had been falsely convicted. The use of DNA evidence profiling in criminal investigations can bring benefits to society by helping police to solve current crimes and cold cases.

However, in the last fifteen years, DNA collection by law enforcement has changed dramatically and is now routinely being used for a multiplicity of purposes that pose significant privacy and civil rights concerns to every citizen.

The federal government and all fifty states have created permanent collections of DNA taken from ever-widening categories of persons and are subjecting these collections to routine searches. Some believe that racial and ethnic disparities that exist in the U.S. criminal justice system are reflected in these databases. For example, while African Americans are only 12 percent of the U.S. population, their profiles constitute 40 percent of the federal database (CODIS).

At the same time, a stunning array of techniques has emerged enabling lab technicians to glean information from DNA that goes well beyond the mere identification of a person. Many argue that law enforcement's use of these tools to search, profile and store the DNA of those who have not been convicted of a crime, without a court order or individualized suspicion, has already exceeded reasonable constitutional protections.

The lack of ethical guidelines for forensic DNA practices in the United States has implications far beyond just the American citizenry. Today, sixty countries worldwide are operating forensic DNA databases and at least twenty-six countries plan to set up new DNA databases. The vast majority of these countries are looking to the United States as a model. Yet, as the United States expands its DNA collection practices, Europe is starting to move in the opposite direction. In 2008, the European Court of Human Rights issued a decision holding the UK's DNA data collection practices were a violation of the right to privacy. European countries are beginning to implement the decision through national legislation.

In contrast to current debates in Europe, expansions of the uses of DNA by law enforcement in the United States are generally occurring in a policy vacuum and then being justified retroactively by a limited number of solved crimes aided by DNA data. Aside from the fact that these cases appear to be the exception rather than the rule, what is not revealed by these stories is the larger picture of the steady erosion of privacy that accompanies the shifting purpose of DNA's use by law enforcement from one of identification to surveillance. Continued use of these techniques and practices outside the arena of judicial oversight and without the application of ethical guidelines demands a rigorous debate about the government's intrusion into the lives of innocent people.

These chapters explore the human rights concerns in the United States and abroad with the unchecked expansion of forensic DNA databases.

Statisticians Not Wanted: The Math behind Cold-Hit DNA Prosecutions

BY KEITH DEVLIN

Keith Devlin, Ph.D., D.Sc., F.A.A.A.S., F.A.M.S, *is a co-founder and Executive Director of Stanford University's H-STAR Institute, a co-founder of the Stanford Media X research network, and a Senior Researcher at the Center for the Study of Language and Information.* *This article was in part adapted from the author's September 2006 column in MAA Online, the monthly online newsletter of the Mathematical Association of America. This article originally appeared in* GeneWatch, *volume 19, number 6, November-December 2006.*

What statistics should be presented in court so that a jury can fairly evaluate the significance of DNA profile evidence when the defendant was first identified by a "cold-hit" search? That is to say, when a person is made a suspect because of a match between his or her DNA at a crime scene and DNA on file in an existing DNA database, what does that information alone tell us about their guilt or innocence? And who is best qualified to make that decision, statisticians or the courts?

These were precisely the questions that led me (a professional mathematician) and a number of other scientists, statisticians, mathematicians and scholars to cosign an amicus letter filed on July 24, 2006 with the California Supreme Court in the case of *People v. Michael Johnson*, No. S144821.

We were worried by previous court rulings in this and other DNA cold hit cases in which courts seemed to be taking the position that it was their job to decide which statistics were pertinent to the case, without seeking advice from the relevant scientific community, in this case statisticians. Our letter argued for one thing and one thing only: that in cold hit cases the court should proceed as it does in other kinds of cases where scientific evidence is involved, namely, to seek the testimony of the appropriate scientific experts. To our amazement and dismay, the court denied the petition of which our letter was a part.

Behind our letter was the fact that it is not at all obvious what exactly is the appropriate statistic to use in a cold hit case. The figure favored by the FBI and many prosecutors is the so-called "random match probability" (RMP), computed by multiplying together the probabilities of random matches on each of the DNA loci tested. (Forensic DNA testing involves determining the number of base pair repetitions at several given locations, or loci, on a person's DNA.) For example, assuming that the probability that the profiles of two randomly chosen, unrelated individuals match on a single, specific DNA locus is one-in-ten (a fairly reasonable assumption), the RMP that two randomly chosen, unrelated individuals have profiles that match on, say, all the 13 loci stored in the FBI's CODIS database is $(1/10)^{13}$ or one in ten trillion.

Like many other scientists, I find this computation troubling, since it depends crucially upon the assumed independence (in the sense of probability theory) of the 13 loci, an assumption for which there appears to be no empirical evidence, and there is even some evidence to the contrary. But that was not the point of our amicus brief. Rather, our focus was on which statistic should be presented in court to enable the jury to weigh the DNA match evidence presented following a cold-hit identification. The FBI, along with the prosecution in the Johnson case, claim that the RMP is still the right one to present, and that the fact that the defendant was first identified by a DNA database trawl makes no difference. I, along with most other mathematicians and statisticians who have considered the matter, have argued otherwise.

To illustrate why the RMP is an inappropriate and misleading statistic in a cold hit case, consider the following analogy. A typical state lottery will have the odds of winning a major jackpot around 1 in 35,000,000. To any single individual, therefore, buying a ticket is clearly a waste of time, since those odds are effectively nil. But suppose that each week at least 35,000,000 people actually do buy a ticket. (This is a realistic example.) Then every one to three weeks, on average, someone will win. The news reporters will go out and interview that lucky person. What is special about that person? Absolutely nothing. The only thing you can say about that individual is that he or she is the one who had the winning numbers. You can draw absolutely no other conclusion. The 1 in 35,000,000 odds tells you nothing about any other feature of that person. The fact that there is a winner reflects the fact that 35,000,000 people bought a ticket—and nothing else. (To put it another way, the 1 in 35,000,000 figure tells you a lot about the lottery—in that sense it is a "relevant" statistic, to use the language the FBI regularly uses in cold hit cases—but, absent other evidence, it tells you nothing about the winner.)

Compare this to a reporter who hears about a person with a reputation of being unusually lucky, goes along with them as they buy their ticket, and sits alongside them as they watch the lottery result announced on TV. Lo and behold, the person wins. What would you conclude? Most likely, that there has been a swindle. With odds of 1 in 35,000,000, it's impossible to conclude anything else in this situation.

In the first case, the long odds alone tell you nothing about the winning person, other than that they won. In the second case, the long odds tell you a lot.

To my mind, a cold hit measured by RMP is like the first case. All it tells you is that there is a DNA profile match. It does not, in of itself, tell you anything else, and certainly not that that person is guilty of the crime.

On the other hand, if an individual is identified as a crime suspect by means other than a DNA match, then a subsequent DNA match is like the second case. It tells you a lot. Indeed, assuming the initial identification had a rational, relevant basis (like a reputation for being lucky in the lottery case), the long RMP odds against a match could be taken as conclusive. But as with the lottery example, in order for the long odds to have any weight, the initial identification has to be before the DNA comparison is run (or at least demonstrably independent of it). Do the DNA comparison first, and those impressive-sounding long odds may be totally meaningless. It simply reflects the size of the relevant population, just as in the lottery case.

But if the RMP is not the right figure to use, what is? In 1989, the FBI urged the National Research Council to carry out a study of the matter. The NRC issued its report in 1992. Titled, *DNA Technology in Forensic Science,* the report is often referred to as "NRC I". The report's main recommendation regarding the cold hit process is given on page 124:

> The distinction between finding a match between an evidence sample and a suspect sample and finding a match between an evidence sample and one of many entries in a DNA profile databank is important. The chance of finding a match in the second case is considerably higher. . . . The initial match should be used as probable cause to obtain a blood sample from the suspect, but only the

> statistical frequency associated with the additional loci should be presented at trial (to prevent the selection bias that is inherent in searching a databank).

In part because of the controversy the NRC I report generated among scientists regarding the methodology proposed, and in part because courts were observed to misinterpret or misapply some of the statements in the report, in 1993, the NRC carried out a follow-up study. That study, *The Evaluation of Forensic DNA Evidence* (often referred to as "NRC II"), was published in 1996. NRC II's main recommendation regarding cold hit probabilities is: "Recommendation 5.1. When the suspect is found by a search of DNA databases, the random-match probability should be multiplied by N, the number of persons in the database."

The statistic NRC II recommends using is generally referred to as the database match probability, or DMP. NRC II's reasoning is essentially the same logic as I presented for my analogy with the state lottery.

Since two reports by committees of acknowledged experts in DNA profiling technology and statistical analysis came out strongly against the admissibility of the RMP, one might have imagined that would be the end of the matter and that judges in a cold-hit trial would rule in favor of admitting either the RMP for loci not used in the initial identification (à la NRC I) or else the DMP (à la NRC II), but not the RMP calculated on the full match.

However, not all statisticians agreed with the conclusions of the second NRC report. Most notably, Dr. Peter Donnelly, Professor of Statistical Science at the University of Oxford, took a view diametrically opposed to that of NRC II. According to Donnelly,

> I disagree fundamentally with the position of NRC II. Where they argue that the DNA evidence becomes less incriminating as the size of the database increases, I (and others) have argued that in fact the DNA evidence becomes stronger . . . The effect of the DNA evidence after a database search is two-fold: (i) the individual on trial has a profile which matches that of the crime sample, and (ii) every other person in the database has been eliminated as a possible perpetrator because their DNA profile differs

> from that of the crime sample. It is the second effect, of ruling out others, which makes the DNA evidence stronger after a database search . . .

Donnelly advocated using a different statistic, which, while generally close in value to the RMP, results from a very different calculation. Donnelly's proposed calculation was considered by NRC II and expressly rejected.

With the experts disagreeing in such a fundamental way, it is scarcely a wonder that judges have become confused as to what number or numbers should be presented as evidence in court.

Two things the statisticians do agree upon, however, is that a DNA profile match following a cold hit search is most definitely not the same as one carried out after a suspect has been identified by other means, and (hence) the calculation that should be performed after a cold hit search should not be the same as the one carried out in other circumstances.

In the *Johnson* case, however, the Court of Appeals of the State of California Fifth Appellate District, in an opinion issued on May 25, 2006, declared: "In our view, the means by which a particular person comes to be suspected of a crime—the reason law enforcement's investigation focuses on him—is irrelevant to the issue to be decided at trial, *i.e.,* that person's guilt or innocence." The court continued a short while later: ". . . the fact that here, the genetic profile from the evidence sample (the perpetrator's profile) matched the profile of someone in a database of criminal offenders, does not affect the strength of the evidence against appellant . . . The fact appellant was first identified as a possible suspect based on a database search simply does not matter."

By totally misunderstanding the one issue on which all the experts agree, the court could hardly have gotten things more wrong. The court is *so* wrong on that issue that their ruling must surely be overturned in due course. But what of the disagreement between the experts regarding which reliability statistic should be used in a cold hit case?

In my view, it is unwise in the extreme (and, as far as I know, inadmissible in court) to allow evidence and base convictions on disputed science. Thus, until the statistics community reaches consensus regarding the appropriate scientific procedure to use, the safest approach would seem to be to adopt a procedure that is free from controversy.

One possibility would be to follow NRC I, taking advantage of much improved DNA testing technology, and extend the match process to more than 13 loci. Such a move would more than compensate for the increase in the accidental match probability, however it is calculated, which results from a cold hit search. Another option would be to follow the logic of NRC II, and use the DMP (and only the DMP) in court. Since the DMP is generally more favorable to the defendant than is the RMP, this procedure would be to risk erring on the side of avoiding false convictions. However, with the current magnitude of the DNA databases (around 3,000,000 entries in CODIS), the figure quoted in court would still be astronomical. Neither approach necessarily makes matters any easier for the defendant in a cold hit case. But at least the court would not be acting in ignorance of, let alone counter to, scientifically established fact.

Genetic Privacy: New Frontiers

BY SHELDON KRIMSKY AND TANIA SIMONCELLI

Sheldon Krimsky *is currently President of the Board of the Council for Responsible Genetics and Lenore Stern Professor of Humanities and Social Sciences at Tufts University.* **Tania Simoncelli** *was formerly the Science Advisor to the American Civil Liberties Union and a member of the CRG Board of Directors. She currently works for the Food & Drug Administration. This article originally appeared in* GeneWatch, *volume 20, number 5, September-October 2007.*

On January 5, 2006, a little-noticed piece of legislation entitled the "DNA Fingerprint Act of 2005" was signed into law by President George W. Bush, greatly expanding the government's authority to collect and permanently retain DNA samples. These ninety-nine lines of text, introduced initially by Senator Jon Kyl [R-AZ], slipped virtually unnoticed through the halls of Congress, buried in the back of the broadly popular, 284-page Violence Against Women Act (VAWA) reauthorization bill. Notwithstanding the lack of public reaction and policy debate, this new law raises extraordinary questions for the future of civil liberties. Among other provisions, it grants the government authority to obtain and permanently store DNA from anyone who is arrested as well as non-U.S. citizens detained under federal authority.

This change in the federal DNA databanking law is emblematic of a new era in forensic DNA – one that is fraught with serious civil liberties and privacy concerns and may ultimately do little to make people safer. While DNA testing was initially introduced into the criminal justice system as a method of developing supplemental evidence to be used in convicting the guilty or freeing the innocent, in the last fifteen years this has changed. The federal government and all fifty states have created permanent collections of DNA taken from ever-widening categories of persons and subjecting these collections to routine searches. At the same time, a stunning array of techniques have emerged allowing lab technicians to glean information from DNA that goes well beyond the mere identification of a person, while the ability to detect and process minute amounts of DNA has steadily increased as costs have declined.

Law enforcement's use of these tools to search, profile and store the DNA of those who have not been convicted of a crime, without a court order or individualized suspicion, has already exceeded reasonable constitutional protections. In particular, a number of new genetic techniques and practices are providing law enforcement unprecedented access into the private lives of innocent persons by way of their own genetic data. These include, but are certainly not limited to: 1) searching for partial matches between crime scene evidence and DNA banks to obtain a list of possible relatives for DNA analysis ("familial searching"); 2) constructing probabilistic phenotypic profiles (descriptors like hair, skin, or eye color) of a perpetrator from DNA collected at a crime scene; and 3) surreptitiously collecting and searching DNA left behind on items like cigarette butts and coffee cups.

This essay explores each of these developments and their implications for civil liberties. We argue that the availability and use of these techniques seriously violates the reasonable expectations of privacy held by law-abiding citizens regarding their DNA. Developing technology, rather than constitutional analysis and informed public decision making, is driving the expansion of DNA databanks. Neglected to date has been a responsible national debate leading to an understanding of the issues and resulting in a societal consensus about the variety of uses of DNA discussed in this paper. To help advance the discussion, we urge that policies on DNA-forensic technologies need to calibrate the proper balance of civil liberties and law enforcement needs. We argue that clear national guidelines are needed to set standards for what governmental authorities, as well as private companies and individuals, may and may not do with DNA. We hope to provide a context for re-assessing these and other practices that raise serious civil liberties concerns. Finally, we briefly suggest what some of those guidelines should be.

Familial searching

"Familial searching" of databases is one of the new methods of creating suspects in the absence of a "cold hit," *i.e.,* matching a crime scene sample with one in a forensic database. Familial searching is premised on the notion that siblings and other closely related individuals have more genetic material in common than non-related individuals. Current methods of familial searching involve generating a list of possible relatives of the owner of DNA picked up at a crime scene by performing either a "low stringency" profile search to look for "partial matches" between crime scene evidence and offender profiles or by conducting a "rare allele" search. Close relatives of those matches are then tracked down and asked to "voluntarily" provide a DNA sample.[1]

In 1973, three women were murdered in South Wales. Twenty-nine years after the crimes were committed, the police submitted crime-scene stains to the United Kingdom's National DNA Database (NDNAD). When no full matches were found, a low stringency analysis indicated that the DNA partially matched the DNA profile of a man named Paul Kappen. Police surmised that someone in Kappen's family was the murderer, leading them back to Paul Kappen's father, Joseph, who had since died. British law enforcement authorities obtained DNA samples from the Kappen family, including Paul Kappen's mother and his siblings. The close match between the crime scene and family DNA profiles was sufficiently credible for the police to obtain a warrant to exhume the body of Joseph Kappen. His DNA was an exact match with the crime scene DNA. The case was solved by posthumous familial searching.

Familial searching has been employed in the United Kingdom in at least 20 criminal investigations.[2] In the United States, the practice was quite limited until 2006 by a policy adopted by the FBI prohibiting the release of any identifying information about an offender in one state's database to officials in another state unless the offender's DNA was an exact match with the DNA evidence found at the scene of a crime. Last summer, however, the FBI changed its policy in response to a request from Denver authorities who found a close match between evidence taken from the scene of a rape and a convicted felon in Oregon, indicating that he was a potential relative of the actual perpetrator. The interim policy, effective July 14, 2006, allows for states to share information related to "partial matches" upon FBI approval.[3] This has opened up the floodgates for using CODIS (the Combined DNA Information System is the FBI's national DNA database) in conjunction with familial searching.

Familial searching raises a series of troubling civil liberties issues. First, if practiced routinely, it effectively expands the database to a whole new category of innocent people whose private genetic data may be mined even though they themselves are not suspects in any criminal case—those who happen to be relatives of convicted offenders or others whose DNA data is kept in government databases. Family searches may also reveal information that family members prefer to keep private; for example, an offender might name someone as a parent or child who turns out to be genetically unrelated to them. In addition, there are a host of unanswered procedural questions associated with how the police might follow up on leads provided by partial matches. A low stringency search (generally defined as a match of 8 to 12 alleles out of the usual 13 to 15) can generate tens, hundreds, or even thousands of partial matches (and these will continue to grow as the databases grow). A partial match only indicates that there is some possibility that a relative of that person could have DNA that fully matches the crime scene evidence—the probability that the partial match is useful depends both on the number of alleles that are found to match and on their respective rarity in the population. As such, the police might be tempted to knock on the doors of hundreds or thousands of individuals, in the event that they do not have further evidence to narrow down their initial list of partial matches. Assuming that a partial match is not sufficient evidence for compelling a relative to provide a DNA sample via a court order, what happens if those individuals refuse to provide a sample? What is the fate of the samples collected? Will they be destroyed if that person is excluded from the crime? Will there be a temptation on the part of law enforcement to follow people around to get their DNA surreptitiously, when a court warrant cannot be obtained because there is insufficient evidence of individual suspicion?

Phenotypic DNA profiling

In an even more disturbing trend, some law enforcement agents have tried to construct phenotypic profiles of the suspected perpetrator based on analyses of the DNA found at a crime scene.[4] In a murder investigation in Louisiana, for example, a relatively new method of DNA analysis was employed to predict the "ancestry" of the alleged offender as 85 percent Sub-Saharan African and 15 percent Native American. The company that performed the analysis, DNA-Print Genomics, has been aggressively marketing the service to police departments, investigators and agencies.[5] The company has also recently started offering to law enforcement agencies a genetic test to infer eye color.[6]

A blood stain left at a crime scene could be subjected to many other tests for genetic conditions, more than a thousand of which are currently available. Law enforcement might use this information in an attempt to narrow down a pool of suspects. For example, suppose a blood stain were to be sent off to a private lab for a battery of genetic tests. The lab finds that the DNA of the stain contains the two genetic mutations associated with Gaucher disease, a metabolic disorder that causes a buildup of fatty substances in the spleen and liver and results in fatigue and bruising easily. Law enforcement might then try to get a list of names of all the people receiving enzyme replacement treatments for Gaucher at the neighboring hospital. Would they be given those names? If so, under what circumstances? Do people on such a list become suspects for a crime simply because they might have a pre-disposition to a certain health condition?

Under the Health Insurance Portability, Accountability Act (HIPAA), a person's DNA information and tissue samples are protected. However, HIPAA contains a broad exception that allows for disclosure of Protected Health Information to law enforcement officials, not only in compliance with a court order or grand jury subpoena but also in response to an administrative subpoena, summons or civil investigative demand. It is worth noting that all of these are legal instruments that can be issued without judicial review.[7] Broad administrative discretion is given to those with stewardship over health information at the hospitals to determine how to respond to written requests from law enforcement for patient records. HIPAA also allows health care providers to disclose to law enforcement, upon request, a broad array of identification information, including name, address, social security number, blood type, date of treatment and a physical description. Federal HIPAA guidelines should be tightened to protect the privacy of medical information, especially in cases where court warrants are not issued, to ensure uniformity in the interpretation of the policy.

There is also an obvious temptation on the part of law enforcement to analyze crime scene DNA in order to make predictions about the physical, behavioral or medical conditions of the alleged perpetrator. This temptation will likely increase over time. Already, claims have been made that genetic factors have been found that are associated with sexual orientation, intelligence, addictive behavior and aggression. Even if such associations are unsound, law enforcement will be tempted to use them so as to generate profiles of suspects from the DNA, such as: "Likely to be a tall, African American, homosexual male, with high intelligence, a propensity for addiction, and recessive for sickle cell anemia." Even if there were reliable population-wide probabilistic inferences from genotype to phenotype, confounding factors would make these inferences questionable for any individual.

This trend is likely to continue with the advent of gene chips, or DNA microarrays, such as those that have been developed by Affymetrix.[8] These gene chips enable researchers to access information on thousands of genes simultaneously. At the same time, scientists have developed DNA sequencing devices as small as ten centimeters in diameter, while the "Personal Genome Project" seeks ultimately to make it affordable for people to sequence their own, individual genome.[9] [10]

Surreptitious DNA collection

In 1974, a woman was raped and stabbed in Buffalo, New York. A 60-year-old man was recently arrested and charged with the crime. The police did not have enough evidence to obtain a warrant for his arrest. Instead, they followed him around, picked up his DNA after he spat on the sidewalk and compared it to the 30-year-old crime scene sample.[11]

This is the latest of an increasing number of known examples where police have collected DNA from individuals surreptitiously and without warrants supported by evidence amounting to probable cause. In another case, currently under review by Washington State's Supreme Court, the police employed a ruse in order to get their suspect, John Athan, to provide them with a DNA sample. Posing as a law firm, the police sent Athan a letter, asking him to join a lawsuit aimed at recovering overcharges in traffic fines. When they received a return letter from him, they lifted his DNA from the dried saliva where he had licked the envelope.[12]

These cases suggest the following question: What does it mean to live in a world where one has to assume that DNA shed on a continual basis might at any time be picked up, extracted and analyzed for information that could lead to one's arrest or conviction, to behavioral profiling or to the even more-attenuated identification of family members as crime suspects?

The primary argument asserted by law enforcement to justify surreptitious DNA searches is that the DNA is "abandoned."[13] In other words, an individual who "abandons" her DNA no longer has any privacy interest in that DNA.

This argument is problematic on a number of counts. First, "abandoned" implies a knowing intent to part with an item. People abandon items they no longer wish to own or carry around. But DNA is not so much abandoned as it is inadvertently and continually shed from people's bodies in the form of skin cells, saliva and hair samples. Short of walking around in the world in a plastic bubble suit, it would be virtually impossible to refrain from "abandoning" DNA in public.

Shedding DNA is not like leaving garbage at the curb. When people leave garbage on the street, they have come to anticipate that someone might rummage through it. They expect that the private information that might be contained in letters or bills can be accessed virtually by anyone who might come into contact with that garbage, which is why many people choose to shred important documents before discarding them. However, DNA cannot be "read" or even seen unless it is collected and then analyzed by sophisticated, expensive equipment. The privacy interest associated with DNA comes into play not in the form in which it inadvertently left the body, but instead when it is analyzed for the information contained in it. And certainly there is no mechanism for "shredding" the DNA that continuously gets released from the human body.

Police point to individual success stories in solving crimes as a way of justifying surreptitious DNA collection as a "clever investigation technique." But allowing police to take DNA without a person's knowledge or consent opens the door to mass DNA collections of anyone vaguely suspected, or even to those who are perfectly law-abiding and suspected of no criminal activity. Individuals would have no way of contesting this collection or use of their DNA. This scenario becomes increasingly worrisome when coupled with developments in behavioral genetics: weak or unreliable genetic markers for aggression or addiction could provide justification for identifying individuals who, it is believed, will commit a crime, and placing them under surveillance or social control.

Expanding databanks and the efficacy of solving crimes

The techniques and practices discussed above go a considerable distance to undermine the privacy of individuals. At the same time, it is possible that people are being asked to sacrifice their privacy for a process that may ultimately do little for criminal justice. In the case of the databanks, while law enforcement tends to boast large numbers of "cold hits" or "investigations aided," so far there has not been a single, peer-reviewed study that demonstrates the true effectiveness of the databanks.[14] While the prevailing notion with respect to these databanks is "the bigger the better," it is worth noting that the ability to use DNA in crime solving is limited by the ability to collect uncontaminated and undegraded DNA at a crime scene, not by the number of people in the databank. As the databanks expand to people convicted of minor offenses or merely arrested, the chances that any given profile in the database will help resolve a future crime apparently diminish. In the United Kingdom, the enactment of arrestee testing in 2004, which has corresponded with a ballooning of the UK database from 2 million to 3 million profiles (including those of more than 125,000 people never charged with any crime), has actually corresponded with a slight decrease in matches with crime scene evidence.[15]

Likewise, DNA dragnets have proven to be highly ineffective. In a study conducted [in 2004] by the University of Nebraska, only one of eighteen dragnets conducted in the United States was found to have led to the actual perpetrator, and this was a dragnet that involved only 25 people who were all staff at a nursing home where repeated sexual offenses were taking place.[16] In other words, a small pool of suspects already existed. Some dragnets have even been found to interfere with crime solving. For well over a year, police had the DNA of the individual who was ultimately charged with the murder of Cristina Worthington, a well-known socialite and fashion writer. The DNA had not been tested, however, because law enforcement officials were busy collecting DNA from about a thousand innocent individuals in hopes of using that DNA to solve the crime.[17]

In the case of familial searching, it is perhaps too soon to tell how helpful this technique could be for law enforcement. But with the problems inherent in DNA dragnets and surreptitious DNA sampling, it is likely that only the successes will be made public. Law enforcement officials are unlikely to publicize the failures, dead ends or number of people who are investigated without their consent or knowledge.

An over-reliance on these practices could well undermine law enforcement. Some law enforcement officials have expressed concern that the tremendous resources funneled into building and expanding forensic DNA banks are channeling away money that should be put into following up on investigational leads or placing police officers on the streets.[18] In addition, crime laboratories all over the country are plagued by extraordinary backlogs resulting from the heedless expansion of the databanks. In February, the California Commission on the Fair Administration of Justice, a bipartisan panel of criminal justice experts and practitioners, released an emergency report that documented enormous backlogs of about 160,000 untested DNA samples in California's state lab arising from the expansion of California's databank to all felons.[19] This backlog is expected to increase exponentially when Proposition 69's arrestee testing provision comes into effect in 2009, when an additional 450,000 samples will be eligible for collection each year.

Backlogs can have tragic outcomes. As the California panel reported, "Delays of six months or more have become the norm" in analyzing rape kits in the state. In one case, a rapist attacked two more victims, including a child, while his DNA awaited analysis. Backlogs can also increase the chance of errors in DNA analysis, labeling or interpretation as lab analysts are pressured to cut corners to meet their workload.

Such errors have already resulted in known miscarriages of justice. As a result of an error made by an analyst at the Houston Crime Lab, Josiah Sutton spent nearly five years in prison, starting at the age of 16, for a rape he could not have committed. In another case, a 26-year-old man faced life in jail and was incarcerated for over a year because the Las Vegas police crime lab mistakenly switched the label on his DNA sample with that of his cellmate.[20] Timothy Durham of Tulsa, Oklahoma, spent four years in prison on the basis of a misinterpreted DNA test, despite having 11 witnesses who placed him in another state at the time of the rape he was convicted of.[21]

The more that DNA is relied upon to create suspects where there are none, the more vulnerable it will be to abuse. Already, several instances have been reported where criminals have planted or tampered with DNA evidence, or paid others to take DNA tests in their stead as a way of confusing investigators or evading prosecution. Prisoners have also been overheard coaching each other on how to plant biological evidence at crime scenes and how to avoid leaving their own DNA behind. Recently, four men in Massachusetts were indicted on charges of DNA tampering for allegedly attempting to switch identity bracelets when having blood drawn for a DNA sample while in custody.[22]

Finally, we will likely see increasing hostility among the public as law enforcement engages in DNA screens that impute suspicion based on neighborhood, vague physical descriptions or racial characteristics, or familial relations. Ultimately, people may be unwilling to cooperate with law enforcement in helping to resolve a crime where these practices become more routine and the rules as to whether and under what circumstances their DNA may be collected and used remain unclear.

Conclusion and recommendations

We can hardly blame law enforcement for wanting to use DNA in any way possible to solve crimes. At the same time, privacy of one's DNA is completely undermined if law enforcement is permitted to use backdoor methods of DNA collection and to examine DNA for any and all information about a person, including their personal characteristics and familial characteristics and connections.

Expansions of the uses of DNA by law enforcement are generally occurring in a policy vacuum and then being justified retroactively by a limited number of solved crimes aided by DNA data. Aside from the fact that these cases appear to be the exception rather than the rule, what is not revealed by these stories is the larger picture of the steady erosion of privacy that accompanies the shifting purpose of DNA's use by law enforcement from identification to surveillance. Continued use of these techniques and practices outside the arena of judicial oversight and without the application of ethical guidelines should spark a rigorous debate about the government's intrusion into the lives of innocent people.

Once the information inscribed in DNA is considered private, then it follows that this principle should be embedded in the policy debate so that it can assist us in establishing an appropriate balance between law enforcement and civil liberties. That principle of balance should guide where and when DNA technology may be used by law enforcement. We offer the following basic recommendations as to how to achieve that balance:

1. Informed consent should be required before law enforcement takes or tests the DNA of a person who has not been convicted of a crime. Surreptitious taking, testing or storing of DNA from suspects or their relatives is a violation of a person's privacy and should be prohibited.
2. Absent valid consent, a court order based upon probable cause should be required for the taking of an individual's DNA. Such DNA should be compared only with the DNA from a crime scene for which that person is a suspect.
3. DNA databanks should be limited to DNA profiles from persons who are convicted of serious crimes. All those presumed innocent do not have

a diminished right to privacy and therefore should not have their DNA included in a forensic DNA databank.

4. Offender biological samples should be destroyed so that the encoded information cannot be mined for purposes beyond identification (such as investigating potential gene-behavior associations).
5. Crime scene samples should be analyzed only for purposes of identification. Law enforcement should generally be barred from looking for rare alleles that are associated with genetic diseases or other traits that are not central to identification.
6. The Genetic Nondiscrimination Act of 2007 should be passed and then amended to provide protections and rules for law enforcement. Otherwise, just as people have been hesitant to undergo genetic testing for fear that their information will be used against them by insurance companies or future employers, so will they fear that law enforcement will mine their medical records for their DNA.[23]
7. A court order based on probable cause should be required for law enforcement to be given access to anyone's medical records for genetic data. The rules protecting medical information in HIPAA and the current broad exemption provided to law enforcement should be tightened in light of emerging interests in health data for forensic uses.

Differential Trust in DNA Forensics

BY TROY DUSTER

Troy Duster, Ph.D., *is a Chancellor's Professor of Sociology at UC Berkeley and professor of sociology and director of the Institute for the History of the Production of Knowledge at New York University.* *This article is adapted from "Explaining Differential Trust in DNA Forensic Technology: Grounded Assessment or Inexplicable Paranoia," which appeared in the Summer 2006 edition of the* Journal of Law, Medicine and Ethics. *This article originally appeared in* GeneWatch, *volume 20, number 1, January-February 2007.*

"What you see depends on where you stand." –Albert Einstein

The aftermath of the 1965 Watts rebellion, bookended by the dramatic images of burning cities and rioting and looting that continued for several days in Detroit and Newark, created two sharply contrasting images of the police in the American public's consciousness. The contrast was distilled in two bumper stickers: "The Police Serve and Protect the Community," and "The Police are a Brutal Occupying Alien Force." Depending upon the community where one lived, the number of vehicles displaying either message was quite predictable.

Similar feelings of trust and mistrust exist toward the American medical community. There is an abundance of evidence of the mistrust of clinical medicine among the US black population in the wake of revelations in 1971 of what came to be known as medical experiments on black men.[1] The Tuskegee Syphilis Study has become so much a part of folklore among African Americans that few will be surprised to learn that it continues to shape how many express deep suspicions of the medical profession.

A similar, perhaps even more deeply held, suspicion exists among African Americans about the criminal justice system in general and the police in particular. In late 1999, and for nine months of ensuing testimony, it was revealed how police in the Rampart division of the Los Angeles Police Department planted drugs and guns on defendants—mainly African Americans and Latinos—then testified in court, under oath, that they had found these items.[2] These machinations came to light only because a police officer working in a special unit of Rampart (Community Resources Against Street Hoodlums, or CRASH) began testifying against his fellow officers while awaiting re-trial on charges of stealing impounded cocaine. The officer, Rafael Perez, testified that he and other police officers had planted guns on suspects, fabricated drug evidence and lied in arrest reports. As a result, more than 120 criminal defendants had their convictions vacated and dismissed, and more than $42 million dollars has been paid in civil settlements.[3]

In the last ten to fifteen years, major police corruption scandals have come to light in Dallas, New Orleans, Philadelphia and Chicago. In Dallas, police framed 39 Latinos and had them deported, by planting what the police testified to be cocaine on them. This turned out to be powdered wallboard gypsum.[4] Also of interest is the infamous Tulia, Texas, drug bust, where a corrupt police officer jailed and then helped convict nearly three dozen people by planting drugs and then testifying against them. These convictions were later overturned when the governor pardoned 35 persons, and the police officer was indicted for perjury.[5]

A few more examples should suffice to begin to fill in the contours of a suggestive national pattern, a mosaic of dotted lines that can be connected to provide at least a plausible account of suspicion. In the early 1990s, in Philadelphia's 39th police district, five officers pled guilty to setting up suspects, bribing witnesses, and planting evidence—resulting in the vacating of more than fifty convictions and an investigation of another several thousand arrests. Lynn Washington, legal scholar and editor of the *Philadelphia New Observer*, recognized deeper issues involved in this case, stating, "What's most disturbing about the Philly corruption is that the DA knew what the cops were up to but tolerated their use of planted evidence because it boosted conviction rates."[6] New York City police were rocked by a similar scandal, when sixteen officers of the Bronx 48th precinct were arrested and indicted "on charges ranging from falsifying evidence to stealing weapons and money from illegally-raided apartments."[7]

Such accounts set the context for a discussion about DNA, genetic technologies, and crime fighting as it relates to the African American and Latino

American communities. Without this discussion, we are left wondering how it is possible that some people see DNA evidence as definitive, while others maintain strong skepticism—that DNA technology, no matter how definitive, may not be used fairly in a criminal justice system that is tainted and sometimes corrupted. Thus, African Americans and Latinos in the poorest neighborhoods in our major cities are far more likely to approach DNA evidence with a general mistrust, for reasons including those described above. Indeed, it is possible to provide more reason for this skepticism by describing the basis for the different perspectives on just how "definitive" DNA testing has come to be perceived. On the one hand, there is the possibility of exoneration of someone convicted when analysis of the crime scene DNA does not match that of the person convicted. On the other hand, there is the arrest and conviction of a person not previously a suspect, when there is a match between that person's DNA and the DNA found at a crime scene (known as a "cold hit"). We will take a closer look at both kinds of cases.

If DNA is the only evidence against the accused in the larger context of the framing scandals just described, we can see how some will fear the considerable abuse potential by rogue police officers doggedly committed to obtaining convictions. That is, if police can plant cocaine and guns on those whom they later testify against, and obtain a conviction, they can surely plant DNA. The legitimacy of the criminal justice system rests primarily on the fair application of laws. Who (or what part of society) would believe that police would actually plant DNA evidence, and even if they did, can DNA evidence ever stand alone without other circumstantial evidence? Before we turn to the differential trust issue, let us explore the nature of DNA technology and the claims made for its degrees of certainty.

The strong case for strong claims of DNA evidence

In 2003, Darryl Hunt was exonerated and released from prison in North Carolina after serving eighteen years for a rape and murder that had occurred in 1984. New analysis showed that his DNA did not match that left at the crime scene.[8] Investigators hoped they could find a match for that DNA sample from the North Carolina convicted offender databank, which includes DNA from 40,000 individuals. Comparing the data from the 1984 crime scene against the 40,000 available DNA profiles, no perfect matches were found. However, the closest single match was to Anthony Dennard Brown, who matched on 16 of the possible 26 alleles. Alleles associated with Short Tandem Repeats (STR) in DNA sequences are inherited in a way that the most likely explanation for a near perfect match is that the DNA evidence sample belongs to a close relative of the individual whose DNA profile is available. However, "while this information may prove to be useful to law enforcement investigators, relatively little has been done to establish the level and kind of similarity between evidence and non-matching database profiles sufficient to justify investigation of an individual's relatives."[9]

Because Anthony Dennard Brown matched only 16 of 26 alleles, he was not a suspect. But this high proportion of matching alleles immediately cast suspicion on his close relatives—most particularly on his brother, Willard Brown. Police followed Willard Brown and confiscated a discarded cigarette butt of his for DNA testing. The laboratory found a perfect match at all the 13 loci (all 26 alleles). Willard Brown was arrested, charged, confessed, pled guilty, and was convicted. He is now serving a life sentence plus ten years.

While Great Britain's Police Forensic Science departments routinely perform what they call "familial searches" of the sort just described, in the United States there is a wide variety of state policies regarding familial searches. In the U.S., federal law bars the FBI from using DNA information from all but perfect matches. However, New York and Massachusetts encourage familial searches, authorized by specific state statutes. Dan Krane, one of the leading experts in DNA forensic technology, notes that, when they are permitted, the thresholds of similarity that must be cleared before relatives are investigated tend to be ambiguously defined and described in terms such as matches needing to "be very, very close" (Virginia), "appear useful" (California) or be at 21 or more out of 26 alleles (Florida).[10-11]

The next case involves DNA as the sole piece of evidence that resulted in a conviction, thirty-five years after the crime was committed. Moreover, the way in which this case unfolded suggested that the laboratory that conducted the DNA analysis most probably made a mistake. On March 20, 1969, Jane Mixer, a University of Michigan law student, was shot and strangled. There was no evidence of sexual assault, nor was any semen from the perpetrator left at the scene. Thirty-five years later, in November, 2004, the police arrested Gary Lieterman, a sixty-two-year-old man whose DNA was in the database because of a fluke. Several years earlier, Lieterman had undergone neck surgery. He developed an addiction to pain killers and forged a physician's signature to obtain them on one occasion. In Lieterman's only brush with the law, he was remanded to drug treatment. His record was supposed to be expunged after his treatment was finished. However, his DNA was left in the database. This is not uncommon.[12]

In 2003, DNA samples of two different men were allegedly found on the victim's pantyhose. One of the samples was Lieterman's; the other sample was from someone who would have been four years old in 1969, the time of the murder. It turned out that this then-four-year-old was in the database because, in 2003, he murdered his mother.

Dan Krane, as an expert witness for the defense, pointed out in his testimony what should have been obvious: that this was an instance of lab contamination. The same lab was handling the two murder cases at the same time. Although the prosecutor maintained that the DNA of the erstwhile four-year-old at the crime scene was not a mistake, after only four hours of deliberation the jury voted to convict Lieterman, who is now serving a life sentence.[13]

Police departments' organizational imperative.[14]

Consider "P," who is arrested and charged with burglary. There have been a number of other burglaries in this police precinct. The arresting officers see a pattern to these burglaries and decide that the suspect is likely to have committed a number of those on their unsolved burglary list. Thus, it sometimes happens that when "P" is arrested for just one of those burglaries, the police can "clear by arrest" fifteen to twenty crimes with that single arrest. "P" will be considered a "repeat offender," even though there may never be any follow-up empirical research to verify or corroborate that "P's" rap sheet accurately represents his crimes.

But only if one is doing close-up observation of police work can crimes be proven to have a pattern.[16] And yet, if social theorists take the FBI Uniform Crime Reports as a reflection of the crime rate, with no observations as to how those rates were calculated, they will make the predictable error of assuming that there are only a very small number of persons who commit a large number of crimes. This kind of bureaucratic decision making generates a theory of "a few bad apples," where both the criminological theory and the policy decision lead one to look for the "kind of person" who repeatedly engages in this behavior. In fact, the long rap sheet is frequently generated by the imperative to "clear by arrest."

The U.S. prison population has undergone a dramatic shift in its racial composition in the last thirty years. The convergence of this shift with the social trend of re-defining race in terms of DNA will present challenges at many levels—from the attempted re-inscription of race as a biological or genetic category to attempted explanations of a host of complex social behaviors. That challenge can only be met by doing what the social researchers of a previous generation did with police work, namely, going to the very site at which those data are generated.

Unreasonable search and "abandoned" DNA

To understand the historical and political contexts of the right of citizens outlined in the Fourth Amendment "to be secure in the persons, houses, papers, and effects, against unreasonable searches and seizures," we must go back to the period when the British crown ruled the colonies. In this period, according to Chapin, an officer of the crown, armed with only the most general warrant for collecting taxes, could break down the door of a person's home, enter, search for taxable goods, and seize whatever items appeared not to have been taxed, items that were called "uncustomed goods."[17]

Today, the Fourth Amendment does not specify what constitutes "a reasonable search," but most courts have interpreted the definition to hinge on the government's requirement to obtain a search that is "warranted." However, even without a warrant, a search is sometimes permissible. In issuing a warrant, the state must balance the "government's special needs" against the individual's right to personal privacy. "Special needs," for example, encompass the safety of airline passengers, so the courts have ruled that it is permissible to test pilots for alcohol and other drugs, likewise bus drivers, train and subway operators, and so forth. Police also need to have other grounds for a warrant, such as the "suspicious behavior" of a suspect.

But a search cannot include a general dragnet of those who exhibit no suspicious behavior—unless, of course, that suspicious behavior is being part of a population group that is thought to contain the likely suspect. In such a circumstance, there is the "limited privacy expectation," where the courts have ruled that ex-convicts have fewer protections of the expectation of privacy. In *Griffin v. Wisconsin*, the Court upheld a warrantless search.[18] The DNA Act of 2000 provided funds for states to expedite the admission of DNA evidence of crimes without suspects. The lower courts have so far withstood challenges to this, and the Supreme Court has yet to rule on it.[19] In late December 2005, President Bush acknowledged that he had authorized warrantless electronic eavesdropping on American citizens in the wake of 9/11.[20] It requires little speculation of which socially designated groups would be singled out for such invasions of their privacy. That is, just being in a suspect group becomes "suspicious behavior." Eavesdropping is aimed at those groups thought to be most likely to "harbor terrorists" or to be in contact with terrorists. The analogy in the criminal justice system is the emerging practice of DNA dragnets. This is a police tactic in which all "likely suspects" in a wide geographical area around a crime scene are asked to provide a DNA sample in order to exclude them as suspects.

DNA dragnets

DNA dragnets originated in England and are most advanced in Europe and the United Kingdom. The first DNA dragnet was conducted in Leicester, England, in 1987. Two teenage girls were raped and murdered in the same area, and police requested voluntary blood samples from more than 4,500 males within a certain radius of the crime scene. When a man asked a friend to submit a DNA sample in his place, he immediately became a prime suspect and turned out to be the killer.[21] Germany is the site of the largest DNA dragnet ever conducted. In 1998, the police collected samples from more than 16,000 people and finally matched the DNA of a local mechanic to the sample collected at the crime scene of the rape-murder of an eleven-year-old.[22]

While the United States has only conducted about a dozen DNA dragnets, they are notable in their focus on specific racial groups. San Diego was among the first jurisdictions to conduct the practice during the investigation of a serial killer in the early 1990s. The suspect was African American, and the DNA of more than 750 African Americans was tested. In 1994, Ann Arbor, Michigan, police obtained nearly 200 samples from African Americans in the hunt for another serial rapist and murderer. In both the San Diego and Ann Arbor cases, the suspect was apprehended and convicted for committing another crime unrelated to the reason for the dragnet.

Then, in 2004, Charlottesville, Virginia, had a racially-driven dragnet that generated a controversial response from civil liberties groups that ultimately convinced the police to ultimately revise and restrict the dragnet strategy.[23] A serial rapist had been active in the Charlottesville area from 1997 through 2003, frustrating police investigations at every turn. DNA evidence linked the rapist to at least six assaults. From a number of leads, the police believed the rapist to be African American, and so, in late 2003, the chief of police initiated a project to obtain saliva samples from 187 men, 185 of whom were black (the two others were Latinos).[24]

However, two students at the University of Virginia who refused raised the issue of what constitutes voluntary submission of a DNA sample. In so doing, they brought pressure on both the university and the local black community to take a position. The Dean of the University of Virginia's Office of African American Affairs organized a forum to discuss the situation, drawing national media attention. At one point, the Dean said, "Because the suspect is black, every black man is suspect." In mid-April of 2004, the police chief suspended the dragnet and restricted its use to a much narrower use of police discretion, based more on whether a suspect resembled a composite profile than on his or her race.

In addition to a DNA dragnet, police have another tactic available to them, as revealed in the Willard Brown case. After a person has become a suspect, the police can literally follow that person around and collect samples of their "abandoned" DNA. In a recent paper, Elizabeth Joh addresses the apparently limitless capacity to pursue this technique for acquiring DNA samples: Deciding whether DNA might ever be "abandoned" is important, because as it is interpreted now, abandoned DNA provides the means to collect genetic information from anyone, at any time. The rules of criminal procedure law appear to pose no restrictions on this kind of DNA collection by the police.[25]

In the Lieterman case cited above, the defendant was convicted of murder on the basis of a single DNA sample putatively left at the crime scene. Since everyone constantly "abandons" their DNA, it is quite easy to obtain samples from anyone. So the question remains, why would anyone suspect officers of the criminal justice system to commit fraud or perjury, or to plant evidence? The answer, sociologically, is that this "anyone" is not a random person, but rather, is predictably located in the very social groups that have been the subjects of unwarranted searches and seizures and of systematic victimization by police corruption.

Moreover, the "anyone" who expresses a belief that the police would not engage in such activities, except as "rogue cops," is also predictably more likely to reside in middle and upper class communities where the police tend to be thought of as those who "serve and protect." In communities in which the police are experienced as a brutal occupation force, antipathy to the whole system of criminal justice can be expected. This was the logic of the Dallas prosecutor's office, which, as part of its staff training, instructed all staff prosecutors to preemptively challenge African Americans as potential jurors in death penalty cases. "A Dallas County district judge testified that, when he had served in the District Attorney's Office . . . his superior warned him that he would be fired if he permitted any African Americans to serve on the jury."[26]

Ancestral informative markers—identifying race from inside the body

In the last decade, using technologies of molecular genetics, remarkable claims are being made in the scientific literature, for example that it is possible to "estimate" the race of a person by looking at specific markers in the DNA.[27] The social implications of this reach far beyond personal recreational usage, in which a person could submit a DNA sample and "discover" the percentage of ancestry that derives from Europe, Sub-Saharan Africa or the Asian continent.[28] Forensic applications of this technology are being touted and marketed as the direct consequence of a successful intervention in a sensational serial rape-murder case. In early 2003, police in Baton Rouge, Louisiana, had been unsuccessful in their attempts to identify a serial rapist-murderer after interviewing over a thousand white males who fit what one witness described as the likely suspect. A tissue sample was analyzed by a company that claimed it could discern that the suspect (based on DNA analysis) had 85 percent African ancestry.[29] When the prime suspect was apprehended, he turned out to be an African American, and the company that did the test has since been advertising its success on its website, and has attempted to market its expertise to police departments around the country. Yet another claim about the capacity to use DNA to identify race appeared in 2005, in the *American Journal of Human Genetics*. Hua Tang and colleagues wrote, "genetic cluster analysis of the microsatellite markers produced four major clusters, which showed near-perfect correspondence with the self-reported race/ethnic categories."[30] This was followed by Stanford University's release of the following statement to the press: "A recent study conducted at the Stanford Medical School challenges the widely held belief that race is only a social construct and provides evidence that race has genetic implications."[31]

Expanding DNA databases while homing-in on putative racial markers

In 1994, Congress passed the DNA Identification Act, authorizing the FBI to establish a national DNA database, the Combined DNA Index System (CODIS). Even as late as in the mid-1980s, most states were only collecting DNA samples from sexual offenders, but within a decade all fifty states were contributing DNA samples to the CODIS system on a wide range of felons and had the capacity to inter-link state databases.

In just three years, the CODIS database has grown from a total of nine states cross-linking "a little over 100,000 offender profiles and 5,000 forensic profiles," to include thirty-two states, the FBI, and the U.S. Army, linking "nearly 400,000 offender profiles and close to 20,000 forensic profiles." States are now uploading an average of 3,000 offender profiles every month.[32]

The expansion of the databases is inevitable. At the state level, in 2004, California voters passed Proposition 69, permitting collection and storage of DNA from those merely arrested for a select number of crimes, joining Louisiana, Texas and Virginia.[33] Also, the federal DNA Fingerprint Act of 2005 authorizes the collection of DNA from citizens arrested for qualifying federal offenses and from non-U.S. individuals who are merely "detained" under federal authority.[34]

As the numbers of profiles in the databases increase, researchers will propose providing DNA profiles of specific offender populations. Twenty states authorize the use of databanks for research on forensic techniques.[35] Tania Simoncelli has analyzed the civil liberties implications of the federal DNA Fingerprint Act of 2005, passed as part of the reauthorization of the Violence Against Women Act.[36] In addition to the expansion noted above, this legislation expands CODIS further still, allowing states to upload DNA profiles from anyone whose DNA samples are collected under applicable legal authorities and eliminating prior protection that prevented the collection and storing of DNA profiles from arrestees who have not been charged.[37]

All of this points to an ever-expanding national forensic DNA database. Coupled with an increasing commitment to use DNA to estimate racial identity so as to reduce the number of suspects needed in a mass-screening program, we can see the ingredients for what has been called a "deadly symbiosis."[38]

Sixty-two percent of all prisoners incarcerated in the United States are either African American or Latino, while these populations comprise only about a quarter of the nation's entire population.[39] Police scandals in major cities show a remarkable pattern of abusive treatment of the members of the poorest communities of African Americans and Latinos. Most significantly, planting evidence has become a common theme in many of these abuse scandals. We must add to this mix the increasing tendency for the molecular re-inscription of race, with "ancestral informative markers" as a product now being sold (lobbied, and hyped) to police departments by the inventors of this technology, the emergence of DNA dragnets (which will inevitably become racially focused in a society in which race creates stratification) and a national DNA forensic database that is expanding to include even those merely arrested. However, none of this seems to generate suspicion or caution in one part of our society, yet in another part of society, agitation, distrust and suspicion are palpable. How to explain these sharply different responses to developments in DNA technology and forensics? Rather than reducing the answer to either misguided mistrust or paranoia—or, alternatively, explaining blind trust in the technology as naiveté—it is probably the better analytic strategy to invoke Einstein's insight, that "what you see depends on where you stand."

The Potential for Error in Forensic DNA Testing

BY WILLIAM C. THOMPSON

William C. Thompson *is a professor in the Department of Criminology, Law and Society and the School of Law at the University of California, Irvine. His research focuses on the use and misuse of scientific and statistical evidence in the courtroom and on jurors' reactions to such evidence. He occasionally represents clients in cases involving novel scientific and statistical issues. This article originally appeared in* GeneWatch, *volume 21, number 3-4, November-December 2008.*

Promoters of forensic DNA testing have, from the beginning, claimed that DNA tests are virtually infallible.[1 2] In advertising materials, publications and courtroom testimony, the claim has been made that DNA tests produce either the right result or no result.[3] This rhetoric of infallibility took hold early in appellate court opinions, which often parroted promotional hyperbole.[4] It was supported when the National Research Council, in the second of two reports on forensic DNA testing, declared "the reliability and validity of properly collected and analyzed DNA data should not be in doubt."[5] It was further reinforced in the public imagination by news accounts of post-conviction DNA exonerations. Wrongfully convicted people were shown being released from prison, while guilty people were brought to justice by this marvelous new technology. With both prosecutors and advocates for the wrongfully convicted using it successfully in court, who could doubt that DNA evidence was in fact what its promoters claimed: the gold standard, a truth machine?[6]

The rhetoric of infallibility proved helpful in establishing the admissibility of forensic DNA tests and persuading judges and jurors of its epistemic authority.[7] It has also played an important role in the promotion of government DNA databases. Innocent people have nothing to fear from databases, promoters claim. Because the tests are infallible, the risk of a false incrimination must necessarily be nil. One indication of the success and influence of the rhetoric of infallibility is that, until quite recently, concerns about false incriminations played almost no role in debates about database expansion. The infallibility of DNA tests has, for most purposes, become an accepted fact—one of the shared assumptions underlying the policy debate.

I argue that this shared assumption is wrong. Although generally quite reliable (particularly in comparison with other forms of evidence often used in criminal trials), DNA tests are not now and have never been infallible. Errors in DNA testing occur regularly. DNA evidence has caused false incriminations and false convictions and will continue to do so. Although DNA tests incriminate the correct person in the great majority of cases, the risk of false incrimination is high enough to deserve serious consideration in debates about expansion of DNA databases. The risk of false incrimination is borne primarily by individuals whose profiles are included in government databases (and perhaps by their relatives). Because there are racial, ethnic and class disparities in the composition of databases, the risk of false incrimination will fall disproportionately on members of the included groups.[8 9]

I will discuss major ways in which false incriminations can occur in forensic DNA testing, including coincidental DNA profile matches between different people, inadvertent or accidental transfer of cellular material or DNA from one item to another, errors in identification or labeling of samples, misinterpretation of test results and intentional planting of biological evidence. I will also discuss ways in which the secrecy that currently surrounds the content and operation of government databases makes these issues difficult to study and assess. I will conclude by calling for greater openness and transparency of governmental operations in this domain and a public program of research that will allow the risks discussed here to be better understood. A coincidental match between different people who happen to share the same DNA profile is one way a false incrimination can occur. To understand the likelihood of a coincidental match, it is important to understand what a DNA profile is and how DNA profiles are compared. Forensic laboratories typically "type" samples using commercial test kits that can detect genetic characteristics (called alleles) at various loci (locations) on the human genome. The test kits used in the United States generally examine the 13 short tandem repeats STR loci selected by the FBI for CODIS, the national DNA database.[10] [Short tandem repeats (or STRs) are repeating sequences of two to six base pairs of DNA.] Some of the newer test kits also examine two additional STR loci.

At each STR locus, there are a number of different alleles (generally between 6 and 18) that a person might have. Each person inherits two of these alleles, one from each parent. Numbers are used to identify the alleles, and the pair of alleles at a particular locus constitutes a genotype. Hence, one person can have a genotype (for a locus called D3S1358) of "14, 15;" while another person has the genotype "16, 17." The complete set of alleles detected at all loci for a given sample is called a DNA profile. When describing DNA profiles, people sometimes mention the number of loci they encompass.

In cases I have reviewed over the past few years, evidentiary samples from crime scenes often produce incomplete or partial DNA profiles. Limited quantities of DNA, degradation of the sample, or the presence of inhibitors (contaminants) can make it impossible to determine the genotype at every locus. In some instances the test yields no information about the genotype at a particular locus; in some instances one of the two alleles at a locus will "drop out" (become undetectable). Because partial profiles contain fewer genetic markers (alleles) than complete profiles, they are more likely to match someone by chance.[1] The probability of a coincidental match is higher for a partial profile than for a full profile.

A further complication is that evidentiary samples are often mixtures. Because it can be difficult to tell which alleles are associated with which contributor in a mixed sample, there often are many different profiles (not just one) that could be consistent with a mixed sample. Because so many different profiles may be consistent with a mixture, the probability that a non-contributor might, by coincidence, be "included" as a possible contributor to the mixture is far higher in a mixture case than a case with a single-source evidentiary sample.

The risk of obtaining a match by coincidence is far higher when authorities search through thousands or millions of profiles looking for a match than when they compare the evidentiary profile to the profile of a single individual who has been identified as a suspect for other reasons. As an illustration, suppose that a partial DNA profile from a crime scene occurs with a frequency of 1 in 10 million in the general population. If this profile is compared to a single innocent suspect, the probability of a coincidental match is only 1 in 10 million. Consequently, if one finds such a match in a single-suspect case it seems safe to assume the match was no coincidence. By contrast, when searching through a database as large as the FBI's National DNA Index System, which reportedly contains nearly 6 million profiles, there are literally millions of opportunities to find a match by coincidence. Even if everyone in the database is innocent, there is a substantial probability that one (or more) will have the 1-in-10 million profile. Hence, a match obtained in a database search might very well be coincidental. Consider that among the 6 billion or so people on planet Earth we would expect about 600 to have the 1-in-10-million DNA profile; among the 300 million or so in the United States we would expect to find about 30 people with the profile. How certain can we be that the one matching profile identified in a database search is really that of the person who committed the crime?

A number of states have recently begun conducting what is known as familial searches.[11] In cases where a database search finds no exact match to an evidentiary profile but finds a near match—that is, a profile that shares a large number of alleles but is not identical—authorities seek DNA samples from relatives of the person who nearly matches in the hope that one of the relatives will be an exact match to the evidentiary sample. In several high-profile cases familial searches have identified suspects who were successfully prosecuted.[12] The key questions raised by familial searches, from a civil liberties perspective, are how often they lead to testing of innocent people—*i.e.,* people who do not have the matching profile—and how often they might falsely incriminate innocent people through coincidental matches. Familial searching may increase the number of people falsely incriminated by coincidental matches because it increases the effective size of the population subject to genetic monitoring. The larger the effective size of the database, the greater will be the likelihood that one of those innocent people will be identified.

People have been prosecuted based on cold hits to partial profiles. Defendants in cold-hit cases often face a difficult dilemma. In order to explain to the jury that the incriminating DNA match arose from a database search (in which the government had thousands or millions of opportunities to find a matching profile), the defendant must admit that his profile was in the database, which in many states entails admitting to being a felon, a fact that might otherwise be inadmissible. Courts in some cold-hit cases have, at the urging of defense counsel, opted to leave the jury in the dark about the database search in order to avoid the implication of a criminal record. Jurors are told about the DNA match but are not told how the match was discovered. The danger of this strategy is that jurors may underestimate the probability of a false incrimination because they assume the authorities must have had good reason to test the defendant's DNA in the first place. In other words, jurors may mistakenly assume the DNA test compared the crime scene sample to the DNA of a single individual who was already the focus of suspicion (a circumstance in which the risk of a coincidental false incrimination is extremely low) and not realize that the defendant was identified through a cold hit (a circumstance in which the risk of a coincidental false incrimination is much higher).

My argument is that jurors' evaluations of the case as a whole may be inaccurate if they are not told the match was found through a database search. I am suggesting that jurors will assume (incorrectly) that the DNA evidence confirms other evidence that made the defendant the subject of police suspicions and hence will underestimate the likelihood that the defendant could have been incriminated by coincidence. This is a process that, in my view, puts innocent people who happen to be included in a database at risk of false conviction.

When DNA evidence was first introduced, a number of experts testified that false positives are impossible in forensic DNA testing. Whether such claims are sinister or not, they are misleading because humans are necessarily involved in conducting DNA tests. Among the first 200 people exonerated by post-conviction DNA testing were two men (Timothy Durham and Josiah Sutton) who were convicted in the first place due partly to DNA testing errors. In both cases a combination of technical problems in the laboratory and careless or mistaken interpretation of the test results produced misleading DNA evidence that helped send innocent men to prison for many years.[13] False DNA matches have come to light in a number of other cases as well.[14][15]

One cause of false DNA matches is cross-contamination of samples. Accidental transfer of cellular material or DNA from one sample to another is a common problem in laboratories and it can lead to false reports of a DNA match between samples that originated from different people. In addition, accidental cross-contamination of DNA samples has caused a number of false "cold hits."

A second potential cause of false DNA matches is mislabeling of samples. The best way to detect labeling errors is to obtain new samples from the original sources and retest them, but this safeguard is not always available. Evidence at crime scenes is typically cleaned up (and thereby destroyed) once samples are taken, and the original samples are sometimes exhausted during the initial round of testing. Retesting is rarely done, even when samples are available.

Routine duplicate testing by forensic laboratories is another possible safeguard, but it too is rarely done.

A third potential cause of false DNA matches is misinterpretation of test results. Laboratories sometimes mistype (*i.e.,* assign an incorrect STR profile to) evidentiary samples. If the incorrect evidentiary profile happens to match the profile of an innocent person, then a false incrimination may result. Mistyping is unlikely to produce a false match in cases where the evidentiary profile is compared with a single suspect, but the chance of finding a matching person is magnified (or, more accurately, multiplied) when the evidentiary profile is searched against a database.

The ability of criminals to neutralize or evade crime control technologies has been a persistent theme in the history of crime.[16 17] There are anecdotal reports of criminals trying to throw investigators off the track by planting biological evidence. When such planting occurs, will the police be able to figure it out? Will a jury believe the defendant could be innocent once a damning DNA match is found? I have strong doubts on both counts and, consequently, believe that intentional planting of DNA evidence may create a significant risk of false incriminations.

Do innocent people really have nothing to fear from inclusion in government DNA databases? It should now be clear to readers that this claim is overstated. If your profile is in a DNA database you face higher risk than other citizens of being falsely linked to a crime. You are at higher risk of false incriminations by coincidental DNA matches, by laboratory error, and by intentional planting of DNA. There can be no doubt that database inclusion increases these risks. The only real question is how much. In order to assess these risks, and weigh them against the benefits of database expansion, we need more information.

Some of the most important information for risk assessment is hidden from public view under a shroud of governmental secrecy. For example, the government's refusal to allow independent experts to examine the (de-identified) DNA profiles in offender databases is a substantial factor in continuing uncertainty about the accuracy of frequency estimates (and hence the probability of coincidental matches). I believe there is no persuasive justification for the government's insistence on maintaining the secrecy of database profiles, so long as the identity of the contributors is not disclosed. The government's refusal to open those profiles to independent scientific study is a significant civil liberties issue.

Can DNA "Witness" Race?

BY DUANA FULLWILEY

Duana Fullwiley, Ph.D., *is an Associate Professor in the Department of Anthropology at Stanford University. She is the author of* The Enculturated Gene: Sickle Cell Health Politics and Biological Difference in West Africa. *This article originally appeared in* GeneWatch, *volume 21, number 3-4, November-December 2008.*

On August 11, 2004, an African American man named Derrick Todd Lee was convicted for the first of a series of murder and rape cases in south Louisiana. In the early 2000s, seven women in the Baton Rouge area had been violently murdered by a serial killer.[1] Lee's eventual convictions were largely based on his Y-chromosome STR DNA profile that matched DNA from samples found on the serial victims' bodies. Before this, however, Lee's DNA underwent a specific genetic analysis that attempted to place him within one of four continental racial groups. Thus, Lee was the first person in the United States to be identified as a possible suspect by an unconventional DNA test that racially profiled his DNA left at a crime scene.

The technology that purported to read Lee's race in his DNA is trademarked as "DNAWitness." The name is not accidental. Its inventors at DNA Print Genomics Inc. want to convey the idea that this technology itself embodies the power of the 'expert witness' through literal genotypes, or "base-calls," of the perpetrator's specific DNA nucleotide pairs. Forensic analysis with "DNAWitness" is, quite simply, a comparison of a sample of unknown origin with a panel of genetic markers called Ancestry Informative Markers, or AIMs.

The basic process of an AIMs analysis consists of a comparative exhibition of varying autosomal coding markers and their relative frequencies in four world populations. The goal of this specific iteration of the AIMs test, packaged only for forensics as DNAWitness, is to infer the aggregate of phenotypes associated with any one racial category in the United States. Such an inference is based on the extent to which the anonymous sample expresses allelic variations of markers comprised in a panel that is thought to differ in people from the continents of Africa, Asia, Europe, and (pre-Columbian) America.[1] In the case of the south Louisiana serial killer, DNAWitness yielded 'ancestry estimates' that the perpetrator's genetic makeup was 85 percent sub-Saharan African and 15 percent Native American. The Louisiana task force's previous search for a 'Caucasian' male was thereafter deemed to be potentially off the mark. The suspect, as deduced by DNAWitness, was most likely a 'lighter skinned black man' as inferred from probabilistic ancestry percentages revealed in the perpetrator's DNA.

I examine the use of DNAWitness to determine the prospective race of a suspect in order to provide evidence to law enforcement for narrowing a suspect pool. I argue that DNAWitness falls short of legal and scientific standards for trial admissibility and eludes certain legal logic concerning the use of racial categories in interpreting DNA. DNAWitness can offer vague profiles in many cases and has a wide margin of error that too often absorbs what might be understood to be important aspects (*i.e.,* substantial percentages) of ancestral heritage and of a forensic 'racial profile.' Moreover, this technology's individual ancestry estimates are highly vulnerable to social and political interpretations of phenotype and may be impossible to accurately interpret with a sufficient degree of objectivity, required of both science and law. It is possible, however, that this test may help to predict a range of skin color phenotypes, as was the case for Lee, since many of the AIMs are skin and hair pigmentation alleles.

The AIMs technology, (again, packaged with different names depending on the market and client) as manufactured by DNAPrint Genomics, is specifically designed to assess allelic frequency differences of coding DNA, or Single Nucleotide Polymorphisms. This is important, since markers that the test makers interpret as 'African' or 'European,' for example, are also found in other world populations that differ from the prior continental referent populations (African, European, Native American and Asian) used by the company in both name and geographic location. This is to say that differences in ancestry profiles may be due to evolution, gene flow, genetic convergence, or genetic drift. The presentation of DNAWitness test results demonstrates no attempt to distinguish between these different mechanisms of locus possession in individuals or in groups. Direct and unique ancestry (gene flow) is but one among several mechanisms that might explain shared sequence variation among and between racialized individuals. The simple description of a certain frequency, or set of frequencies, as 'African' ancestry may constitute a false designation of 'racial type,' while, conversely, it might not. The fact that there is no gold standard for this technology (a specific proprietary test) should make the legal community pause before lauding its potential success and eventual adoption on a broad basis.

From the outset, before evaluating scientific criteria for admissibility in a trial setting, it must be clarified that DNAWitness has not been used at the trial stage but rather at the pre-trial stage as prospective information for investigating officers. Nonetheless, it is critical to consider the scientific standards for legal admissibility to shed light on the ways in which this technology may actually do harm in the courtroom, since its scientific shortcomings can be easily identified with regard to admissibility rules. Furthermore, holding this technology to accepted legal standards with regard to 'expert' use of science and technology will also allow us to better understand DNAWitness's problematic role in the legal setting at any stage.

Legal precedent would have us focus on three federal cases to determine how scientific merit constitutes the rules for admissibility in a court of law: Daubert v. Merrell Dow Pharmaceuticals, General Electric & Co. v. Joiner, and Kumho Tire Co., Ltd. v. Carmichael. Issues of a)"reliability," b)"scientific validity," and c) whether techniques "can be tested" and "falsified" are of critical concern. As stated in Daubert v. Merrell Dow, "scientific methodology today is based on generating hypotheses and testing them to see if they can be falsified; indeed, this methodology is what distinguishes science from other fields of human inquiry."[2] More specifically, a "non-exclusive checklist for trial courts to use in assessing the reliability of scientific expert testimony," provided in Notes to 702, Federal Rules of Evidence, includes:

1. "whether the expert's technique or theory can be challenged in some objective sense, or whether it is instead simply a subjective, conclusory approach that cannot be reasonably assessed for reliability;
2. whether the technique or theory has been subject to peer review and publication;
3. the known potential rate of error of the technique or theory when applied;
4. the existence and maintenance of standards and controls; and

5. whether the technique or theory has been generally accepted in the scientific community."[3]

DNAWitness fails to meet this basic checklist on four out of the five items. (It has been subjected to peer review, as AIMS, in several research studies for means other than inferences of racial phenotype.)

Notwithstanding that these Federal Rules were established for the use of scientific evidence in a court of law independent of DNA testing, they nonetheless hold for all scientific evidence.[4] Effective December 1, 2000, several amendments to the rules, namely with regard to procedure and methods of reliability, made it clear to both the bench and bar "that an attack on the procedure used to test DNA for evidentiary purposes can be an effective challenge to the weight of any DNA evidence admitted."[5] Thus, presenting genetic results in less than exact and recognized ways could prove detrimental to case arguments.

The rise of new genetic technologies in the past two decades has yielded a range of scientific possibilities for the courts. Not all genetic tests perform the same kinds of tasks, and none were instituted without prolonged discussion, debate, and research consensus with regard to their reliability and consistency among scientists and law enforcement.[6] As this brief discussion makes clear, DNAWitness is based on Ancestry Informative Marker technology, or coding SNPs, that are largely shared among individuals and groups for varying reasons—reasons that are neither described nor acknowledged explicitly in the test results offered by DNAPrint. AIMs-based technologies, like DNAWitness, are attempts to model human history from a specifically American perspective to infer present-day humans' continental origins.[7] Such inferences are based on the extent to which any subject or sample shares a panel of alleles (or variants of alleles) that code for genomic function, such as malaria resistance, UV protection, lactose digestion, skin pigmentation, etc. There is a range of such traits that are conserved in, and shared between, different peoples and populations around the globe for evolutionary, adaptive, migratory, and cultural reasons. To assume that people who share, or rather co-possess, these traits can necessarily be 'diagnosed' with a specific source ancestry is misleading. Not only will siblings often share the same profile—or not—but individuals from all four 'parental' continental groups offered up by the model could feasibly share similar profiles—or not. As a forensics market version of the AIMs technology,

DNAWitness may offer precise mathematical ancestry percentages, but the accuracy of that precision remains debatable.

At best, this technology is an experimental modeling tool that hopes to mimic recent American human history as it reconstructs four racial types through an artificial homogenizing of markers found with relatively higher frequencies on some continents and lower frequencies on others. As compelling a tool as DNAWitness may seem, investigators should require that DNA analyses used in the serious proceedings of law be falsifiable, reliable, and thoroughly vetted. Anything less would prove irresponsible if incorporated into criminal investigations.

Forensic Genetics: A Global Human Rights Challenge

BY JEREMY GRUBER

Jeremy Gruber, J.D., *is President of the Council for Responsible Genetics. This article originally appeared in* GeneWatch, *volume 24, number 5, August-September 2011.*

DNA testing was initially introduced into the criminal justice system of the United States, one of the first countries to adopt its widespread use, as a method of developing supplemental evidence to be used in convicting violent felony offenders or freeing the innocent on a case-by-case basis. A 1992 report on New York State's original DNA database legislation stated that it would be limited to: "murderers and sexual offenders because DNA evidence is more likely to be uncovered in homicides and sexual attacks than in other crimes. And sexual offenders . . . often are recidivists."[1]

Such a characterization is almost quaint by today's standards. Over just a short time, function creep has overcome the system and the balance between the legitimate needs of law enforcement and individual rights has been lost. In the last fifteen years, forensic DNA collection and the resulting databases have changed dramatically, with DNA collection by law enforcement around the globe now routinely being used for a multiplicity of purposes that pose significant privacy and civil rights concerns for every citizen with little public debate and few safeguards to protect against possible adverse effects. In the United States, for example, the federal government and all fifty states have created permanent DNA databases taken from ever-widening categories of persons and subjected these collections to regular searches. The result is that the United States now maintains the largest DNA database in the world, with the Combined DNA Index System (CODIS) holding over 9 million records as of 2011[2] (the United Kingdom's National DNA Database (NDNAD) is of similar size, giving it the unfavorable distinction of having the largest percentage of its population recorded on a national DNA database). This has occurred despite the fact that DNA is far different from other methods of identifiction, such as fingerprints. It is a window into an individual's medical history and that of their entire family.

Law enforcement now routinely uses these tools to search and profile citizens convicted of even petty crimes, and collection practices are heavily trending toward the permanent retention of both biological samples as well as profiles from individuals arrested for but never convicted of a crime. At the same time that forensic DNA databases are expanding, a stunning array of techniques has emerged enabling lab technicians to glean information from DNA that goes well beyond the mere identification of a person and are providing law enforcement unprecedented access into the private lives of innocent persons by way of their own genetic data without a court order or individualized suspicion.

Some of these techniques include:

1. Trolling for suspects using DNA dragnets where police take samples from the public.
2. Comparing partial matches between DNA evidence and profiles in databanks to obtain a list of possible suspects from their relatives ("familial searching").
3. Constructing probabilistic profiles (including but not limited to race) of perpetrators from DNA collected at a crime scene.
4. Surreptitiously collecting and searching DNA left behind on items like cigarette butts and coffee cups.
5. The creation of local "offline" forensic DNA databases.
6. Dismissal of petty offense arrests in return for "voluntarily" joining a DNA database.

Many of the same problems that routinely plague criminal justice systems are reflected in these practices, including racial disparities in arrests and convictions. For example, while African Americans are only 12 percent of the U.S. population, their profiles constitute 40 percent of the federal database (CODIS). The lack of ethical guidelines for forensic DNA practices in the United States has implications far beyond just the American citizenry, however. Governments around the world are looking to the United States as well as the United Kingdom for guidance.

Today, 56 countries worldwide are operating forensic DNA databases and at least 26 countries, including Tanzania, Thailand, Chile, and Lebanon, plan to set up new DNA databases.[3] A number of countries, including Australia, China, Israel, and New Zealand are actively in the process of expanding their databases. And a number of other countries, such as Bermuda, the United Arab Emirates, Uzbekistan and Pakistan are even proposing including their entire populations on their databases. DNA databases around the world vary widely on issues ranging from access and consent to retention of both DNA samples themselves as well as the computerized profiles created from them. All of them share one common trait, though: a lack of sufficient privacy and human rights safeguards.

The growth of forensic DNA databases worldwide is often characterized as the natural response to public demands in each respective country. But the alarming rate of creation and expansion of such databases, with little public input and discussion, has been anything but piecemeal.

Public and private entities from the US as well as the United Kingdom are actively promoting DNA databases, often portrayed as technical solutions to high crime rates. The UK's Forensic Science Service (FSS), for example, has directly contracted with foreign governments including the UAE. Meanwhile, the Department of Justice FBI Laboratory has worked with over 29 countries to plan and create their databases (running the FBI-developed CODIS system) including promoting international agreements and authorizing legislation. Even individual US elected officials are actively promoting DNA databases abroad; Denver District Attorney Mitchell Morrissey, for example, regularly consults with foreign governments on their DNA databases.

The promotion of forensic DNA databases is by no means limited to government. A diverse private industry has developed to directly contract with foreign governments to build and maintain such systems, including offering policy recommendations largely modeled on United States practice. Over the last eleven years, Life Technologies has advised over 50 foreign governments and states on forensic DNA legislation, policy and law and regularly makes promotional presentations to foreign countries. In 2009, for example, the government of Japan standardized its DNA collection and analysis for the country's forty-seven prefecture laboratories using Life Technologies DNA testing systems. Life Technologies continues to provide support to Japan's National Police Agency. Such services are by no means limited to those countries that request them, either. In 2009, the Bermuda government signed a million-dollar multi-year contract with Florida-based firm Trinity DNA solutions to set up and run their database after salesmen for the company had approached Bermudan officials engaged in a comprehensive marketing plan.

Efforts to share DNA data between countries have expanded just as rapidly, allowing for data sharing across borders with little oversight. In Europe, data-sharing agreements established through the European Union (such as the PRÜM DNA Search Network) have begun this process.

In the United States, the National Institute of Justice's International Center promotes information sharing among similar Institutes worldwide.

Certainly the single largest contributing factor to DNA sharing across borders is Interpol. Interpol, the largest international police coordinating organization, has facilitated plans to create and maintain an international DNA database since Resolution No. 8 of the 67th General Assembly[4] in 1998 in Cairo. This resolution endorsed the encouragement of international cooperation on the use of DNA in criminal investigations and consequently the Interpol DNA unit was established. The objective of this unit is to: "provide strategic and technical support to enhance member states' DNA profiling capacity and promote widespread use in the international law enforcement environment."[5]

Through its DNA Gateway and G8/Interpol Search Request databases, Interpol is able to collate DNA profiles provided by member states and make these available to investigators globally where a match has been recognized. Interpol has 184 member countries and currently forty-nine members are making use of the DNA Database to varying degrees. Despite over ten years of operation, Interpol's databases still do not have clear operating procedures; for example, there is no transparent system for removing DNA profiles nor are there standards for comparing requests between member states with sophisticated and not so sophisticated forensic practices.

In addition to simply maintaining the database, the Interpol DNA unit is also urging harmonization of technical standards and methods across member states when creating DNA profiles from DNA samples. A number of countries use different operational practices when treating genetic data. For instance, countries have different practices on the number of markers used in their national systems (which poses obvious issues of accuracy). Advisory groups have been formed in Europe and elsewhere tasked with improving harmonization of forensic DNA methods to allow for the ease of sharing data across national boundaries (such as the ISO Standards for DNA Database Exchange).

At the same time that the growth of forensic DNA databases is exploding, there has been little public discourse on the privacy and human rights concerns they raise; nor has there been any domestic or international effort to create standards reflecting such concerns by those international bodies that are promoting information sharing. That may be changing. The Forensic Genetics Policy Initiative, a collaboration of GeneWatch UK, Privacy International and the Council for Responsible Genetics, seeks to achieve a direct impact on the human rights standards adopted for DNA databases across the world. The project aims to build global civil society's capacities to engage in the policy-making processes on the development of national and international DNA databases and cross-border sharing of forensic information and to protect human rights by setting international standards for DNA databases. An appropriate middle ground between the legitimate needs of law enforcement and a respect for individual rights is achievable. We must start now.

54

Twenty Years of DNA Databanks in the U.S.

BY SHELDON KRIMSKY

Sheldon Krimsky, Ph.D., *is Chair of the Board of Directors and a founder of the Council for Responsible Genetics. He is the Lenore Stern Professor of Humanities and Social Sciences in the Department of Urban & Environmental Policy & Planning at Tufts University. This article originally appeared in* GeneWatch, *volume 24, number 5, August-September 2011.*

Forensic DNA databanking in the United States began in 1990 as a pilot program serving fourteen states and local communities after an earlier start in Britain. The FBI's goal in developing the national Combined DNA Index System (CODIS) linking all the state DNA databanks was to collect the DNA of convicted violent felons and recidivist sex offenders. Within two decades several trends can be identified:

There has been an expansion of the categories of individuals whose forensic DNA samples are deposited into CODIS, extending from convicted felons and recidivist sex offenders to undocumented immigrants and misdemeanants who have neither been charged with nor convicted of a crime.

Courts have continued to rule that forcibly taking blood samples or other sources of DNA from a suspect on the mere chance that incriminating evidence might be found violates the individual's 4th Amendment protection. The 4th Amendment of the Constitution provides that "the right of the people to be secure in their persons, houses, papers, and effects, against unreasonable searches and seizures, shall not be violated, and no warrants shall issue, except on probable cause." Thus the forcible taking of one's DNA is a breach of one's privacy in the U.S. legal system and requires the state to have a probable cause or an overriding interest. In American jurisprudence, when suspicion is low and invasiveness is high, the 4th Amendment protection is generally high. In contrast, as suspicion grows and invasiveness diminishes, the protection against the invasion of privacy by law enforcement is diminished. Law enforcement agents can obtain a court order for forced blood samples. As DNA identification no longer requires a blood sample—a cheek swab will do—its intrusiveness into the body has dropped propitiously, and with it 4th Amendment protection. In the U.K., mouth swabs and hair samples were reclassified from first being considered "intimate" to "non-intimate." Under British law non-intimate samples can be taken without a person's consent from anyone arrested for a recordable offense and/or detained in a police station.

States and local police jurisdictions that legally obtain a person's DNA for a forensic profile to be entered into a database typically also retain the person's biological sample, which contains intimate information about a person's genotype.

Most courts have ruled that police can obtain a person's DNA surreptitiously without a warrant, even as the person has an expectation of privacy for the information in the DNA left on a discarded object. Ironically, while a warrant is required to acquire the DNA from a person, the police are free to follow a person around or use a ruse to obtain one's DNA when there is no probable cause and no warrant.

DNA can be used either to implicate or exonerate an individual accused or convicted of a crime. The power of DNA in exoneration is more powerful than its power in conviction. When the DNA doesn't match, as in a rape, it is highly improbable that the incarcerated person is guilty of the crime. When the DNA does match in a violent crime, the evidence may be strong, especially when other evidence links the suspect to the crime, but there are other hypotheses which could explain a false positive match, such as contamination. Over 250 prisoners have been exonerated for their crimes based on the probative post-conviction value of DNA evidence. Prosecution with DNA is the role of the state and federal police; exoneration with DNA is the role of nonprofit organizations, such as the Innocence Project, which operate on philanthropy. Increasingly, states have recognized that prisoner claims of actual innocence have been thwarted by lack of access to the crime scene DNA, which could possibly exonerate them. More states are providing falsely convicted prisoners with compensation for their false imprisonment.

While privacy of one's DNA has become increasingly valued and protected in medical genetics, in the workplace, and for people seeking health care insurance—as a result of passage of the Genetic Information Non-Discrimination Act (GINA)—the opposite trend can be found in the criminal justice system. The courts are more likely to view forensic DNA profiles as they do fingerprints: simply a means of identification. Moreover, because DNA can be taken from an arrestee by a cheek swab, the courts have lowered the bar on personal intrusiveness, thus expanding circumstances where police can obtain a DNA sample without a warrant.

GINA, passed in 2008, prohibits access to an individual's genetic information by insurance companies making enrollment decisions or employers making hiring decisions. Whether collected by police or by insurance companies, an individual's DNA can reveal inherited genetic disorders, predispositional disease states, parental linkages, ancestral identity, sibling connections, familial disease patterns and environmental and drug sensitivities. According to the American Society of Human Genetics, "Genetic information, like all medical information, should be protected by the legal and ethical principle of confidentiality." Bioethicist George Annas noted: "It is useful to view the DNA molecule as a medical record in its own right for privacy purposes." Many, but not all, of the privacy issues associated with DNA databanks would be resolved if the biological sample were destroyed once the forensic DNA profile was obtained. In Germany and Japan, biological samples are routinely destroyed once a successful DNA profile is made. In Germany, profiles are destroyed upon acquittal or discontinuance of criminal proceedings.

Areas where law and policy remain to be resolved include "abandoned DNA," "familial searching," "arrestee DNA" and retaining DNA profiles of those not convicted of crimes.

In 2007, *The New York Times* (April 2) ran a story titled, "Stalking Strangers." Author Amy Harmon wrote:

> They swab the cheeks of strangers and pluck hairs from corpses. They travel hundreds of miles to entice their suspects with an old photograph, or sometimes a free drink. Cooperation is preferred, but not necessarily required to achieve their ends... The talismans come mostly from people trying to glean genealogical information on dead relatives. But they could also be purloined from the living, as the police do with suspects. The law views such DNA as "abandoned."

Helena Kennedy, chairperson of the Human Genetics Commission in the UK, commented after Parliament passed a law honoring the privacy of a person's DNA to anyone outside of law enforcement: "Until now there has been nothing to stop an unscrupulous person, perhaps a journalist or a private investigator, from secretly taking an everyday object used by a public figure—like a coffee mug or a toothbrush—with the express purpose of having the person's DNA analyzed. Similarly, an employer could have secretly taken DNA samples to use for their purposes." This is the first law of any country with a DNA databank that honors the expectation of privacy by prohibiting people who are not legitimate members of law enforcement from analyzing so-called "abandoned DNA."

Familial DNA searching allows police to explore the family members of someone who came up as a close but not exact match between a crime scene DNA sample and their profile on a DNA databank. The issue as yet unresolved is what privacy considerations can be given to family members of a person who has his or her forensic profile on a databank, when there is no probable cause. To what extent can police troll members of the population for evidence of guilt when all they have is a "familial DNA resemblance?"

In a July 12, 2010, editorial, *The New York Times* raised civil liberties and civil rights concerns over familial searching: "Hundreds of people could fall under suspicion simply because they are related to someone in the criminal DNA database. Because blacks and Hispanics are disproportionately represented there, a first-time black offender has a better chance of having his DNA lead to a familial match than does a first-time white offender."

Japan and Germany do not allow familial searching. In the United States, individual states can pass their own familial searching laws and issue regulations for their use. Four states—California, Colorado, New York and Florida—have such laws. Only Maryland has categorically banned familial searching. At the very least, familial searching, which usually involves acquiring the DNA of family members to determine if there is an exact match, should be limited by a court warrant.

About 11 states have passed laws allowing police to obtain DNA forensic profiles of arrestees who have not been charged or convicted of a crime. In November 2004, California voters passed Proposition 16, which amended the "DNA Fingerprint, Unsolved Crime and Innocence Protection Act." According to Proposition 16, persons arrested or charged with any felony could be subject to warrantless seizure of their DNA. The arrestee provisions of the 2011, struck down the arrestee provision of Proposition 16. The majority wrote: "[T]he DNA Act, to the extent it requires felony arrestees to submit a DNA sample for law enforcement analysis and inclusion in the state and federal DNA databases, without independent suspicion, a warrant or even a judicial or grand jury determination of probable cause, unreasonably intrudes on such arrestees' expectation of privacy and is invalid under the Fourth Amendment of the United States Constitution." This is the first major court decision that questions the extension of DNA databanking to arrestees.

With no suspects in a murder case, police have sometimes resorted to "DNA dragnets" in communities small enough to initiate a voluntary program of DNA collection from all men between certain ages. While such dragnets have not proved very successful in tracking down the perpetrator, police add the forensic profiles they collect to the state databank, which enters it into CODIS. The people who provide the DNA do so to exclude themselves, thereby narrowing the field of suspects. But they do not expect that, after being excluded as a suspect in the crime, their DNA will be under constant surveillance and remain on the national forensic network of felons when they were not arrested or charged with a crime. Most states do not guarantee to DNA dragnet volunteers, who are excluded as suspects, the removal of their forensic profile and the destruction of their biological sample.

In the Cape Cod community of Truro, Massachusetts, police began collecting DNA samples from nearly 800 male residents within three years after the murder of fashion writer Christa Worthington. While the DNA dragnet did not yield the murder suspect, the samples and forensic profiles of those men who had volunteered their DNA remained on the databank. After a lawsuit filed on behalf of 100 men who had volunteered their DNA, by 2008 police had only "returned" the DNA of one man. This illustrates the need for uniform rules on returning the voluntarily submitted DNA of innocent people to law enforcement in community DNA dragnets. It should be part of the informed consent process that voluntary DNA of innocent people is taken off the databank.

The recent California Appeals Court decision suggests that many of the unresolved questions pertaining to forensic DNA will see their day in court, if not in the legislature.

The UK DNA Database: The Founder's Effect

BY HELEN WALLACE

Helen Wallace *is the Executive Director of* GeneWatch UK. *This article originally appeared in* GeneWatch, *volume 24, number 5, August-September 2011.*

Britain's DNA database is the oldest in the world, established in 1995. Originally intended to retain the DNA profiles of persons convicted of serious crimes, in 2001 the New Labour government led by Tony Blair began to transform it from a criminal database to a database of suspects. Two changes in the law led to DNA being collected routinely on arrest for any recordable offence and held until the suspect reached 100 years of age, regardless of whether or not they were convicted of, or even charged with, any offence.

Britain's low age of criminal responsibility (ten years old in England and Wales) and the wide range of offences classified as recordable led to people being added to the database for such alleged minor crimes as pulling each others' hair or hitting a police car with a snowball. Although young black men were disproportionately affected, people added to the database came from all sectors of society, including a white grandmother arrested for theft when she allegedly failed to return a football kicked into her garden.

What was originally a populist policy became increasingly contentious as ordinary families found themselves or their children with permanent records on police, DNA and fingerprint databases, even when the police openly acknowledged they had done nothing wrong. The new legislation was implemented in England and Wales, and later in Northern Ireland, but the Scottish parliament blocked similar changes in 2006, arguing that indefinite retention of innocent people's DNA was unacceptable.

During the debate in Scotland, a senior police officer claimed that 88 murders had been solved as a result of allowing the retention of innocent people's

DNA in England and Wales. Similar figures were later repeated by Prime Minister Gordon Brown, but these claims were exposed as spurious. The false claims were based on a police estimate of the number of matches that had been obtained between DNA swabbed from murder scenes and unconvicted individuals on the database. The number of matches was not verifiable but, more importantly, matches are not solved crimes and many such matches are with the victims or with passersby. To date, a decade after expansion of the database began, the police have yet to identify a single example of a murder that would not have been solved if innocent people were taken off the database.

The UK statistics show that as the database has ballooned in size, there has been no increase in the number of crimes detected using DNA. This is presumably because many of the people added to the database over the past decade are at very low risk of committing the types of crimes for which DNA evidence may be relevant. The common misconception that a bigger database is better therefore needs to be rethought. A focus on analysing DNA from crime scenes, and on thorough traditional policing, has proved much more effective than testing and storing DNA profiles randomly from arrested persons who are not suspects for a crime involving DNA.

The law in England and Wales allowing retention of DNA and fingerprints following acquittal or when charges were dropped was challenged in the courts. However, the English courts ruled repeatedly that there was no interference with people's right to privacy. The legal breakthrough came in December 2008, when the European Court of Human Rights made a unanimous ruling in the case of *S. and Marper v. the UK* that the UK law on DNA was clearly in breach of the European Convention on Human Rights. The UK government was forced to hold a consultation and adopted a new law allowing the retention of DNA and fingerprint records from unconvicted persons for six years after arrest, a position regarded by many as still in breach of the Convention.

Following the UK elections in May 2010, the new Coalition Government committed to rolling back this legislation: a new bill is currently being debated by parliament that would make the law in England and Wales similar to Scotland's. Although aspects of the bill require improvement, about a million innocent people's records are expected to be taken off the database. In addition, all biological samples will be destroyed once the DNA profiles stored on the database as a string of numbers have been obtained from them, providing an important additional safeguard to prevent access to sensitive genetic information.

One of the most disturbing aspects of the UK DNA database expansion is the way in which it was intended to set a precedent for DNA databases around the world. Lobbying for the changes in the law came not from the police but from commercial interests, which wished to profit from analysing the DNA of every citizen in every country. In 2003, the UK watered down UNESCO ethical guidelines for genetic databases, which originally stated that DNA from innocent people should be destroyed at the end of an investigation. In 2005, the US adopted the Violence Against Women Act, mirroring the UK law at a federal level. The Act allows the uploading of DNA profiles to the federal database on arrest, rather than on charge or indictment, and removes the burden from the state to remove the DNA profile of someone who is later acquitted or whose charges are dismissed. In 2006, the UK Forensic Science Service (FSS) signed the first of a series of contracts with the United Arab Emirates, with the aim of putting the entire population on a DNA database. Other countries, such as South Africa, became the targets of lobbying to expand their databases along the same lines as the UK law. A draft bill in South Africa was dropped following the decision by the European Court of Human Rights, and more careful consideration is now being given to the need for better safeguards.

DNA is undoubtedly a useful tool for solving crimes; during an investigation it can exonerate the innocent as well as help to convict the guilty. But databases of individuals' DNA profiles, linked to sufficient personal information to allow the police to track them down, raise important issues about human rights. The same technology that can track a criminal through their DNA can also be used by abusive regimes to track political opponents or identify their relatives. If a DNA database is not secure, unauthorized access could reveal the identities of individuals who may be hiding for good reason on witness protection schemes, fleeing from an abusive relative, or working as an undercover officer. In countries where sex outside marriage is a criminal offence, women may be put at risk by familial searches of the database for a suspect's relatives, because such searches can reveal non-paternity.

These risks are not offset by increased benefits as DNA databases grow in size. Larger databases increase the risk of errors and false matches, which can occur by chance or through poor laboratory practice, and hence risk miscarriages of justice. And, as the UK data shows, expanding DNA databases can deliver very little benefit, as a large proportion of most populations is very unlikely to commit serious crimes for which DNA evidence might be relevant.

The question remains what kind of precedent the UK's experience will set for other countries: one in which unfettered expansion of such databases is the norm, or one in which some important lessons are learned from some serious mistakes. There is a choice between allowing narrow vested interests to set the agenda for DNA database expansion or having a balanced debate of the pros and cons, taking full account of the impacts on civil liberties and human rights.

Presumed Innocent? The Confused State of U.S. DNA Collection Laws

BY MICHAEL T. RISHER

Michael T. Risher, J.D., *is an attorney at ACLU of Northern California. This article originally appeared in* GeneWatch, *volume 24, number 5, August-September 2011.*

In 2009, two of our nation's largest criminal justice systems—the federal government's and California's—began a massive new program of seizing DNA from individuals who had not been convicted of any crime, without a search warrant or any judicial oversight. The synchronous timing was mere coincidence, but the effect has been that a number of courts are now examining the constitutionality of the practice and reaching conflicting conclusions. This article will summarize the history of constitutional challenges to warrantless DNA seizure and the current litigation dealing with this contentious and rapidly evolving issue.

History

The Fourth Amendment to the U.S. Constitution prohibits "unreasonable" searches and seizures. This generally means that the government must obtain a search warrant before it conducts a search, although there are many exceptions to this warrant requirement. When laws were first enacted that required persons who had been convicted of serious crimes to provide a DNA sample for analysis and inclusion in CODIS, they were immediately challenged in court on the grounds that the mandatory extraction and analysis of their DNA constituted a search (just as, courts had held, did mandatory drug testing), and that the government should therefore have to obtain a warrant to do it. The courts universally—and quite properly—accepted the first part of this argument, holding that DNA sampling implicated both the physical integrity of the body and personal privacy and is therefore a search. But the courts were initially split on the question of whether this meant that the government needed a warrant. Some judges believed that because DNA contains so much personal information a warrant should be required; others believed that the fact of a criminal conviction reduced a person's privacy interest so much that it was reasonable to insist that convicted felons provide DNA samples without a warrant. After all, a person serving a prison sentence has essentially no privacy rights; even a person who is granted probation or is released from prison on parole has vastly reduced privacy rights and must allow the police to search his or her house without a warrant or even a reason to suspect any wrongdoing. The latter view eventually won out, and every appellate court to consider the issue has held (often over strongly worded dissents) that the government may require people convicted of felonies to provide DNA samples for inclusion in CODIS.

Mandatory DNA collection from persons who have been arrested but not convicted presents a very different question, and appellate courts have reached different opinions about its legality. People who have merely been arrested because a single police officer has reason to believe they may have committed a crime have much stronger privacy rights than have people who have been convicted of a felony following a trial or a guilty plea; and the government's interest in collecting DNA from people who may well be innocent is much less than it is in taking DNA from convicted murderers and rapists.

The first two appellate decisions on the question were decided in 2006 and 2007 but reached conflicting conclusions. The Minnesota Court of Appeals held that forcing arrestees to provide DNA samples violated the Fourth Amendment, while the Supreme Court of Virginia held that it did not.[1] Also in 2007, a federal trial court held that arrestee testing violated the Fourth Amendment in an opinion that the court did not even publish. Then, in May 2009, an Arizona state trial court ruled that a law requiring minors to provide a DNA sample upon arrest violated the Fourth Amendment.[2]

The federal and California laws brought a new wave of litigation over the practice, again with varying results. The first two cases were federal prosecutions in which a criminal defendant was ordered to provide a DNA sample as a condition of being released on bail pending trial, as required by a recently amended federal law. In May 2009, a federal judge in California upheld the statute in *United States v. Pool;* six months later, in *United States v. Mitchell,* a federal judge in Pennsylvania came to the opposite conclusion, holding that the law's requirements violated the Fourth Amendment's prohibitions against unreasonable searches and seizures.

The losing side in each case took an appeal to the United States Court of Appeals in San Francisco and Philadelphia, respectively. The way these two appellate courts treated the cases in itself illustrates how seriously the courts are taking this issue. In general, the United States Court of Appeals uses three-judge panels to hear and decide cases; in very rare cases the court may decide that the whole court (or a larger panel) will rehear a case that was already heard by one of these panels, a procedure known as rehearing *en banc*.[3] In *Mitchell*, the court apparently thought that the issue was so significant that the entire court decided to hear the case without even allowing a three-judge panel to decide the matter, something that has rarely if ever occurred before. As discussed below, in July 2011, a slim majority of the court upheld the DNA collection.[4] And in *Pool*, after the initial panel upheld collection in a 2-1 decision, the Ninth Circuit also decided to rehear the case *en banc*.[5]

Meanwhile, other courts were hearing challenges to California's law, which requires that anybody arrested on suspicion of a felony—which can include simple drug possession, knowingly writing a bad check, or entering a store planning to shoplift a pack of gum—provide a sample, regardless of whether they are ever even charged with a crime. This law affects a huge number of people; some 300,000 people are arrested in California every year on suspicion of a felony, and approximately 100,000 of them are never convicted of anything. The first of these challenges is a civil class action challenge to the law brought by the ACLU of Northern California. The named plaintiffs in the suit were all arrested for felonies, three of them at political demonstrations, and one when he tried to return camera equipment that had been stolen from the federal government. Three of them were never even charged with a crime, and one of the demonstrators was charged but the case was dismissed. But all four were forced to provide DNA samples simply because of the arrest. They asked the federal district court in San Francisco to issue a preliminary injunction to stop the state from taking DNA samples from arrestees without obtaining a search warrant, but the court refused to do so.[6] That case is now on appeal to the Ninth Circuit, the same court that is considering the *Pool* case. It is impossible to know when the court will issue its opinion in either case.

At the same time, Californians who had refused to provide a sample upon arrest were being prosecuted in state court; one of them went to trial, was convicted, and took an appeal. And in August 2011, the California Court of Appeals in *People v. Buza*[7] held that California's arrestee-testing law violated the Fourth Amendment.

The *Mitchell* and *Buza* appellate opinions

Thus, within the space of a few weeks, two appellate courts reached very different conclusions about the constitutionality of mandatory DNA sampling. Although the cases' holdings are technically not in conflict—*Mitchell* upheld taking DNA from a person only after a grand jury had already issued an indictment, whereas *Buza* dealt with people who had only been arrested—the opinions take starkly contrasting approaches to the broader issue of when the government can force a presumptively innocent person to provide a DNA sample without a warrant.

The *Mitchell* majority (9 of 14 judges) first acknowledged that DNA sampling and analysis constituted two separate searches for the purposes of the Fourth Amendment, searches that violate the Fourth Amendment unless they are reasonable. It also acknowledged that our DNA contains a huge amount of personal information; but it then discounted the seriousness of the intrusion that these searches caused, for two different reasons. As to the initial sampling, it reasoned that the physical intrusion was minor, less than the intrusion caused by a blood draw or mandatory urinalysis, for example, both of which have been described by the Supreme Court as being only minimally intrusive. As to genetic privacy, the court opined that people who provide samples have little to worry about: federal law prohibits the government from using the samples for anything other than criminal-identification purposes, and the profile that is generated by the DNA analysis involves only a small portion of "junk" DNA, which contains little private information. On the other side of the balance, the court held that the government has a strong interest in identifying arrestees, a concept that the court said could include not just knowing who it had arrested but also whether he or she has committed any uncharged crimes. The court believed that just as the police take fingerprints from people they have arrested and then use those prints not just to identify who they have but also to see whether those prints have been found at the scenes of unsolved crimes, so they should be able to use DNA for both of those same purposes.

The *Buza* court took a very different view, one that displayed a much clearer understanding of how the government is using DNA and how that may threaten privacy. First, the court rejected the analogy to fingerprinting, both because fingerprints contain none of the personal information that our DNA does (most courts hold that fingerprinting is not even a search for Fourth Amendment purposes) and because fingerprinting of arrestees had become a routine part of our criminal justice system long before the courts started taking the Fourth Amendment rights of arrestees seriously (and, the court might have added, long before computerized databases made it possible to use fingerprints in the way that DNA is now being used). It also properly rejected the government's assertion that the police are using arrestee DNA sampling to identify arrestees—meaning to determine who they have arrested; the law requires that the police identify an arrestee (through fingerprints) before they even take a DNA sample, and since the state takes about a month to process an arrestee's DNA sample it could not even use a sample for identification purposes. Instead, as the *Buza* court made clear, the only reason the government is taking DNA from arrestees is to try to implicate them in uncharged crimes. And although solving crime is certainly a legitimate governmental interest, the courts have long made clear that, outside of emergency situations, the Fourth Amendment prohibits the police from conducting mass, warrantless searches just on the hope that they will uncover evidence of a crime.

The future

As opinions by intermediate appellate courts, both the *Buza* and *Mitchell* opinions will not be the last word on this topic. Since *Buza* struck down a statute, the California Supreme Court may well decide to take up the case, and the issue of whether the Fourth Amendment allows the government to seize DNA from all arrestees will almost certainly end up in the U.S. Supreme Court within the next few years. It is impossible to predict how the issue will ultimately be resolved. But one thing is clear: courts looking at the issue of when the government should be allowed to seize, analyze, and stockpile our genetic blueprint should do so based on a clear and complete understanding of the interests involved on both sides, rather than relying on assumptions about DNA databanks or unsupported claims that it is or is not useful for particular purposes. This will require that they take a hard look at the privacy interests involved, the efficacy of taking DNA from all arrestees rather than waiting for the criminal justice system to sort out who is guilty and who is not, and also why the government is taking DNA from arrestees. A decision that will have such an impact on the genetic privacy of so many Americans must be made based on evidence and science, not on speculation and mere assertions.[8]

Can a DNA Dragnet Undermine an Investigation? A Case Study in Canada

BY MICHEAL VONN

Micheal Vonn *is a lawyer and Policy Director of the British Columbia Civil Liberties Association. This article originally appeared in* GeneWatch, *volume 24, number 5, August-September 2011.*

The 2007 conviction of notorious serial killer Robert William Pickton did not see the end of the decades-long story of murdered and missing women in British Columbia. The police investigation into reported disappearances of women from Vancouver's Downtown Eastside has been widely criticized and is now the subject of a controversy-plagued public inquiry. Meanwhile, women and girls continue to disappear in Northern BC along Highway 16, dubbed the "Highway of Tears."

Earlier this year, the Royal Canadian Mounted Police (RCMP) in Prince George collected approximately 600 "volunteer" DNA samples from taxi drivers, on the theory that a Prince George cabbie might be one of the last people to have contact with one of the missing women. On the basis of reports, it does not appear that the decision to seek voluntary DNA samples from taxi drivers was particularly well-considered. A contact at Emerald Taxi, who spoke to the media on condition of anonymity, put it this way: "At that time, someone said a cab driver might know something, so they called us all in for an interview, and then said, 'Oh, how about a DNA sample, too?' which was a surprise."

The surprise requests came with the unsurprising DNA dragnet catch-22: the drivers could "volunteer" their DNA *or* immediately find themselves on the list of suspects. As one driver said, "They made it sound like if I didn't, that could cause problems." Taxi company managers and drivers told the media that many of the drivers initially refused to submit a sample, and many initially told the RCMP that they would not provide a sample without a warrant; but threatened with becoming a suspect, most capitulated.

News outlets across the country covered the DNA dragnet story and the BC Civil Liberties Association (BCCLA) was asked to comment on the "privacy issues"—specifically, to comment on whether there *were* any privacy issues given that the RCMP had promised the samples would be destroyed after the volunteer-suspects had been "eliminated." But what started out as a conversation about the national DNA databank system and relevant legislation quickly turned to questions about whether the choice to employ a DNA dragnet could be a tactical blunder for the investigation. The anger of the cab drivers, not least because members of the public had started to voice fears about taking taxis because the police focus on the taxi companies obviously meant that one of the drivers was a serial killer, could only mean that the police were effectively alienating the very people that they were hoping would have useful information for them. In deciding to coerce the drivers into providing DNA samples, the police might be effectively shutting the door on precisely the community cooperation they need for the investigation.

There are at least eighteen disappearances that have occurred along the highway that cuts through Prince George, which is the urban center of Northern BC. It is impossible that members of the community don't have information that could prove valuable to these investigations. But the RCMP in general and the Prince George police in particular are suffering from a deep distrust by members of the public. In 2002, a provincial court judge in Prince George was convicted of sexually assaulting underage aboriginal girls in the community. A number of allegations were also made against local RCMP officers by the girls involved and their social workers but no charges were ever laid and the code of conduct investigations into the allegations against ten RCMP officers were mysteriously allowed to exceed the limitation period without resolution. The police accountability scandals have continued ever since, with RCMP finding no wrongdoing in their members repeatedly tasering a 'hog-tied' prisoner who subsequently died, refusing to impose internal discipline on an officer convicted of causing bodily harm for an assault that broke bones in the victim's face, and failing to reprimand officers who were found by a Provincial Court to have destroyed or concealed cell block surveillance footage of an incident in which a man held in jail said he was tasered by the RCMP more than 20 times. When the BCCLA conducted a public forum on policing in Prince George in the summer of 2010, participants reported a lack of trust in the police. As one participant said, "It's hard to tell your kids to trust the RCMP when you have a hard time trusting them."

To date, the DNA sweep through the Prince George cabbies has generated no reported investigative leads and a great deal of anger at the perceived aggression of the police. We don't know yet what will prove to be the critical pieces of information and evidence that solve these cases. But especially in a context where the relations between the police and the community are already fraught, it is important to consider how distrust that is exacerbated by a controversial dragnet is going to affect the ability of people to come forward with information. How much small-town policing, or big-town policing for that matter, is reliant on a relationship of trust between the police and the community? How terribly old-fashioned and no-tech that sounds. Certainly in the Pickton murders many members of the community had critical information and many who came forward were not listened to. It's possible that what unlocks the cases along the Highway of Tears will have more to do with listening to members of the community than profiling their DNA. In fact, it may prove that a DNA dragnet, in some cases a forensic tool, could also prove a liability in an investigation.

58

Forensic DNA, the Liberator

Interview with Peter Neufeld

Peter Neufeld *is the co-founder and co-director of the Innocence Project. This interview originally appeared in* GeneWatch, *volume 24, number 5, August-September 2011.*

The Innocence Project is a U.S.-based organization dedicated to exonerating wrongfully convicted people through DNA testing and reforming the criminal justice system. Peter Neufeld co-founded the Innocence Project with Barry Scheck in 1992.

GeneWatch: What role do forensic DNA databases play in the Innocence Project's work?

Peter Neufeld: They play an interesting role. There are occasions where we get a DNA test result on a material piece of evidence from a crime scene that would exclude our client, but prosecutors still resist motions to vacate the conviction. In some of those cases, what then tipped the balance in our favor was that the profile of the unknown individual [whose DNA was found at the crime scene] was run through a convicted offender database and a hit was secured. Once we were able to identify the source of the semen or the blood as somebody who had a criminal record and who had no explanation for his DNA being there, we were then able to secure the vacation of the conviction for our client.

GeneWatch: Many of the proponents of the expansion of forensic DNA databases have pointed to the usefulness of DNA as a tool for exonerating those who have been wrongly convicted, and some actually use the Innocence Project's work as an argument in favor of expanding forensic databases. Would you endorse that?

Neufeld: I wouldn't do a blanket endorsement. I think what we have to do in all these situations is weigh the advantages and downsides, and it depends on the expansion. For instance, expanding the database to other convicted felons, is a reasonable tradeoff because the people have been convicted of a felony.

We would not be in favor of expanding the database to arrestees. We certainly would not be in favor of the kind of gray-area databases that have been established, for instance throughout New York, where local medical examiners' offices or crime laboratories create their own databases outside the state system, which includes profiles of people who volunteered to give DNA so they could be excluded [from an investigation]. Those people should not be included in databases, but expanding it to other felons seems like a reasonable compromise.

GeneWatch: What about familial searching—running evidence from a crime scene against a database and looking for near-matches that could be family members of the source? What kind of a role do you see that playing?

Neufeld: There hasn't been a lot of familial searching—yet. There has been a lot of talk about it. The Innocence Project itself does not have a position at this time on familial searching. I can tell you that, personally, I'm very worried about it. I see it as a further encroachment on privacy and civil liberties, and with very little bang for the buck.

There's another issue we have to think about that is just over the hilltop: namely, these phenotype profiles that the FBI is trying to develop. Let's say there's blood found at a crime scene and they don't get a hit in the database, but they can determine through looking at a whole bunch of markers that there's a higher probability that the source of the blood has blond hair, for instance, or has blue eyes, or some other physical attribute. Then they will try to use this information to focus on a particular subpopulation in the community. That is something that could be extremely dangerous and that people should be vigilant about.

GeneWatch: It sounds like something that could lead to more people being wrongly convicted.

Neufeld: Well, not necessarily *convicted*, but we certainly will see a situation where lots of completely innocent people will be harassed by police. They could be stalked, they could have their reputations compromised; all kinds of things could happen. For instance, there was a serial murder investigation in Louisiana where certain people declined to have their DNA profiled. The prosecution went to court to seek an order [to obtain their DNA], their names and addresses were put in local newspapers, and they were the object of all kinds of rumors. They turned out to be completely innocent, but that's what happened.

There are all those kinds of problems lurking out there, and people have a tendency not to take them seriously enough, thinking, "Well, it's all about public safety. No one has anything to fear." When folks get lackadaisical like that, that's when there can be terrible adverse consequences for the whole community.

GeneWatch: In proving someone's innocence, particularly someone who has already been convicted, how essential is it to locate the perpetrator?

Neufeld: Well, you know, it shouldn't be. When there's a rape of a 90-year-old woman and semen is recovered, the fact that our client is excluded [on the basis of the DNA evidence] should be enough; but lately we've seen an uptick in prosecutors who are doing everything they can to fight a DNA exclusion, which is a presumptive proof of innocence. In some of those cases, it takes a CODIS hit to carry the day. Actually, in some cases even a CODIS hit isn't enough. We have two cases right now in Illinois, for instance, one where five kids confessed and one where three kids confessed, 19 and 16 years ago. In those two cases, we've identified the real perpetrator through DNA typing, and nonetheless the prosecutors are taking the absurd position that the confessions are reliable, all these kids are guilty, and in one case the convicted rapist, who is 25 years older than the 14-year old girl who was killed, happened to be a necrophiliac. So once you start seeing those kinds of really offensive inculpatory attempts to explain away the DNA exclusion, you realize that CODIS hits are going to be very, very useful, although not always conclusive. They may be conclusive in the eyes of the public and rational thinking people, but they're not necessarily conclusive in the eyes of a recalcitrant prosecutor.

GeneWatch: When people argue that forensic DNA databases should be expanded in order to help exonerate people who were wrongly convicted, it seems that argument often has less to do with the work that the Innocence Project does than just saying, "The way we will help people who have been wrongly convicted is by finding the actual perpetrator." Is that even enough in itself?

Neufeld: It's not just that. You're right in the sense that the quantum of evidence that I would deem sufficient to exonerate somebody may be less than what a prosecutor will require before he throws in the towel. Unfortunately, there are a number of judges who will be very deferential to the assertions and absurd speculative theories of certain prosecutors, for a variety of political reasons. We simply have to be aware of that, and we have to be able to respond. Our response takes two different directions: One, we will try to generate additional proof of innocence wherever possible, and sometimes that entails trying to identify the real perpetrator. At the same time, we will shine a bright light on the kind of absurd speculations of prosecutors trying to defend a conviction that should be immediately overturned.

Let's get one thing straight, though: There's no question that there would be fewer wrongful convictions if there was a universal DNA databank. The reason we don't have a universal DNA databank is that probably a majority of the population is strongly opposed to it for privacy and civil liberties reasons—not unjustifiably. Consequently, since there isn't a critical mass to support that kind of police action, the police are trying to do what they can to generate databases of more vulnerable populations. If you create an arrestee database, there's going to be a disproportionate number of people of color in that database. They may not have the same clout, electorally speaking, as upper-middle-class white people who don't want to have their DNA on file.

Why DNA Is Not Enough

BY ELIZABETH WEBSTER

Elizabeth Webster *is Publications Manager at the Innocence Project. This article originally appeared in* GeneWatch, *volume 24, number 5, August-September 2011.*

After twenty-seven years of wrongful imprisonment, Thomas Haynesworth was released in March on his forty-sixth birthday at the request of Virginia Governor Robert McDonnell. The Virginia Attorney General and two of the Commonwealth's attorneys support his exoneration. So why is the Innocence Project still fighting to prove his innocence?

Haynesworth's troubles began in early 1984, when a serial rapist began terrorizing women in Richmond, Virginia. Police apprehended Haynesworth, an 18-year-old with no criminal record, after one of the victims spotted him on the street and identified him as her attacker. His photo was shown to victims of similar crimes. Ultimately, five victims identified him.

Haynesworth protested, saying that he was innocent, but the eyewitness evidence compelled the juries. Haynesworth was convicted of two rapes and one attempted robbery and abduction. "I thought they were going to see that they made a mistake and correct it," he says. "It's been 27 years, and I'm still waiting."

Haynesworth was sentenced to 74 years in prison, which might have been the end of the story, if not for a lab technician named Mary Jane Burton. While Virginia courts and police agencies routinely lost or destroyed evidence, Burton took the extraordinary effort of saving cotton swabs and other evidence samples in her notebooks. Had Burton followed lab policy and returned all of the samples to the investigating agencies, all evidence in these cases would have been gone forever. The blood type testing, or serology, that Burton performed was not nearly as probative as DNA testing would later become.

The Innocence Project and others pushed for a review of Burton's case files. In response, then-Governor Warner launched a massive DNA review of convictions. Burton died in 1999 and never learned of the tremendous impact of her work; so far, six wrongfully convicted Virginians were proven innocent because of her practice of preserving evidence. The review also led to DNA tests in Haynesworth's case.

The DNA testing cleared Haynesworth in one of the rape convictions. Moreover, the Department of Forensic Science matched the sperm sample to the genetic profile of a convicted rapist named Leon Davis. DNA testing on a second rape that Haynesworth was charged with, but not convicted of, also cleared Haynesworth and pointed to Davis. Biological evidence from Haynesworth's other two convictions, however, does not exist. In one of the convictions, documents show that evidence was destroyed. There never was any biological evidence available in the other conviction.

Profile of a perpetrator

Davis was suspected of committing at least a dozen rapes in Richmond and Henrico County in 1984, and he is currently serving multiple life sentences for those crimes. Davis and Haynesworth lived in the same neighborhood; they resembled each other and were sometimes mistaken for each other.

In Haynesworth's first letter to the Innocence Project in 2005, he writes: "There is an inmate named Leon Davis who is in prison for some of the same things I'm charged with, and he was living down the street from me in Richmond, Virginia. . . . I will bet my life this is the man who committed these crimes. . . . Just get my DNA tested, and you will see I'm innocent of these crimes."

All of the crimes were perpetrated within the same one-mile radius and all shared the same modus operandi. If DNA testing proved that Davis committed two of the crimes, it follows logically that he committed them all.

Haynesworth now waits for a hearing with the Virginia Court of Appeals. Until then he is on parole and is a registered sex offender. He leaves his mother's house only to report to his job as an office technician with the Office of the Virginia Attorney General, Ken Cuccinnelli. Without a writ of innocence or a pardon, Haynesworth cannot be cleared.

Accepting DNA's limitations

In 1992, the newly founded Innocence Project began taking cases from prisoners with claims of innocence whose cases were suitable for post-conviction DNA testing. By 2000, 67 people had been exonerated, and the organization was swamped with letters from prisoners seeking assistance. This trend has not slowed in 19 years. Today the organization receives over 3,000 letters a year and nearly 300 people have been exonerated.

Yet the total number of wrongful convictions surely surpasses this. Many wrongfully convicted prisoners have no DNA evidence to test. Over 20 percent of the cases closed by the Innocence Project since 2004 were closed because evidence had been lost or destroyed. Despite diligent searching, there just isn't a Mary Jane Burton in every state. Evidence preservation laws have become more commonplace, but many jurisdictions, even major metropolitan areas, still have hopelessly outdated paper-based systems.

If DNA evidence had existed in all three of Haynesworth's convictions, it could easily have proven his innocence of those crimes. In a sexual assault case in which a single perpetrator attacks a stranger, and consent is not an issue, the results are easily interpreted. When a real perpetrator, like Leon Davis, can be identified through a DNA database hit, it not only exonerates the innocent but also solves the crime.

As it stands, investigators and attorneys had to look carefully at the pattern of crimes in Haynesworth's cases to see if they were likely committed by the same perpetrator. Such cases have resulted in exoneration before. But what about those cases that are not suitable for DNA testing at all? Very few cases involve physical evidence that could be subjected to DNA testing, even among violent crimes.

Unlike any other type of evidence, DNA testing can conclusively prove innocence (or guilt) to an unprecedented degree of scientific certainty. But a system that depends on DNA testing alone to protect the innocent is a failed system. DNA illuminates the flaws in the criminal justice system; it does not eliminate them.

Those flaws include eyewitness misidentification; improper use of the forensic sciences or reliance on outdated or invalidated forensic methods; false confessions, admissions, and even guilty pleas; jailhouse informant testimony; and more. Simple, cost-effective reforms—improving police lineup procedures, for example, or mandating that all interrogations be recorded—can reduce the rate of wrongful convictions, and, by extension, assist in the apprehension of real perpetrators. Some states have been slow to adopt these reforms.

Before DNA exonerations became common, criminal justice professionals struggled to understand the implications of the technology. DNA testing trumps all other types of evidence, whether it's a witness who swears that he could never forget a face, a co-defendant's confession, or a hair analyst who claims to have a match. Admitting that the system has devastated an innocent person's life is never easy. The only way to truly reconcile the loss is to learn from the wrongful conviction and make sure it never happens again.

The Innocence Project continues to advocate for Haynesworth's exoneration on the remaining two cases. With the complete support of law enforcement in Virginia, we hope that day comes soon. When it does, there will be those who say that the criminal justice system works. They will celebrate the power of DNA to find justice. But 27 years of wrongful imprisonment is not justice, and DNA cannot erase those years. While Haynesworth was behind bars, the true perpetrator continued to commit brutal crimes; additional women were harmed. DNA cannot erase what happened to those crime victims. But if the criminal justice system learns from these errors and continues to adopt reforms that will prevent future injustice, then DNA can be rightfully thanked for leading the way.

The CSI Effect

BY ROBERT A. PERRY

Robert A. Perry, J.D., *is Legislative Director of the New York Civil Liberties Union.*
This article originally appeared in GeneWatch, *volume 24, number 5, August-September 2011.*

On July 7, 2001, the Las Vegas Metropolitan Police Department issued a video that explained how a mix-up in the department's crime lab led to the conviction and imprisonment of an innocent man.[1]

The video presentation provided a detailed, step-by-step replay of what was essentially a documentary exposé of the manner in which apparently mundane human error can confound the science of forensic DNA—a technology that is perceived by many, including many scientists, to be all but infallible.

The perception is validated by compelling news accounts in which a DNA specimen is the magic bullet, so to speak, that leads inexorably to the perpetrator. Media accounts of criminal forensics read like a script from an episode of the American television series, *CSI: Crime Scene Investigation*.

Under optimum circumstances DNA is an extraordinarily precise forensic tool. However, the collection and analysis of DNA evidence is a human endeavor. It is susceptible to human fallibility—and venality—and this scientific fact has not been given sufficient consideration in the rush to create ever larger DNA databanks.

There is a growing body of research that demonstrates that error can and does occur with extraordinary frequency. Professor William C. Thompson has observed that the problem is widespread; that it occurs even in the best labs; and that forensic scientists have downplayed the scope of the problem and its import.[2] Thompson and his colleagues warn of "limitations or problems that would not be apparent from the laboratory report, such as inconsistencies between purportedly 'matching' profiles, evidence of additional unreported contributors to evidentiary samples, errors in statistical computations and unreported problems with experimental controls that raise doubts about the validity of the results."[3]

The problem is not mere negligence—mislabeling and cross-contamination of DNA samples and computational mistakes—but also unconscious bias and something more sinister: the intentional misrepresentation of DNA analyses.

The routine use of forensic DNA in criminal investigations and prosecutions poses novel and complex challenges to the integrity of the criminal justice system. A former district attorney in Manhattan put the issue this way:

> DNA databanks do help apprehend dangerous criminals (and thereby prevent crime). But most people aren't violent criminals and never will be, so putting their DNA on file exposes them to risks that they otherwise wouldn't face. First, people who collect and analyze DNA can make mistakes. . . . Second, people can be framed by the police, a rival or angry spouse. Third, DNA is all about context; there may be innocent reasons a person's DNA is at a crime scene, but the police are not always so understanding.[4]

A major study by the *British Journal of Criminology* finds that at a certain point there are diminishing returns to society, in terms of public safety, from the inclusion of DNA samples in a massive databank. The report raises concerns about the "tactical" use of DNA when interviewing suspects, and about police officers with insufficient knowledge of forensics resorting to DNA evidence in lieu of rigorous detective work.[5]

And, of course, these issues also pose a formidable challenge to the integrity of forensic science.

In a 2009 report, the National Academy of Sciences issued a sweeping critique of the nation's crime labs in the United States, observing that forensic scientists with law enforcement agencies "sometimes face pressure to sacrifice appropriate methodology for the sake of expediency."[6] Other researchers have observed that lab technicians typically fail to take measures to "blind" themselves to the expected outcome of their analysis. Studies of this phenomenon indicate that when faced with ambiguity in the information presented in a DNA sample, crime lab analysts will often fit their interpretations to support the prosecution's theories.[7]

The proponents of an ever-expanding DNA databank dismiss such concerns, challenging the skeptic to demonstrate a wrongful conviction caused by error or abuse related to the use of DNA evidence. But this defense poses the wrong question. The pertinent question is this: Are there sufficiently rigorous and independent quality assurance procedures to ensure that error and abuse will be discovered; and if it is, will rigorous case review follow, along with sanctions and discipline if negligence or wrongdoing are involved? The answer is no.

If the foregoing is not sufficient to give pause, consider that twenty-five states require persons arrested to provide a sample for inclusion in a databank. Sixteen states now authorize law enforcement to investigate the family members of an individual whose DNA does *not* match crime-scene evidence but is a near match. The scientific rationale for this practice is that a "partial match" between a crime-scene sample and the DNA of someone in the state's databank may implicate a blood relative of that individual. This practice is aptly named: familial searching.

Consider this thought experiment: What is the race of the individual wrongfully convicted in the Las Vegas Police Department video? African American. To the extent error or fraud taint criminal investigations, the consequences—including wrongful prosecution and conviction—will be borne overwhelmingly by persons of color. The criminal justice system is not race-neutral. The gross racial disparities in the population incarcerated for drug offenses are but one example of this fact.

So what's to be done? What is a reasonable, prudent response to the emergence of this CSI counter-narrative?

First, it is incumbent upon policy makers to undertake a rigorous and objective reexamination of the assumptions that have informed government's creation of an ever-expanding DNA databank. The scientific community must play a central role in framing the terms of this debate.

Second, policy makers must create a robust regulatory model that dictates rigorous quality assurance protocols; standards for the accreditation and review of lab performance; methodologies for evaluating the outcome and resolution of criminal investigations that involve forensic DNA; and procedures for protecting the privacy and due process rights of defendants.

A 2009 Supreme Court ruling reflects perhaps a growing awareness that advances in forensic science are no guarantee of justice or fairness. In that case (affirmed in a 2011 ruling) the court held that lab reports may not be used at trial as evidence of a person's guilt unless the analysts who created the reports testify in court, subject to cross-examination. This requirement, Justice Antonin Scalia observed, "is designed to weed out not only the fraudulent analyst, but the incompetent one as well." The right to confront witnesses would be required by the Constitution, Justice Scalia concluded, even "if all analysts possessed the scientific acumen of Mme. Curie and the veracity of Mother Theresa."[8]

PART VI

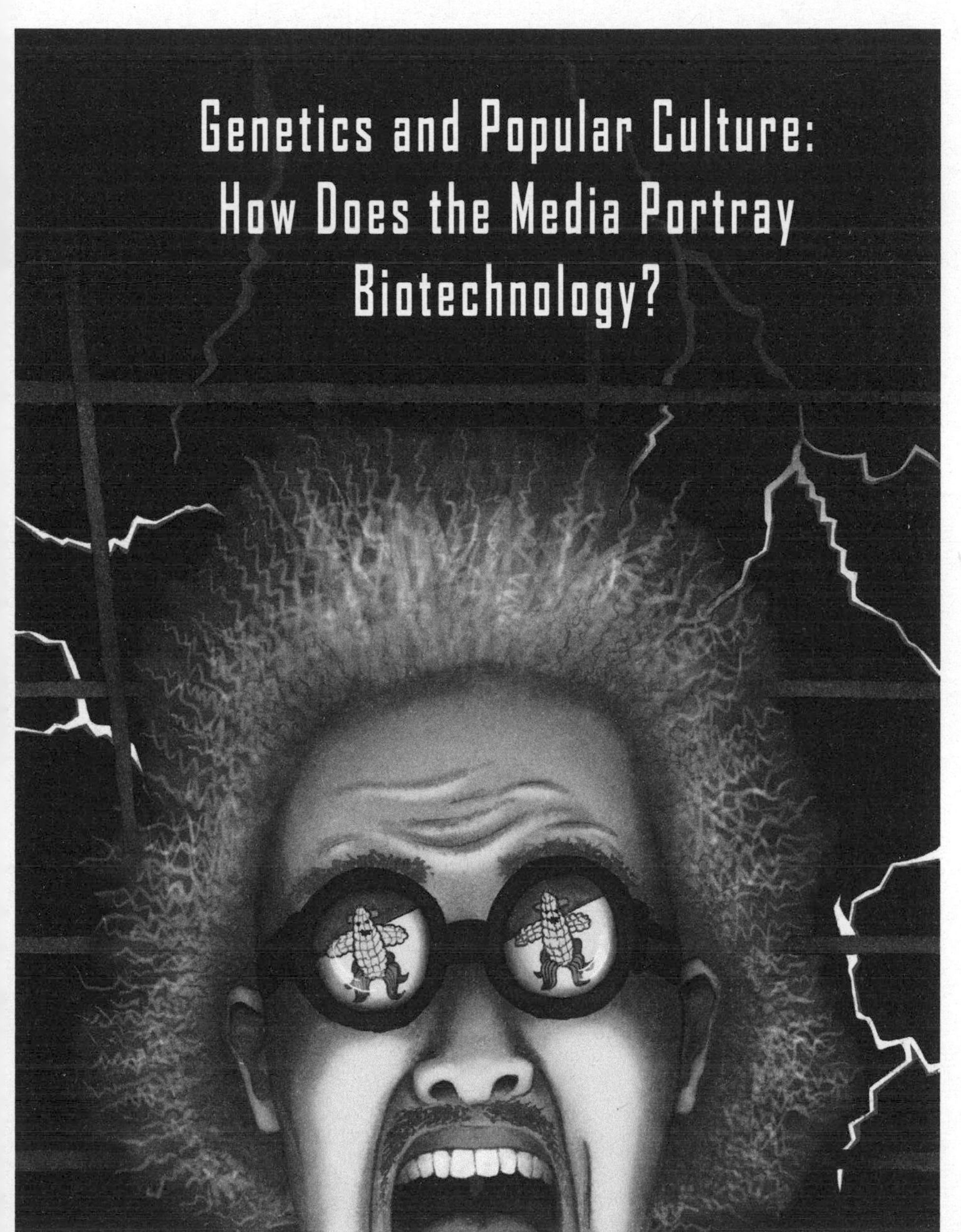

Artist: Sarah Kim

We can all think of ways that genetic science (or a rough interpretation of it) has been portrayed in popular culture. We've seen it as thematic elements driving the plots in sci-fi movies from *The Fly* to *Gattaca*, on television in shows like *Andromeda*, and in books like Mary Shelley's *Frankenstein* and Aldous Huxley's *Brave New World.* Companies like DNA 11 are pioneering the application of genetic science in the creation of personalized custom art. *Orchids,* a play about reproductive technology by Canadian playwright, physician, and educator Jeff Nisker, brings such issues to the stage. Films like *The Convicted* and theater like *The Exonerated* as well as numerous TV programs under the theme "crime scene investigation" (or CSI) explore forensic DNA in criminal investigations.

We see genetic science translated for the layperson in the news or documentaries, or packaged as a consumer product by direct-to-consumer genetic testing companies. If the science is accurate, it is a great way to get people educated and involved in science. If not, it can misinform and create an exaggerated sense of the power of genetic science.

We may find it more difficult, however, to think of ways that these representations of genetics or genomics have in turn impacted the science itself—but they have. They help us determine what scientific achievements have already happened, what is likely to happen, and what is highly improbable. The way genetic science and technologies are presented in popular culture affects our perceptions and informs our attitudes toward them and the importance society—including policymakers and scientists themselves—places on certain research or technologies. As Priscilla Wald points out in her chapter, "Scientists read newspapers and popular fiction, too." But most of all, popular culture makes genetic science fun and accessible, encourages thought and provokes discussion.

From music to the visual arts, film and literature, the following chapters explore genetics in popular culture and the resulting influence on science itself.

How Art and Literature Can Contribute to Genetic Policy

BY LORI ANDREWS

Lori Andrews *is a mystery writer and professor of law at Chicago-Kent College of Law where she is Director of the Institute for Science, Law and Technology. This article originally appeared in* GeneWatch, *volume 22, number 6, November-December 2009.*

A few years ago, a boutique opened in a trendy shopping area in Pasadena, California. The store, Gene Genies Worldwide, said it offered "the key to the biotech revolution's ultimate consumer playground." Its products were new genetic traits for people who wanted to modify their personalities and other characteristics.

The store was filled with the vestiges of biotechnology—petri dishes and a ten-foot model of the ladderlike structure of DNA. Brochures highlighted traits that studies had purportedly shown to be genetic: creativity, conformity, extroversion, introversion, novelty-seeking, addiction, criminality, and dozens more.

Shoppers initially requested one particular trait they wanted to change in themselves or their children, but once they got into it their shopping lists grew. Since Gene Genies offered people not only human genes but ones from animals and plants, one man surprised everyone by asking for the survivability of a cockroach.

The co-owners, Tran T. Kim-Trang and Karl Mikhail, were thrilled at the success of their endeavor, particularly since none of the products they were advertising was actually yet available. Despite their lab coats, they were not scientists but artists attempting to make a point, striving to serve as our moral conscience. "We're generating the future now in our art and giving people the chance to make decisions before the services actually become available," said Mikhail.

At the time I met Kim-Trang and Mikhail, I was chairing the Ethical, Legal and Social Implications (ELSI) advisory committee to the Human Genome Project. The HGP was providing grants totaling 3 to 5 percent of its budget to economists, philosophers, anthropologists, law professors, sociologists, physicians, molecular biologists and others to anticipate the social issues that would arise with genetics. But for decades, artists, novelists, and poets had already begun exploring these issues in greater depth than any short-term ELSI grant could allow.

Beyond assessing the complex and often surprising impact of genetic technologies, novels and art works can provoke discussion. When people read about a genetic development in *Science* or even *The New York Times*, they are often reluctant to challenge what they read. They think their lack of a Ph.D. in molecular biology makes their comments irrelevant. In contrast, people feel comfortable addressing the same issues when they are presented in a novel or a work of art. Artist Bradley Rubinstein creates digital photos showing children whose eyes have been digitally replaced with dog eyes, giving people a starting point for an intensive discussion of the morality and advisability of genetic enhancement.

Nancy Kress's novel, *Beggars in Spain*, allows people to understand the social context in which genetic enhancement would occur. Kress anticipates what society would look like if some children were genetically engineered not to need sleep. Unlike a university press release or biotech company annual report extolling the virtues of such a development, her book analyzes what might happen, given human emotions, social stratification, and the economics of access to technologies. As might be predicted, the "enhanced" children in the novel learn much faster and ultimately attain better jobs, making other people jealous of them. "Normal" people begin to sell items from factories with the We Sleep logo, even when the products are shoddier than those the genetically-engineered people create. (Shades of "Buy American.") But the people who don't sleep are also jealous of those who do—and take drugs to have the chance to dream.

The arts not only transform us individually, but they also transform us socially. Literature, paintings, photographs, and poems have documented social injustices and challenged and sometimes changed the social structure. Following Dorothea Lange's publication of one of her photos, "Migrant Mother," in the *San Francisco News*, the U.S. Government allocated $200,000 to establish a migrant camp for homeless workers. Upton Sinclair's novel *The Jungle* provides a disturbing indictment of the meat-packing industry, leading to the passage of both the Pure Food & Drug Act and the Meat Inspection Act.

In my own work, I've begun to explore how to weave popular culture into public policy debates. As a midterm project in my classes on genetics policy at Princeton and at Chicago-Kent College of Law, I asked students to read a science fiction book of their choice dealing with genetics. Their assignment was to analyze how close we were to using the technology at issue in the book; to assess the impacts of the technology on individuals, families, society, and the legal system and to determine whether the problems raised in the book could be handled by existing laws or whether new laws were needed.

The students were surprised at how quickly science fiction had become science fact (or at least science hype), with one or more actual doctors or companies doing research on or offering the technology that had been described in each science fiction book. And the students found creative ways to use existing laws to handle some of the problems, which we then applied in actual *pro bono* legal cases.

Since novels seemed to provide a way to stimulate discussion of genetics issues, four years ago I began writing a series of mysteries with a geneticist main character, Dr. Alexandra Blake. I chose to write about issues that are here and now, rather than writing science fiction. In the first novel, *Sequence* (published in 2006), I addressed issues of genetic privacy. A Navy lieutenant is asked to provide DNA when a murder occurs in a naval base library. He refuses, even though he is innocent, because he is gay and is concerned that the Navy might test his DNA for the alleged "gay gene."

In *Sequence*, I also wrote about a government official who tried to personally profit by selling to a biotech company the tissue samples that were part of the Armed Services DNA bank. After the book was published, the U.S. Congress began investigating a government official who had personally profited from providing tissue samples to a pharmaceutical company. Those samples had been collected by other researchers who had convinced Alzheimer's patients and healthy volunteers to provide tissue samples, including spinal fluid collected through a spinal tap, so that they might find biological markers associated with Alzheimer's and develop a cure.

When one of the other researchers went to the freezer to get the samples for research, Dr. Trey Sunderland, the chief of NIMH's Geriatric Psychiatry Branch, told her that the samples had been destroyed when a freezer malfunctioned. In actuality, Dr. Sunderland had provided the pharmaceutical company Pfizer with over 3,000 tissue samples and associated clinical data. Pfizer paid Dr. Sunderland about $285,000 for supposed consulting fees and approximately $311,000 for lectures and travel expenses. He was ultimately indicted and pled guilty.

In my latest mystery, *Immunity* (published in 2008), I point out problems with the regulation of human research, but I also take on some of the concerns about criminal DNA databases. One of the suspects in the book is Native American, but traditional forensic databases do not contain a wide-enough selection of Native Americans in the population database to indicate whether a certain allele is rare or common. The fact that a suspect's DNA matches at 9 or 13 loci doesn't mean much if many other Native Americans have that exact same genetic profile at those alleles.

My intention in writing mysteries was to smuggle in a few policy issues within a traditional thriller. But I hadn't anticipated that my fiction would inform my legal work. Some of my legal projects were inspired by research I undertook for the novels on the problems of regulating research on monoclonal antibodies and the potential use of mandatory quarantine for a pandemic. Exploring genetics in fiction has also given me a new view on the ethical issues I routinely handle in my legal work. In *Sequence*, the protagonist surreptitiously tests her boyfriend's DNA to see if he's the killer. I'd fight against that in real life, but writing her character made me see how easily a person could be pulled into a seemingly unethical action.

Fiction and art can help expand the discussion of genetic policy issues to a larger audience. They can empower readers and viewers to discuss the pros and cons of genetic technologies. By portraying the larger social context in which technologies are adopted and addressing who loses and who benefits, they can also help chart the appropriate regulation of genetics.

Generations Ahead: Genetic Technologies Meet Social Justice

Interview with Sujatha Jesudason

Sujatha Jesudason *is the former executive director of Generations Ahead and currently the director of the CoreAlign Initiative at the University of California, San Francisco. This interview originally appeared in* GeneWatch, *volume 21, number 1-2, Summer 2008.*

GeneWatch: Can you give us an idea of how Generations Ahead got started?

Sujatha Jesudason: Generations Ahead grew out of the Program on Gender, Justice and Human Genetics at the Center for Genetics and Society (CGS), in 2008. The goals of the program and now the goals of Generations Ahead are to build a national network concerned with social implications of genetic technologies and to build a proactive policy agenda. Our primary goals are to increase the awareness of these complex issues, build the capacity of social justice leaders to participate in the debates and advocate for policies that protect human rights and affirm our shared humanity.

GeneWatch: When Generations Ahead was getting set to become an independent organization, what was the vision for the group? Were there specific goals you felt you would be better able to accomplish?

Jesudason: CGS is an organization that is putting out cutting-edge thinking on these issues. As we started working with social justice organizations, because they were new to these issues, it was a big leap for some of them to get involved in the debates. In California, when we were working on legislation to ensure the health and safety of women giving eggs for research, we found our reproductive rights allies hesitant to get involved in the discussion both because they didn't know that much about the issues and because they didn't feel comfortable with a policy that women shouldn't be compensated. They needed the space and opportunity to explore a variety of positions, and that was hard for us to create under deadline since CGS clearly took a position that women should not be compensated for fear of creating an exploitative market in eggs. Generations Ahead, while deeply rooted in social justice and human rights values, tries to create an open space for different organizations to figure out their policy positions, particularly in dialogue with other allies.

We support social justice groups to make genetic technologies a part of their broader justice agendas and work their way toward policy positions on these complex topics. We still look to CGS for the cutting-edge issues, and we create opportunities for groups to discuss those issues and come up with their positions and plans for action. To create an open space that worked for organizations just taking on these issues, we needed an organization more committed to facilitating dialogue and policy development across different movements. Generations Ahead works closely with reproductive rights, disability rights, racial justice, human rights and LGBTQ rights movements, bringing them together to figure out what a collective policy agenda would look like.

GeneWatch: Can you tell us what you mean by "policy-neutral?"

Jesudason: It's critical that there are organizations like CRG and CGS putting out good policies. We start with those policy proposals and work to build broader alignment around what can be the more winnable policies. Generations Ahead does not start out with a defined policy position. We develop our position with our allies in the context of what they think is workable and winnable.

There's a certain amount of pragmatism in our approach rooted in the complexities of existing policy struggles. For example, the right has pushed for policies around "informed consent" to discourage women from having abortions. While it is definitely the right policy to give women comprehensive genetic counseling in the prenatal context, we have to figure out how that doesn't get used as an anti-choice practice, otherwise reproductive health and rights advocates will not be on board.

GeneWatch: Can you tell us about your typical day—or is there such a thing?

Jesudason: Usually, I start out in meetings with our program directors. We have ongoing conversations about our issues at the intersection of different movements and are constantly refining our strategies across the four different program areas: reproductive rights, racial justice, disability rights, and Lesbian, Gay, Bisexual, Transgender, Queer, Questioning and Intersex rights. We talk about areas of growth: working with young people, working with families, and working with progressive religious organizations. And as you can imagine, all these groups have their own issues and agendas. We are constantly trying to figure out how to grow our organizational capacity so that we can do more outreach and capacity-building for our allies.

I usually have a meeting with a racial justice organization, where we talk about Generations Ahead briefing their staff and members about these genetic issues, and then we explore if there's some collaborative project we can work on together to increase their capacity for engagement. Then I usually have at least one conversation with a funder—both current and potential funders—letting them know about our work, trying to find out how much they have been tracking these issues, and whether they would be interested in funding us. Right now we're trying to get funding for a national racial justice strategy meeting. We're planning to bring together racial justice leaders from across the country for a three-day meeting to talk about race and genetics (DNA databanks, behavioral genetics, race-specific medicines, and ancestry testing) and figure out what work we want to get off the ground together. We're working with the Center for Social Inclusion to co-host this meeting in the next six months. We're doing similar national strategy meetings for disabilities rights and LBGTQ rights.

GeneWatch: Is there an issue that you feel is the single biggest issue Generations Ahead is tackling right now?

Jesudason: I just got back from the CRG conference on DNA databanking and, thinking of the many issues raised, perhaps DNA databank issues are the most pressing right now. One of the challenges we face is that genetic technologies often don't seem pressing to organizations working on educational reform or post-Katrina rebuilding. So it's hard to mobilize people on those issues right away. With the DNA databanks, there's a pressing need for policies: the technology is in place and it's being used right now, so the impact is more immediately felt, as opposed to, say, reproductive cloning.

GeneWatch: When you decide which issues to prioritize, is it a question of relevancy to the movement's goals, chances of success, or something else?

Jesudason: Because we work with multiple movements, we start with: what are the most pressing concerns for those groups: LGBTQI, racial justice, disabilities rights, reproductive rights? Within each movement we have priorities, and we look at the possibilities given the political landscape, and the possibilities in terms of political ripeness—is there enough energy or interest around a set of issues—and strategic leveraging.

For instance, there's been a lot of concern and interest around cloned beef. But from our perspective strategically, that hasn't been an issue that has activated the people we work with because it doesn't speak enough to their core concerns—but I think the DNA databank issue speaks very directly and strategically to what they're already doing.

The way we see our work is that we're not creating whole new sets of issues but that this is an extension of the groups' existing justice-related concerns. For racial justice movements, the DNA databank issue fits in with what they're already doing. For reproductive rights organizations concerned about fertility and infertility, reproductive technologies are an extension of that set of concerns. We're updating political agendas for an age with increasing issues around biotechnology.

GeneWatch: How did you get into this work?

Jesudason: In 2004, I was working for an Asian reproductive rights organization, and I had been invited to a meeting hosted by CGS. That was my first introduction to these issues. I knew the conversations in the reproductive rights movement and nobody was talking about human genetic issues. It was the same in the racial justice movement. If both reproductive rights and racial justice groups were not taking part in the debate now, when the terms of the debate are being set, then at the point of policy making we would end up always fighting defensively. I got involved because I wanted to make sure that social justice movements were an integral part of the policy debate.

Before this—I'm a social scientist, not a life scientist—I hadn't thought about genetic technology as a social justice issue. Now, as I understand these technologies, how they are developed and used, I totally get how socially constructed they are and how important it is to shape how we understand, develop and use them.

Social justice organizations and leaders need to be thinking about science as a social justice issue, particularly for the biological sciences and human biotechnology. How we relate to each other and define family, community and humanity—all of these come into question with human genetics.

GeneWatch: How do you find allies—or coalition members, if that's your term—or how do they find you?

Jesudason: We definitely consider them "allies"—not just our allies but allies that work with each other.

We do most of the finding. We have a movement strategy for each movement that locates key leaders and organizations and what they are working on. We identify where there is compatibility with our approach and our issues, and we go talk to them. We proactively go seek them out and we also do a lot of work to get invited to the conferences and events within these movements. This way people get a chance to hear about our issues and then often afterwards they will approach us and ask if we can come to their organization and give them a briefing.

GeneWatch: It sounds like you travel a lot!

Jesudason: Yes! Yes I do. All of our project directors travel a lot. The social implications for these technologies are really just on the edge of peoples' consciousness, so if we were to wait for groups to come to us, it might be too late. Even when people do start tracking these issues, it's hard for them to find the time and resources to figure out what to do about them organizationally. That's where we provide an important service. We go to them where they are and show them some options of what they can do.

GeneWatch: How does Generations Ahead—and how do you personally—approach "opponents," if at all? Or do you actually classify any groups, corporations, individuals as "opponents"?

Jesudason: We don't tend to think in terms of opponents right now, in part because so much of our work is building the capacity for social justice organizations and leaders. We aren't quite ready or interested in sitting down with whoever the opponents might be yet. The only clear opposition that we do not meet with right now are anti-choice advocates, because we clearly position ourselves within a progressive agenda on women's health—though there are other organizations that are trying to work across that "choice" divide.

GeneWatch: When you're looking at how to get a message out or how to pursue action on a certain issue, are there certain venues or approaches you find are becoming more difficult or promising? Are you noticing a trend?

Jesudason: What I have noticed is that you need to have a lot of organizational infrastructure to truly have an impact on policy either at the state or federal level, and part of that is having resources and relationships. We partner with organizations that already have very robust policy advocacy departments—for instance, Planned Parenthood has a very powerful presence in Washington, D.C., as does the American Association of People with Disabilities and the Center for American Progress.

One of the things we're doing is bringing policy folks from different movements together to develop a set of principles in terms of evaluating policies that are coming up and developing shared messaging and activation strategies. So we see ourselves in policy work rather than grassroots work—and a big part of that is just that we're a small organization.

GeneWatch: Is there such a thing as 'too big' for Generations Ahead's purposes, or do you think there is an optimal size?

Jesudason: There's definitely such a thing as too big for our purpose because of our network-building strategy—we don't want to be one of the "big" organizations, we want to help activate networks of organizations. We are at six staff now, and I can imagine probably the biggest we would be is 12 to 15. We have a network-centric advocacy strategy, relying on a network of organizations and leaders and coalition building as opposed to creating a big organization with a membership base.

GeneWatch: What's your ultimate goal as an activist? Is there a time that you could ever stand back and say "our work here is done?"

Jesudason: We will know our work is done when every social justice organization is working on issues in genetic and reproductive technologies, that they have their own in-house experts on it, and that they see it as an integral part of their work. That is our discrete five-year organizational goal. In the long term our mission statement says: "Generations Ahead brings diverse communities together to expand the public debate and promote policies on genetic technologies that protect human rights and affirm our shared humanity."

Now, in terms of our long-term goal—it's not totally clear what that will look like, in part because we haven't heard from everyone yet on what a social justice agenda on genetic technology would look like, so we don't know yet what that success would look like at the policy advocacy level. We are clear about our values and about our strategies. What this changed world will look like is yet to be determined by all our allies working together to envision a hopeful world in an age of genetics.

Dinosaurs in Central Park

BY ROB DESALLE

Rob DeSalle, Ph.D., *is a curator in the American Museum of Natural History's Division of Invertebrate Zoology and co-director of its molecular laboratories. He is also a Board member of the Council for Responsible Genetics. This article originally appeared in* GeneWatch, *volume 22, number 6, November-December 2009.*

Upon the release of the recent blockbuster movie *2012*, my colleague at the American Museum of Natural History, Neil deGrasse Tyson, sent out to all of the curators at the museum a list of talking points concerning the lack of "reality" of the premise of the movie. Neil is an astrophysicist with a great reputation as a communicator of science, as demonstrated in his many appearances on PBS, but I was immediately puzzled by the need for Neil's pre-emptive strike against a silly movie about the geological and astrophysical demise of our planet. Then I remembered a similar situation I found myself in a decade or so ago, when the second *Jurassic Park* movie was in production and near release. Our genomics labs at the AMNH, about three years earlier, made the claim to have successfully extracted a small piece of DNA from a termite embedded and preserved in amber. Not having read the book or seen the original movie, I wondered: what connection, if any, might this work have to *Jurassic Park*?

As it turned out, everything! The premise of *Jurassic Park* (if you happen to have been in a coma for the last fifteen years), comes from author Michael Crichton's mastery of suspending disbelief. The story is about the reanimation of dinosaurs from DNA isolated from amber-preserved insects that had bitten and fed on dinosaurs. All of a sudden, I found myself inundated with phone calls from reporters of all kinds asking about the feasibility of the premise of the book. While we had reported the isolation of DNA from a long deceased organism, we had not gotten anywhere near the fulfillment of the premise from Crichton's pen, made equally vivid by the movies based on the books through Steven Spielberg's cinematic flair. As responsible scientists, we patiently explained to the media that we wouldn't be seeing dinosaurs prancing around Central Park anytime soon. However, the questioning by the media became so intense and anticipatory of a positive answer, that I was oftentimes openly frustrated. One reporter was so persistent with the hopes that it could be accomplished that he finally prompted me to respond: "Yes, *Jurassic Park* could happen. . . and monkeys could also fly out of my butt." (To explain: back at that time, this was a phrase made popular by Mike Meyers of *Wayne's World* and *Saturday Night Live*.) This comment did not endear me to the Public Relations Department at the AMNH.

While dealing with the press was frustrating, we also realized that this interest in the book and movie was a golden opportunity to teach. Oftentimes it is impossible to get the public even remotely interested in what we do as scientists and here was a situation where the intense interest in the potential science actually became annoying. And so, in anticipation of the release of the second *Jurassic Park* movie in 1997, David Lindley and I wrote a book titled, *The Science of Jurassic Park and The Lost World*. As publishers are wont to do, a subtitle was added to the book: "Or How to Make a Dinosaur." They added this subtitle because my co-author and I were clear throughout the book that neither dinosaurs nor any organism would be reconstructed in the near future from its naked DNA and the publisher felt that the negative response on our part would hurt the sales of the book.

In the book, we focused mostly on the scientific issues raised by the premise of *Jurassic Park*. For instance, the amber we had worked on at the AMNH in 1993 was from Miocene deposits in the Dominican Republic. Any self-respecting geologist would recognize right away that this deposit is only 30 million to 35 million years old, meaning that the amber we had worked with was way too young to yield dinosaur DNA because dinosaurs had all gone extinct 65 million years ago. Ironically, the scenes from the *Jurassic Park* movie where the amber fossils are discovered are also from the Dominican Republic and also incapable of having any dinosaur DNA in them (unless, of course, one considers birds dinosaurs, which actually one should). Right after we published the book, reports started to come out of other labs of older and older DNA fragments being isolated from older amber. Counter to that, reports came out of the British Museum of Natural History that claimed all of the previous studies on amber were faulty and not valid. This latter work prompted most labs to stop working with samples from amber and to hold to the idea that DNA older than 30,000 years could not be retrieved from fossilized tissues, whether it be in rocks or amber. Ancient DNA studies instead focused on subfossil remains (remains not mineralized or amber-preserved) less than 10,000 years old. To date, the oldest subfossil to yield DNA is between 100,000 and 30,000 years. Whether amber preserved fossils will yield up DNA is still debatable, but scientists as a matter of practice have focused on much younger samples.

Our second talking point to debunk the *Jurassic Park* premise usually rested on the general inability, when working with "ancient" tissues, to obtain enough high-quality DNA to reconstruct a genome. And if it were possible to obtain even enough small fragments, it would be computationally impossible to reconstruct, or "assemble" as the genomicists put it, an entire genome. Here, I have to admit, we were wrong. Our suggestion was made well before the announcement of the first draft of the human genome, let alone several recent publications using "next generation" sequencing methods. Our book was finished in 1996 and coincided with Craig Venter's suggestion that shotgun sequencing and assembly of small fragments was the way to go when sequencing a genome. More recently, the 454 and Solexa "next gen" approaches have revolutionized genome level sequencing. It is now possible to assemble genomes from relatively small fragments of DNA, especially if another closely related species is available as a scaffold upon which to do the assembly. The announcements coming from the Max Planck Institute concerning the sequencing of the Neanderthal genome and the announcement of the sequencing of both Venter's and James Watson's genomes are evidence of the power of these "next gen" sequencing techniques.

The third talking point was that even if we did get a whole genome, the insertion of this genome into an egg to get a developing embryo was impossible. We also questioned the ability to produce cross-species embryos using a surrogate species to provide the egg for development. We were partially wrong on this claim, too, as several cross species cloning and surrogacy examples exist. In fact the book was published just after Ian Wilmut and colleagues announced the birth of Dolly, a sheep cloned from somatic tissues. Perhaps the most prominent of the advances in this area after Dolly concerns the birth of Noah, a baby Gaur. Gaurs are extinct in nature and Noah was born from the application of Nuclear Transplantation experiments using frozen, stored Gaur tissue and a cow as a surrogate mother. Noah died soon after his birth but is still regarded a success story in this area of research. Even though success stories exist for this area of research, I would still suggest we are far from being able to inject "naked DNA" into a surrogate egg and have significant development occur.

The fourth argument we would use is that there would be nowhere to keep the beasts. We called this the ecological catastrophe problem and used Barbados, an island big enough to have substantial enough vegetation to support the number of dinosaurs mentioned in *Jurassic Park*, as an example of what might happen. We predicted that a severe ecological collapse would occur even on an island as big as Barbados, and that this might be the most important scientific point to consider in the entire scenario. Crichton and Spielberg did away with the ecological catastrophe problem by having Ian Malcom, the mathematician of the story (played wonderfully by Jeff Goldblum), babble about chaos theory. This problem is bigger than just chaos theory, though, as how we treat our environment is immensely important. An entire discipline in science has emerged around the pertinent subject of conservation biology. One of the most important subjects that conservation biologists concern themselves with is what they call "naturalness." Naturalness refers to the state of the environment that would be there without the overt human activity we have seen in the past two centuries (or perhaps, before Columbus came to the New World, or perhaps even pre-human civilization). Reintroduction in the real world is an attempt to recreate naturalness in one of these contexts. It seemed strange to us to even attempt to recreate naturalness as it existed in the Cretaceous. Why not instead worry about the species we have driven to extinction while humans have been the predominant organisms on the planet—not necessarily in reanimating them, but making sure other species don't suffer the same fate. We suggested that the billions of dollars that might be spent to create a Jurassic Park would be better used buying land for conservation preserves.

Our final talking point was always an ethical one. We pointed out that just because a technology is there, it doesn't necessarily mean it should be used. The fallacy of the "progress marches on" attitude was our main point. Scientists, we suggested, have a duty to consider the ethical ramifications of their work. If some new technology is developed and we go wild using it, we place everyone at a disadvantage. We can always say that scientists are good at self-policing these kinds of things, but the kinds of ethical questions that are arising from the modern technology have outpaced the knowledge and social experience of scientists. The impact is so extreme that, in my opinion, cogent ethical decisions can only be made by taking a broad ethical assessment approach involving experts from the many areas of science, philosophy, social science, economics and other disciplines.

As scientists, we hope that our work has an enduring impact on science and society. The *Science of Jurassic Park* episode of my career actually has had a huge impact on how I view science and the world. I look at the opportunities created by the scientific silliness of a well-crafted story like *Jurassic Park* as a great educational opportunity. When the public has an interest in something like dinosaurs and it opens up an opportunity to teach something, as scientists and educators we need to grab that bull by the horns and use it to the utmost.

I also learned that the general public often gets swept away by what scientists do. They are sometimes dazzled by the technology so much that their imaginations run wild. In general, most of the lay public want to see "monkeys fly out of our butts," but part of the job of being a scientist is to ensure that the public understands our science.

Finally, I learned some hubris about science and society from thinking about the feasibility of *Jurassic Park*. All of the scientific arguments that *Jurassic Park* won't work pale in comparison to the ethical arguments. Not considering the ethics of our science before doing it is like putting a 65-million-year-old dinosaur in Central Park.

Myth, Mendel, and the Movies

Interview with Dr. Priscilla Wald

Dr. Priscilla Wald *is a professor of English at Duke University and the author of* Contagious: Cultures, Carriers, and the Outbreak Narrative *and* Constituting Americans: Cultural Anxiety and Narrative Form. *This interview originally appeared in* GeneWatch, *volume 22, number 6, November-December 2009.*

GeneWatch: In your observations of the way science is represented in popular culture, have you found any particular portrayals of genetics or genomics that were especially insightful or dangerous?

Priscilla Wald: One film, actually, is *Journey of Man*. It is a film that seems to me—I don't want to say dangerous, that's a little too strong—but the film is very misleading about what genetics can and can't tell. I think it was making an effort to respond to the critics of some of the work on the Human Genome Diversity Project [which recorded certain genetic sequences of world population groups to detect differences], and it was preparatory to the Genographic Project. Spencer Wells, in the narration, suggests that it would be very easy to pin down somebody's ancestry, and the film does not get into the complexity and the implications. What I've written about is how the narrative of it reinforces racist narratives. I'm not calling anyone a racist, I want to be very clear about that—the film was very explicitly made to be anti-racist—but the logic of the narrative of the film really reproduces the idea that 'civilization' moved out of Africa and 'progressed' westward. And it's a really disturbing film because I don't feel that whoever wrote it was fully in control of the narrative and the impetus of the film, so it takes on a life of its own and I think does not convey the message that they were hoping to convey. Yet I think people watching it wouldn't necessarily pick up the inaccuracies if they didn't know something about genetics.

GeneWatch: And that's the problem, right—that the science can get lost in translation, even if not on purpose?

Wald: That's the real danger. I write in general on depictions of science in mainstream media and popular fiction and film, and consistently the nuances of science get lost and other kinds of stories take over. And I think that in turn affects the science in all kinds of ways. Scientists read newspapers and popular fiction, too; and in applying for funding, certain projects are going to get funded because of the anxiety level that they represent, for instance. Something becomes the new solution: we're going to find the gene for this or that—the gene that codes for schizophrenia, say—even if it's way oversimplified. When policy decisions get made, how much do the policy makers or the funding agencies necessarily know about these things? So the way people feel about the topics has a huge impact that we can't even chronicle.

But to answer your first question, the other film that I just taught that is very problematic is *Gattaca*. It's a fun film, but basically what it does visually and narratively—and I don't think this was in Andrew Niccol's or anybody's mind—is consistently set up genetics and genomics against religion, against the family, against justice. The message of the film is 'there is no gene for the human spirit.' A lot of scientists really like it. They see it as making the point that we shouldn't fall into genetic determinism, and that their research doesn't mean that someone's character can be reduced to their genes. But the logic of the film is anything but, especially visually. Everything is associating genetics with sterility, with the idea of living in a laboratory, with things that are dehumanizing, and I think it does a real injustice to the science. And then people get it into their heads. At one point the film says, "Whatever possessed my mother to put her faith in God instead of the local geneticist?" And you see a crucifix as he's giving that voiceover narration. It's telling us that there's an opposition between genetics and religion, and those are the kinds of messages that I find really disturbing.

So on the one hand there's something like *Journey of Man*—which both of my children saw multiple times in middle school and high school—really perpetuating false information about what genomics can and can't tell us about who we—"we," in quotation marks—are, and where we came from; and then there's something like *Gattaca* setting up genomics as this sterile, anti-human kind of science. Those, to me, are the two bookends, and equally dangerous.

A film that I think is really interesting, actually, is *X-Men*. You think of it as cartoonish, but it's really about bias and human nature and the complexity of being gifted. It doesn't fall into the naïve categories of good guys and bad guys. It's a very interesting film that explores what it means to be gifted.

GeneWatch: It seems that science fiction often actively undermines the romanticizing of science. Are there exceptions to this that we're missing—any scientific utopias?

Wald: I would say that *X-Men*—or another interesting one, the TV show *Heroes*—is quite favorable in its depiction of science. Dr. Xavier is a brilliant scientist, and some of the main characters in *Heroes* are research scientists. They are very well-intentioned and actually doing some very important work, presumably.

As for utopias . . . I've read so much science fiction, but nothing is coming immediately to mind, and maybe it's because I'm much more interested in perspectives that are about the complexity of something, not all bad or all good. If I see it going all in one direction I tend to go somewhere else.

I would also say a novel that's really interesting—certainly not utopian—is *Darwin's Radio* [by Greg Bear]. I think it's one of the most brilliant explorations of what happens when a widely held scientific theory is challenged. Another example—which is speculative fantasy, so it's not really exploring the science directly but more the issues coming out of the science—is Octavia Butler's *Xenogenesis*, or *Lilith's Brood* trilogy. She's writing in the late 1980s, and all of the things that are coming out of the Human Genome Project are things that she was exploring in really intricate and unresolved ways in that trilogy.

For some people, science fiction needs to be truly engaging with the literal science; for me it's often the speculative fiction that is actually most deeply engaged with the science, because it's getting at the deep issues that are informing the research.

GeneWatch: Do you see any marked similarities or fundamental differences in the way that genomics is represented in mainstream media versus science fiction?

Wald: I don't draw a huge distinction there, but I'm also looking for scenarios in popular culture that are being plucked out of the news. The popular depictions that interest me have tended to be amplifiers of some of the deep issues being brought up in the media—a sort of magnifying glass that takes some concept and really extends it into an ongoing scenario that allows us to really tease out the implications that are implicit in the stories that are emerging from the media.

GeneWatch: You have been researching the idea of genomics as creation myth. Where do you see that happening?

Wald: I'm writing a whole book on that, but I'll try to give you a very quick version. As for the 'where,' the obvious place is the DNA ancestry testing going on. In *Journey of Man*, at one point, the book that Spencer Wells wrote from the script of the film—the film was first—he actually called it a creation myth or a creation story. At one point he says, "We have our own creation stories, too." And I think that's a very astute point. I don't know how he meant it, but he's referring to Western scientists, and I think it's exactly right.

I don't mean to be at all disparaging or dismissing when I use the term 'myths.' I think it is a mistake to associate myths with "primitive" culture. I think that everybody has myths, and I think myths are essential to our existence. Myths are stories about social groups and group identities; they give us purpose and give our lives a kind of poetry. They are also very hard to challenge, because we believe them on a very deep level, and they are kind of invisible. So when I call genomics a creation myth, I don't mean it to be dismissive. I'm certainly not anti-science or anything like that. At the same time, I think the science is getting incorporated into new origin stories that we are telling ourselves.

We have new ways of thinking about our interconnectedness at the same moment that we have a kind of pushback from an effort to have a science that is telling us that we can trace our ancestry, and we can go back to seeing our differences as well as our similarities.

So the surface message of people like Spencer Wells is that we're all the same, we're all mixed, there's more genetic similarity within groups than between groups and so forth; but at the same time, the message of the research is about going back and separating these strands of where we come from. And my question is: why are we doing this at this time? I don't mean that in a paranoid sense—I think there could be good reasons—but I think it's an important question to ask.

Part of what I'm arguing in my book is that whenever there is a convergence of a real challenge to previously accepted definitions of human beings coming out of both science and social thought, you tend to get new creation stories. I see that happening in the eighteenth century with the idea of natural rights (the narrative that accompanies the "Rights of Man"); I see it happening in the Victorian period around Darwinian evolution; and I see it happening after World War II, at the dawning of genetic research, which is also, and I think this is not irrelevant, the moment at which science fiction comes into its own as a genre. I think science fiction, deeply and theoretically, is engaging exactly those issues: the changing definitions of human beings and their relation to creation stories and myths.

GeneWatch: You talk about myths being invisible in a way, and it's possible to see how genomics can be invisible, too, but at the same time, is it less of a myth because genomics can claim this scientific grounding?

Wald: I think all myths do. Well, not all myths, obviously . . . but in early cultures I don't think there's such a difference between myth and science. If you think of science as an effort to have a systematic observation of the world, myths are also an effort at explanation. Currently, we have a certain story about who we are and our origins, which comes from Darwin, and that's based on scientific inquiry. But it still has a mythic dimension.

Our myths are grounded in, we could say, a more sophisticated science than the myths of two thousand years ago, but I think there's an analogy, and it would be arrogant for us not to think that if the world continues, our understanding of the world will be superseded by the understandings of others.

GeneWatch: Because we don't know what the myth will be in a hundred years. We don't know how silly ours might sound.

Wald: Right, but none of them really sounds silly to me in the end; if I understand all of them to be about the poetics of collective identity, they become a lot more profound. So I'm not taking them at face value.

A world without myths would be a very dry world. Myths are beautiful; but they can blind us as much as they can illuminate and enlighten. That's true of any story, any ideology, any narrative or theory. Take something like Gay Related Immunodeficiency. That term, 'GRID,' made it easier for a doctor to identify the early signs of HIV in some people (specifically gay men). It also created terrible and destructive biases that both stigmatized populations in dreadful and counterproductive ways and blinded researchers to the full cause of HIV, which meant the blood supply went uninvestigated for way too long, and people who were not gay men were not suspected of having HIV. It illuminated certain areas and obscured others, and that's true of any theory, any definition. We have too much noise coming in; if we didn't filter, we would see nothing. But we have to be very careful of how we filter.

There are some myths that are more dangerous than others, so what I advocate is just being attentive to the myths and figuring out which ones we want to keep going with and where we want to make changes.

Liberation or Enslavement?

BY ELAINE GRAHAM

Elaine Graham, Ph.D., *is Grosvenor Professor of Practical Theology at the University of Chester in the United Kingdom. This essay is excerpted from* Representations of the Post/Human: Monsters, Aliens, and Others, *Manchester University Press (2009). This piece first appeared in* GeneWatch, *volume 22, number 6, November-December 2009.*

"... the question can never be first of all 'what are we doing with our technology?' but it must be 'what are we becoming with our technology?'"

- Philip Hefner, Technology and Human Becoming[1]

At eleven o'clock on the morning of June 21, 1948, in a workshop in a tiny side street on the campus of the University of Manchester, a team of mathematicians and engineers conducted the world's first successful electronic stored-program computing sequence on a machine they named 'Baby'. Since then, computer-mediated communications have transformed the abilities of their users to store and process information—and, more widely, they have changed leisure habits, the jobs many people do, and the types of machines that fill offices, shops and homes. Together with the identification, half a decade later, of the structure of DNA, the birth of 'Baby' has triggered a technological and cultural revolution.

I want to focus here on the impact of these technologies, the digital and genetic, respectively: not just upon our material and economic existence but upon our very experiences and understandings of what it means to be human.

The "posthuman condition": being and becoming

Reference to the so-called 'posthuman condition'[2] is becoming commonplace to denote a world in which humans are mixtures of machine and organism, where nature has been modified by technology, and technology has become assimilated to form a functioning component of organic bodies. But the impact of digital and genetic technologies is not simply a scientific or ethical issue; it carries a number of deeper, existential implications.

Firstly, the digital and biotechnological age challenges our assumptions about what it means to be human because new technologies are transforming our experiences of things like embodiment, communication, intelligence and disease, even blurring the very distinctions between the 'organic' and the 'technological', between humans, machines and nature. For some, this is not simply a matter of coming to terms with the economic and cultural impact of new technologies, but, as the opening quotation suggests, it also challenges the very nature of human ontology—our being and becoming.

Secondly, it raises questions of the kinds of images, discourses, authorities and narratives that will be decisive in helping us to debate the very question of what it will mean to be human in the twenty-first century. I am particularly concerned with the relative influence wielded by scientific institutions and discourses, by the media, by popular culture and by religion in advancing and adjudicating this question. It's intriguing to note how much debate about the impact of new technologies on Western culture involves a further blurring of supposedly impermeable boundaries: between scientific objectivity (the world of fact) and that of cultural representation (the world of fiction).

As Katharine Hayles argues, "culture circulates through science no less than science circulates through culture"—in itself, perhaps, a postmodern acknowledgement of the constructedness of scientific theories, a contention that science does not report on or portray an a priori reality but is dependent on wider cultural metaphors to weave a narrative about the world. This will often draw on images, analogies, other narratives, to achieve rhetorical effect. Thus, the Human Genome Project is depicted not only as a scientific exercise but as a heroic quest for the 'holy grail' or the 'code of codes'; Francis Bacon's discourse of science rests on an image of the male appropriator seizing nature and forcing her to relinquish her secrets; and witness the continuing potency of the Frankenstein myth in informing public reception of genetic technologies. All of these suggest a creative interplay between science and culture. But similarly, popular culture—science fiction especially—becomes both enthusiastic advocate and critical opponent of science and technology.

Different representations of the 'posthuman' in scientific literature and popular culture may not simply be neutral depictions, therefore, but in fact expressions of deep value judgments as to what is distinctive about humanity and what should be the relationship between humans and their tools and technologies. These questions are perhaps not only about viewing technology as 'savior' or 'servant', but about our own very existence: about the nature of life and death, the potential and limits of human creativity, about humanity's very place in the created order. And so, if the future is one of augmented, modified, virtual, postbiological humanity, what choices and opportunities are implicit in the hopes and anxieties engendered by the technologies that surround us? What understandings of exemplary and normative humanity will be privileged in the process—and who will be included or excluded from that vision? In what follows, I have sketched out four alternative responses to the dilemma of what it will mean to be human in a technological age.

Liberation or enslavement?

Creation out of control

In the mid-1990s, public concern over the effects of genetically-modified crops was accompanied—and fueled—by media references to 'Frankenstein Foods'. Even Monsanto, the multinational biotechnology corporation, saw fit to counter what it regarded as the negative publicity this occasioned.[3] Associations with mad scientists and monstrosity evoke (deliberately?) a reaction of horror and suspicion—of creation turning on its creator who dares to 'play God'. The alliterative quality of the term adds to its impact; but in comparison to more neutral terms like 'genetically modified crops' we are presented already with a powerful set of allusions and associations. Science and popular culture/myth are interwoven in the way these issues are represented.

Note also how theological themes are evoked, however incoherently: the notion of 'playing God' plays on fears that humanity has in some way usurped the creator, which is of course a long-standing theme in cultural engagement with technologies. The subtitle to Frankenstein was, after all, "The New Prometheus"; Mary Shelley evoking the Greek myth of the mortal who stole fire from the gods and who was punished as a consequence.

Dehumanization

Fritz Lang's *Metropolis*, made in 1926, articulates both the wonders of mechanization, but corresponding fears too. Set in the year 2000, the city from which the film takes its name is fatally divided. Whilst the sons and heirs of the factory magnates disport themselves in a rooftop garden of delights, the working populace is enclosed in a subterranean world of unremitting drudgery.

The opening scenes of the film depict 'The Day Shift' as a human collective robbed of individual will or personality. The lethargy of the workers, their movement as one, suggests a fear of the loss of individuality in the era of mass production and central planning. Interspersed with shots of the pumping of pistons and the spinning of cogs and wheels are pictures of a futuristic, decimal clock face, a juxtaposition that communicates how the relentlessness of time, the rhythms of machinery and the imperatives of productivity take precedence over human need.

So the wealth and technological sophistication of the city has its downside, and the workers in their underground labyrinth have become dehumanized by their routine, slaves to the relentless demands of the assembly line. Like the robot Maria, designed to suppress a workers' cult, the underground masses have become automata, driven only by the imperatives of machinery and efficiency. There is also a contrast in the film between the workers' factory workplace and their subterranean home, where Maria the preacher-prophet propagates values of the heart, affectivity and spirituality, in contrast to the dehumanizing forces of industry. This expresses an important theme, therefore, of technology effecting a kind of disenchantment; not only the erosion of something distinctively human, but the loss of some spiritual essence of human nature.

Technocracy

This is a third position, in which technologies are neutral instruments, merely means to an end. It is often allied to an unreconstructed futurism, in which technology solves our problems, grants us unlimited prosperity, guarantees democracy, fulfills our every desire, and is very pervasive in the media and other representations of popular science.

In his book *Visions*, the Japanese-American physicist Michio Kaku predicts a world of microprocessors that will be so cheap to manufacture that we will treat them like so much scrap paper in 2020; there will be 'smart' machines that think and anticipate our needs; by 2050 there will be intelligent robots and by 2100 self-conscious, sentient artificial intelligence. Technology is elevated to mythical status: "The Internet will eventually become a 'Magic Mirror' that appears in fairy tales, able to speak with the wisdom of the human race."[4]

Similarly, in describing his own (temporary) transformation from organic human to cyborg via the implantation of a silicon chip transmitter in his forearm, Kevin Warwick is expansive in his speculations about the potential of such technological enhancement. "Might it be possible for humans to have extra capabilities, particularly mental attributes, and become super humans or, as some regard it, post humans [sic]?"[5] His enthusiasm encapsulates perfectly the vision of those who see the promise of cybernetic technologies as going beyond mere clinical benefits to embrace nothing less than an ontological transformation. Warwick sees no limit to the transcendence of normal physical and cognitive limitations, an achievement that for him signals nothing less than a new phase in human evolution.

Evolution

For some, technologies promise the evolution of homo sapiens from organic to silicon-based life. This perspective is sometimes known as 'transhumanism'. Transhumanists argue that, augmented and perfected by the latest innovations in artificial intelligence, genetic modifications, nanotechnology, cryonics, the human race will be liberated from the chains of poverty, disease and ignorance, to ascend to a better, higher, more superior state: the 'posthuman' condition. With the aid of technological enhancements, human beings can guarantee themselves immortality and omnipotence (Regis, 1990; More, 1998). Machinic evolution will complete the process of natural selection.

The apotheosis of the transhumanist ethos is to be found in a group known as the 'Extropians', their name encapsulating their quest to defy entropy as expressed in human bodily deterioration such as disease and aging. The transhumanist spirit of technological and evolutionary inevitability expels defeatism and negativity, qualities that have no place in the Extropian world. One of their leading gurus, Max More, puts it this way: "No mysteries are sacrosanct, no limits unquestionable; the unknown will yield to the ingenious mind. The practice of progress challenges us to understand the universe, not to cower before mystery. It invites us to learn and grow and enjoy our lives ever more."[6] Yet this is not a human distinctiveness grounded in embodiment or even rational mind per se so much as a set of abstract qualities enshrined in a human 'spirit' of inventiveness and self-actualization: "It is not our human shape or the details of our current human biology that define what is valuable about us but rather our aspirations and ideals, our experiences and the kinds of lives we live. To a transhumanist, progress is when more people become more able to deliberately shape themselves, their lives and the ways they relate to others, in accordance with their own deepest values."[7]

While many of transhumanists' proposed technological developments are yet to be realized, it may be more appropriate to regard transhumanism, like all other posthuman thinking, as another kind of thought experiment, which, like fictional representations of technologized humanity, serve to illuminate and refract deeper hopes and fears. What makes transhumanism such a vivid example of posthuman thinking is the way in which it articulates a particular set of humanist ideals and transposes them into the technological sphere. Transhumanists deliberately harness the aspirations of Enlightenment humanism and individualism as a philosophical underpinning for their endeavors. In its endorsement of human self-actualization unconstrained by fear, tradition or superstition, transhumanism exhibits a secular skepticism toward theologically-grounded values, arguing that these serve to rationalize passivity and resignation in the face of human mortality and suffering.

As to technology, I think we can see how this tendency regards it not as threat but promise. Implants and prostheses, artificial intelligence, smart drugs, genetic therapies and other technological fixes will compensate for our physical limitations, overcoming even the existential challenge of mortality. Technology thus furnishes humanity with the means to complete the next phase of evolution, from homo sapiens to 'techno sapiens'.

Conclusion

So, new technologies promise to enhance lives, relieve suffering and extend capabilities, yet they are often also perceived as threatening bodily integrity, undermining feelings of uniqueness, evoking feelings of growing dependency and encroaching on privacy. But whether advanced technologies are to be regarded as essentially bringing enslavement or liberation will be shaped by implicit philosophical and theological convictions about what it means to be human, what the purpose of humanity is, what will contribute to human flourishing, what threatens human integrity and so on.

The various representations of the posthuman are a tribute to humanity's propensity for constructing new technological worlds but reveal also our tendency to invent other worlds of meaning and value and to invest these creations with diverse hopes, fears and aspirations. For embedded in the various representations implicit in new technologies are crucial issues of identity, community and spirituality: what it means to be human, who counts as being fully human, who gets excluded and included in definitions of the posthuman—and in our understandings of the nature of the God in whose image we have been formed. How we conceive of God will, even in a supposedly secular age, still impinge on the kinds of normative and exemplary models of divine nature and human destiny that fuel our technological dreams.

Exercising some control over the posthuman future will necessitate not just ethical debate, but, I contend, a theological orientation also. That's because, in thinking about the values embodied in these representations, we are effectively asking a theological question: in a digital and biotechnological age, what choices, destinies—and ideologies—has Western culture chosen to elevate as its objects of worship?

Musical Genes

INTERVIEW WITH GREG LUKIANOFF

Greg Lukianoff *is the founder and curator of the Genetic Music Project. This interview originally appeared in* GeneWatch, *volume 24, number 2, April-May 2011.*

The Genetic Music Project is a community art endeavor in which musicians convert genetic code into songs. Greg volunteered pieces of his own code, taken from test results from consumer genetic testing company 23andMe, and uploaded them to the website (www.geneticmusicproject.com) for musicians to use as a starting point. Although musicians are free to get more creative about the way they convert the genome into music, the first few songs on the site assign a note to each of the four nucleotides (A, C, T, and G); pick a FASTA sequence (which uses those four letters as a shorthand way to express a piece of genetic code) associated with a certain trait (such as likelihood of baldness or schizophrenia); and just see what series of notes emerges.

```
TCCCCGCAGA ACTCCTCTGT GCCCTCTCCT CACCAGACCT TGTTCCTCCC
AGTTGCTCCCACAGCCAGGG GGCAGTGAGG GCTGCTCTTC CCCCAGCCCC
ACTGAGGAAC CCAGGAAGGTGAACGAGAGA ATCAGTCCTG GTGGGGGCTG
GGGAGGGCCC CAGACATGAG ACCAGCTCCTCCCCCAGGGG ATGTTATCAG
TGGGTCCAGA GGGCAAAATA GGGAGCCTGG TGGAGGGAGGGGCAAAGGCC
TCGGGCTCTG AGCGGCCTTG GCCCTTCTCC ACCAACCCCT
GCCCTACACTMAGGGGGAGGC AGCGGGGGGC ACACAGGGTG
GGGGCGGGTG GGGGGCTGCT GGGTGAGCAGCACTCGCCTG CCTGGATTGA
AACCCAGAGA TGGAGGTGCT GGGAGGGGCT GTGAGAGCTCAGCCCTGTAA
CCAGGCCTTG CCGGAGCCAC TGATGCCTGG TCTTCTGTGC
CTTTACTCCAAACACCCCCC AGCCCAAGCC ACCCACTTGT TCTCAAGTCT
GAAGAAGCCC CTCACCCCTCTACTCCAGGC TGTGTTCAGG GCTTGGGGCT
GGTGGAGGGA GGGGCCTGAA ATTCCAGTGT
```

GeneWatch: The process starts when you get a genetic test—yours was from 23andMe—and when you get the results back, you select a chunk of the genetic code and assign a note to each of the four nucleotides. What does that first piece of code look like? What does it represent?

Greg Lukianoff: The genetic test is selecting the FASTA sequence for certain genetic markers (see top of image on previous page). What you're getting back is, to the best of my knowledge, the entire FASTA sequence for an allele. So I put that up there, and I indicate what conditions those alleles are associated with, which provides a lot of opportunities for metaphor and tone for the music that people submit.

For instance, when I found out that there was a particular genetic marker for bitter taste perception, I found the FASTA sequence for it, put that up, and sent it to my friend Amy, who is an unbelievably talented country artist. If someone was going to sing a song on the theme of "bitter," who better than a country singer?

I was really surprised that nobody had already made a community art project out of genetic sequencing. I knew that people had been making music out of genetic code—for example, the composer Alexandra Pajak wrote an album based on the sequence of the HIV virus—but given that DNA is part of this wonderful organic system that grows by its own nature, that seems to be perfect for a community art project, something that you just put out into the world and see how it does. Of course, there's an element of *un*natural selection to it—I'm curating the website—but at the same time, I don't know where it's going to end up.

GeneWatch: You provided the code; who writes the music?

Lukianoff: When the site first went up, pretty much all of the music came from friends of mine who were excited about the idea. Having someone in Denmark upload a song was something else. My big hope is that it continues to grow, and that people recognize that you can go well beyond taking the four individual "notes" from the nucleotides (A, C, T and G).

Using the computational brilliance of genetics, you can create more and more complex art. The different nucleotides could stand for pitch, or you could even take a bunch of nucleotide sequences and relate that to the 64 different settings of a Casio keyboard and let the code tell you which instruments to use. So there is a tremendous amount of fun to be had with this, and hopefully it will grow organically.

I'm particularly interested in getting different genres of music, too. I'd really love to get a rock song, for example. I think when people think of 'genetic music,' they would first think of some sort of atonal electronica, but as we've already shown, it can work out in so many musically interesting ways.

GeneWatch: Is there any genre that you think wouldn't work for this project?

Lukianoff: I think you could figure out a way to make any genre work. The only genre I'm not interested in hearing is smooth jazz. Because smooth jazz is terrible.

But with *real* jazz, I actually think that the surprising twists and turns of the FASTA sequences would be perfect for jazz saxophone.

One of the things about the musicality of the genetic code is that it does surprising things. For example, the piece on heroin addiction ["Heroin Addiction" by Liz Wade] is just based on the first ten nucleotides of the FASTA sequence for a gene that is supposed to indicate that you may be more easily addicted to heroin. When she played that, just the first ten nucleotides, it was really haunting. Then it can start to get interesting, where you can stumble into themes. It sounds very traditional, but then it will run maybe one or two notes longer than you would expect. You end up hearing some surprising musical things come out of it.

GeneWatch: Are you coming at this more from the science or music side?

Lukianoff: I wouldn't pick one or the other. I'm sort of a science hobbyist—it's the degree that I never got but always wanted to. So I sort of came at it from the science side rather than from the music side. The inherent musicality of the information behind all life just seemed irresistible.

GeneWatch: Have you learned more about music or genetics?

Lukianoff: I think I've learned a lot about both. There were certainly things I didn't know about genetics, and there were plenty of things I didn't know about music. That's one of the things I'm trying to be open about, that I'm neither a musician nor a scientist!

The difference between this website and others, really, is that we decided to just put it out there and see what happens with it, to encourage people to be creative and apply their own approach—but also to teach everyone, including me, about the science and the music behind it.

GeneWatch: One of the FASTA sequences you posted is supposed to be the genetic marker for "longevity." Are you taking 23andMe's test results with a grain of salt?

Lukianoff: One of the things 23andMe does is indicate the amount of research behind the traits. For some there has been a lot of research and for some there have been only tiny studies. They're good at explaining it, and this is something I suspect I'll eventually be asked; but I try to make clear that I'm not vouching for the accuracy of the information, I just think it's a great starting point.

Back on "The Farm"

BY ROB DESALLE

Rob DeSalle, Ph.D., *is a curator in the American Museum of Natural History's Division of Invertebrate Zoology and co-director of its molecular laboratories and a member of CRG's Board of Directors. This article originally appeared in* GeneWatch, *volume 24, number 2, April-May 2011.*

About a decade ago, I had the great pleasure to spend some time in the studio of a well-known New York City artist who was interested in the then-burgeoning and often over-publicized science of genetic modification of animals and plants.

This inquisitive artist was Alexis Rockman, a painter with a reputation among his colleagues for paintings "depicting nature and its intersections with humanity," and "painstakingly executed paintings and watercolors of the phenomena of natural history."

The interaction was as timely as it was interesting. Alexis knew little about the techniques of genetic modification or genomics, but was—and continues to be—a superb natural history artist. Meanwhile, I had just begun to curate an exhibition on the human genome and genetic technology at the American Museum of Natural History entitled, "Genomic Revolution." Thus began our relationship: an artist and a scientist talking every Thursday afternoon about genetic technology over coffee in his studio.

Some of the topics that Alexis wanted to discuss seemed pretty bizarre. However peculiar the topic, I would first try to explain the technology to Alexis and then would add an extra layer of science once we felt comfortable with the primer and its jargon. Throughout this process, it became immediately obvious to me that, as he was soaking up the material, Alexis was worried that some of the genetically modified versions of animals would impact our natural world.

After about two months of my genomics "tutorial," Alexis dismissed me and began work on a piece of art he was later to call "The Farm." I left the last meeting with some apprehension about the art that might come from our conversations. While I felt that Alexis possessed a firm general understanding of genetic technology, I was—and continue to be—wary of how an artist or an author might take creative license with science.

While walking in the SoHo neighborhood of New York City one fall day in 2000, I looked up at a huge billboard at the intersection of Lafayette and Houston Streets. The billboard stunned me. Erected by an organization called DNAid, it featured Alexis' "The Farm" in all of its glory. Since I had only seen sketches of some of Alexis' ideas, I was blown away by the immensity of the piece, by its vividness and candor. Through his strong understanding of natural history, Alexis strove to create an awareness about the existence of plant and animal ancestral forms among his audience. More specifically, Alexis wanted his audience to understand that all living organisms—not just plants and animals—have ancestral forms. To facilitate this understanding, Alexis painted certain domestic animals while including the "ancestral" versions of them. Hence, we see chickens, swine, cattle, wild mice, and domestic crops in the background of the painting. It is purposefully ironic that each of these domestic forms has its wild form that existed probably at most 20,000 years ago, when domestication began.

In the foreground, a slew of genetically modified organisms lurk near a barb wire fence. All of these peculiar creatures came from Alexis' thinking about the extent and limits of genetic modification. The painting includes an interesting menagerie with plants, like tomatoes, grown in cube-like shapes; a mouse with an ear growing off its back; a rather porcine pig with human organs growing inside it; and a large cow I can only describe as "Schwarzenеg-gerish." As bizarre as the painting's modified organisms look, they were, as Alexis suggests in the description that accompanied the piece, informed by reality.

I thought it might be interesting to look at these four modified organisms a decade later to see how well-informed the artist was in drawing them and what their status is now. Let's start with the geometrically bizarre domestic plants. I recall from our conversations that Alexis was already aware of "Flavr Savr," one company's attempt to genetically modify tomatoes to maintain freshness, but he was particularly taken aback by the possibility of genetically modifying things to change their shape. The square tomatoes in "The Farm" would be much more easily and efficiently packaged. Alas, Flavr Savr went bust around the time Alexis produced the piece, and to my knowledge no genetically modified cube tomato has been produced.

Perhaps the most peculiar animal in the piece is the mouse with a human ear attached to it. I recall that Alexis and I had discussed the potential of using non-human animals as culturing media for human organs. In 2000, this idea was very prevalent in the news, so I didn't label this fascination as "bizarre"; rather, I thought that his questions about the topic were timely and warranted. In fact, the pig with the human organs growing inside it was also a popular news story at that time.

The "earmouse" (also known as the Vacanti mouse) actually did not have a human ear growing out of it. The "ear" consisted of a gob of cow cartilage grown in the shape of an ear. It was produced not by genetic engineering but by inserting a polyester fabric that had been soaked with cow cartilage cells under the skin of an immune-compromised mouse. Pigs as donors of human organs, meanwhile, are something we may see in our lifetime. Some pig organs, including their hearts, are about the same size and have the same general plumbing as human organs, and scientists have suggested that pigs might be a good source of organs for human transplants. Like any transplantation, though, tissue rejection is an important consideration, and some genetic modification of the pigs to overcome rejection would be needed. While this might seem far-fetched—especially when Alexis produced "The Farm," over ten years ago—the possibility has been resurrected as a result of work in 2009 in China producing pig stem cell cultures. Such cultures can be used as an easier method to genetically modify pigs to circumvent the rejection problem. A year later, Australian scientists genetically modified a line of pigs by removing a stretch of a single chromosome in order to alleviate the rejection problem and allow experiments with lung transplants to proceed.

The last animal in Alexis' menagerie is the Incredible Hulk Cow. "Double-muscled" cows do exist, such as the Belgian Blue and the Piedmontese. These breeds have a defective myostatin gene that would otherwise, when expressed normally, slow muscle cell growth in a developing cow. However, these breeds came about through conventional breeding, not genetic modification. To my knowledge, super-muscular cows have not been successfully engineered, although researchers are working on several other bovine genetic engineering tricks, including cows with altered genes to improve the conversion of milk to cheese or engineered for resistance to mad cow disease.

While only one of the bioengineered animals in Alexis's menagerie still exists today—pigs being developed as organ donors—ten years later, Alexis's "The Farm" still makes a compelling statement about technology and nature. We are in the midst of our second generation of genetic modification (the first being the initial domestication of plants and animals), and we have not yet figured out how to proceed. Alexis draws attention to the fundamental novelty of genetic engineering in depicting the progression of domesticated animals as beginning with already-domesticated forms, subtly omitting their wild ancestors.

All of this reminds me of a conversation I had with an uncle who is a farmer in upstate New York. Over a beer in the shade of a barn, we were talking about how farming has changed since he was a young man. During the conversation, I excitedly tried to explain to him the possibility of using plants with a genetically engineered gene involved in stress tolerance. This gene would be linked to a luminescent beacon, so that when the plants were under drought stress, the beacon would glow, telling the farmer to water them. My uncle looked at me incredulously and said, "Now how good of a farmer would I be if I couldn't tell my crops needed water?"

Deflated Expectations

BY EMILY SENAY

Dr. Emily Senay, M.D., MPH, *is currently the medical correspondent for* PBS Need to Know. *She is also an Assistant Professor of Preventive Medicine and a course instructor in the Masters of Public Health Program at Mount Sinai School of Medicine. Prior to joining PBS, she was a medical correspondent for* CBS News *for fifteen years. This article first appeared in* GeneWatch, *volume 25, number 1, January-February 2012.*

The ten-year anniversary of the completion of the Human Genome Project reminded us that genetic and genomic research has yet to fulfill the promise of cures for devastating diseases like Alzheimer's, Parkinson's and cancer. Those promises were wildly overblown but nevertheless made it into the zeitgeist, leaving many discouraged. If the past is the best predictor of the future, what does this mean for genetic technology twenty years on? Will the big breakthroughs come? Will genetic technologies emerge that surprise us all? To help in making somewhat accurate predictions about the future, I thought it might be wise to consult the Ouija board of new technologies: Gartner's Hype Cycle Graph.

Developed to help investors understand how technology matures, Gartner's Hype Cycle begins with the emergence of an important new technology that captures popular imagination and promises to change everything. The new technology is then over-hyped by scientists, journalists and investors, creating the first phase: the peak of inflated expectations. When those expectations are not quickly met, disappointment sets in and the new technology is declared a bust—the trough-of-disillusionment phase. Then, while nobody is paying attention anymore, the new technology begins to yield innovative and available products, most of them unanticipated in the initial hype phase. Everybody is surprised but pretty soon forgets their earlier skepticism and readily adopts the new products. The new thing becomes old hat and the technology enters the phase of the hype cycle known as the slope of enlightenment.

So where is genetic technology on the hype cycle graph? Plotting events since the completion of the Human Genome Map, it is clear we are currently in the trough of disillusionment. Bummer, I know. But if you trust the Hype Cycle Graph, this could actually be good news. If we are in the trough then there is nowhere to go but up. It is during this phase that the few stalwart innovators toil away in obscurity creating early versions of what will eventually be the next big thing that really will deliver. So before too long we could be experiencing an explosion of innovation in genetics and genomics.

But will this explosion include a cure for big problems like cancer? Two great minds think so: Jim Watson, co-discoverer of DNA, and Albert Brooks, comedian and author. Writing in *Cancer Discovery* in November 2011, Watson prophesies that with hard work scientists will be able to use RNAi (an RNA molecule that inhibits gene expression is called RNA Interference or RNAi) to selectively block cancer genes, leading to a cure for many cancers within the next five to ten years. Brooks also predicts a cure in his new novel, *2030*—all cancers, all comers, 100 percent cured. Unfortunately the resulting dystopia is not so appealing. Old people don't die, debt balloons, and a generational war breaks out when the "olds" suck up all the resources. Personally, I don't think either scenario is likely. For starters, this isn't the first time Watson has predicted a cancer cure, and Brooks is famous for his overly negative outlook. Secondly, a cure for cancer would be the most obvious thing to predict, and according to the hype cycle it's usually stuff nobody sees coming that emerges first and takes off. So I expect that curing cancer, Parkinson's or Alzheimer's is going to be tougher, more incremental, and slower than we all want.

So what will emerge? Going out on a limb, I predict "social genetics" will take off much sooner and in a bigger way than currently anticipated and will be driven not by scientists but average folks. I base this prediction on nothing more than recent random conversations with two friends: both educated and savvy. For no particular reason each had their genes mapped by one of those new personal genetic companies. They weren't worried about Parkinson's or breast cancer. In fact, they couldn't really articulate why they spent a couple hundred bucks and sent their spit to California. Maybe they're weird or just early adopters. No matter, though they learned nothing of real medical utility, they couldn't have been more thrilled to share with me their risk of restless leg syndrome, excessive earwax, abdominal aortic aneurysm, etc. It all seemed like TMI, even a little creepy; at best, a novelty. Then it hit me: that's just what I thought when I first heard about a lot of things, like the Internet, email, Google, and Facebook. The genetic genie is out of the bottle! Everybody is going to want their codes cracked—and they are then going to want to share that info with you. Imagine a website called GeneticConnections.com. Upload your code, find long-lost relatives, make common variant friends, or find your perfect genetic mate! Nah. Forget it. It'll never work. I'm sure I'm wrong and Jim Watson is right. I really hope so . . . even if Albert Brooks is right, too!

Culturing Life: From the Laboratory to the Studio

BY SUZANNE ANKER

Suzanne Anker, MFA, *is a visual artist and theorist working at the intersection of art and the biological sciences. She is Chair of the BFA Fine Arts Department, School of Visual Arts.*

Set in accelerating motion by the decoding of the human genome in 2000, many art exhibitions, residencies, and "sci-art" collaborations are making their way onto the international stage. From artists in the lab programs in Switzerland[1] to a Master's Degree in the Biological Arts in Australia[2] to Bio-Labs being constructed in art and design departments at universities and colleges, such intersections continue to attract a growing number of participants. Are artists-in-laboratories doing science or are they employing the tools of science to develop other projects? Although there is no consensus on this subject, artists are developing novel methods that are garnering results, both aesthetically and scientifically. This is not an either/or situation, but rather a Janus-faced example that dismisses conventional discussions about art and science. Expanding on global issues like environmental toxicity, industrialized farming, and population genetics, a techno-scientific undertow is challenging our perceptions, beliefs and desires concerning altering life.

The nexus between the visual arts and the biological sciences is consistent with current discourses about art as a form of research and knowledge production. Brian Degger, from transitlab.org, defines "speculative research" as "a platform for researching and developing interactive, social and biological artworks." In Linda Candy's *Guide to Practice-Based Research,* she cites the "creative artifact as the basis of a contribution to new knowledge."[3]

Employing concepts, processes and materials intrinsic to scientific laboratories, visual artists are producing hybrid works that address the narratives of life and its transformation. Although the term "bio-art" is a contested one as to when and how it came to be, there is sufficient consensus as to what it encompasses. A critical analysis can be made of this practice. Although many of the ideas circumscribed in bio-art reach back to the origins of art itself, such as the chimerical figures in Paleolithic art, the infiltration of novel imaging practices, laboratory methods and software apparatus have expanded the potentialities inherent in aesthetic practice. This brief chapter is an update on the intersections of art and biology that the late Dorothy Nelkin and I formulated almost a decade ago, in our seminal text, "The Molecular Gaze: Art in the Genetic Age."

The bio-art practices currently of interest are not exhaustive in this article but represent the next phase of attitudes and concerns within this practice. Briefly stated, the categorical subjects implicated here are synthetic biology, food science, DIY biology, speculative research and a proliferation of projects that employ microorganisms.

Do-it-yourself biology is a fast-paced practice spurred on by social media. It is a practice in which "hackers" are re-purposing scientific apparatus, exchanging information, and offering "wetlab" workshops outside of traditional academic or institutional settings. Writing in *Wired* Magazine in 2005, Rob Carlson stated, "the era of garage biology is upon us." Log on to diybio.org and see for yourself an international camaraderie erupting. As their website instructs, you are invited to join in on global discussions about techniques and protocols, check on safety regulations and find workshops and other interested parties in your geographical area.

In Brooklyn, New York, one such space, Genspace,[4] is a community-based bio-lab open to the public. Hosting workshops and events, the general public or amateur biologists can learn about microbiological techniques through hands-on practice. Genspace, founded by Ellen Jorgensen and Oliver Medvedik, offers courses and membership to artists, amateurs, students and the general public ranging from synthetic biology to painting with bacteria to workshops in PCR amplification. Both Ellen and Oliver are card-carrying biologists attaining their Ph.D.s in biology at NYU and Harvard, respectively. Other members of the group include artists, computer programmers, journalists and scientists.

Synthetic biology, a discipline involving the application of engineering principles to the living cell has become an interest for many artists. Recently, an exhibition entitled, *Synthethic: Art and Synthetic Biology* took place in 2012 in Vienna.[5] Oron Catts, Paul Vanouse and Joe Davis were some of the artists participating. Paul Vanouse's *Latent Figure Protocol (LFP)* (2011) *(fig 1)* employs electrophoresis as his working methodology and signature icon. Vanouse questions the truth value in "genetic fingerprinting." He points to the facts that banding patterns produced during electrophoresis are influenced by primers and probes affecting the DNA. He creates images from this scientific technique to create cultural symbols like infinity, a zero and one, and a skull and crossbones. Can DNA be used to falsify results, questions the artist?

Joe Davis, a pioneer in bio-art, produced a bacterial radio, fusing utility with biology. His work has been described thus: "*Bacterial Radio* exhibits several bacterially-grown platinum/germanium electrical circuits (crystal radios) on glass substrates. Joe Davis, in collaboration with Ido Bachelet and Tara Gianoulis from Harvard Medical School in Boston, used bacteria altered with variants of a gene from orange marine puffball sponges (*Tethya aurantia*) to plate electronic circuits on Petri dishes and microscope slides. This gene codes for a protein—silicatein—that normally forms *Tethya aurantia*'s glass skeleton, its tiny, glass, needle-like spicules composed of silicon and oxygen. Variants of this gene have now been optimized to plate metallic conductors and semiconductors including germanium, titanium dioxide, platinum and other materials. Here, genetically modified bacteria are embedded in non-conductive materials containing metal salts and then optically induced to plate-specific electrically conductive circuits. These *Bacterial Radios* on display are connected to high impedance telephone headsets, antennae and ground, so that visitors may use them to actually listen to AM radio broadcasts."[6]

Food Science has also become a popular idiom. Although GMO's [genetically modified organisms] have been a source for critique among artists, molecular cuisine, urban farming, vertical gardens *et al.* are now being added to the mix. Roger Buergel, Director of Documenta XII in Kassel, included the molecular cuisine of Ferran Adrià from his reknowned restaurant, *El Bulli,* as an artwork. Employing the ways in which the physical properties of food could be manipulated by additives like agar-agar, sodium alginate and calcium lactate, as well as the use of liquid nitrogen and vacuum sealing, the manipulation of taste and its experience results in a transformed flavor. Another restaurant of particular note is the *MOTE* in Chicago. Taking into account ecological blight, food shortages in third world countries as well as obesity issues in the US, chefs Homaro Cantu and Ben Roche and their staff have come up with novel applications of food preparations far exceeding molecular gastronomy as an e-list enterprise. In their short video on TED,[7] they show how hamburgers, for example, can be made out of corn, oats and the like, eliminating the cow itself. As a form of bio-mimicry, this type of cuisine could be employed by third world nations as well, turning weeds and leaves into food. Lisa Ma, in her "speculative research design"[8] project has worked with joy-stick factory migrant workers to help them gain alternative methods of income when first world countries no longer desire outdated technology. In *Farmification*, she shows how unused land in China can be employed as a resource for these farmers to cultivate their own food when there is no work at the factory.

For Georg Tremmel and Shiho Fukuhara, "bio-hacking" or reverse engineering is a project that explores nature as cultural property. Employing a genetically modified carnation as their subject, they were able to turn a genetically altered purple flower to its original color of white. In addition to "bio-hacking" organisms, microorganisms are also being employed by artists. Philip Ross, for example, is able to construct significantly dense structures out of fabricated fungal bricks. Employing *Ganoderma lucidum* as his organism, the artist envisions fungal material as a possible architectural resource for the future. Although not altering the genetic sequences of this microorganism, the artist is able to alter its physical characteristics and repurpose its function.[9]

Mark Dion and Rick Pell both create installations that address the ways in which genetics is a marker for taxongenomic categories. Mark Dion's installation at the Museum of Natural History in London commemorated the 300th anniversary of Linnaeus' system of the organization of living matter. Employing unorthodox methods of specimen collection, Dion tied fly paper to the roof of his car while assistants held insect nets out of the car windows. Once collected, these specimens were then analyzed genetically only to find that several species, heretofore unknown in London, were discovered. Rich Pell, on the other hand, set up the *Center for Postnatural History* in 2010 as an archive of organisms that have been altered; his collection points to the ways in which we are compelled to reconsider taxonomies in the twenty-first century. Both artists use traditional museum display techniques in their work, to further explore the ways and means knowledge is situated in public spaces.

With the rise of epigenetics, research into "junk" DNA and proteomics, genetic determinism has taken a back seat. Whereas in the 1990s attention was being paid to the internal genetic script of organisms, in this post-genomic era the nature/nurture debate is including more environmental factors to assess changes in intrinsic genomes. From the desiccation of wild fish by genetically altered ones who are considerably larger to studies of twins whose DNA has changed significantly after birth to Craig Venture's artificially constructed organism, we continue to see how the adaptation of ready-made resources is being reshuffled in novel ways. As the genetic sciences continue to unravel in detail the differences in organisms, artists, too, have remained, and prospered, in the dialogue about altering life.

Biology in Science Fiction

BY MARK C. GLASSY

Mark C. Glassy, Ph.D., *is the Chief Executive Officer of Nascent Biologics, Inc. and is on the faculty of the University of California, San Diego in the Mechanical and Aerospace Engineering Department and Bioscience Department. He is the author of* The Biology of Science Fiction Cinema *and* Movie Monsters in Scale.

The most popular art form is the cinema. More people see films than all other art forms, like paintings and sculpture, concerts, plays, etc. Much of society derives its current pop culture concepts from the dialogue, costumes, and social behavior found in films. Science fiction films in particular have elements that hinge on real-life science. Whether the science depicted in a film is correct or not is dependent on many factors. The accuracy of the science will always take a back seat to the action, plot, and budget constraints. Filmmakers oftentimes make up their own scientific facts to suit their particular cinematic needs. In this respect, much of the science in science fiction films is based on popular pseudoscience. As such, more often than not real science gets lost in the translation, and most filmgoers will not pick up the inaccuracies; science fiction films are not necessarily meant to be educational. If popular science influences popular culture, then it follows that popular culture also influences popular science.

All films mirror and reflect something of the society in which they were made, whether they are contemporary, period pieces, or futuristic. In science fiction films the science—or more specifically the biology—also mirrors contemporary thinking and, more importantly, the knowledge and understanding available at the time. For example, genetic engineering was not explored in films until the 1950s since the concept was not yet understood. And, of course, films produced today contain contemporary science that in fifty years will also seem primitive. The science portrayed in films can help us understand how popular culture affects society's science literacy.

Presentation of science in popular culture

Many science fiction films of the 1930s and 1940s were set in dark laboratories with bubbling glass containers and mad scientists conducting glandular and hormonal experiments. These laboratories were often hidden in the scientists' homes, further imprinting the idea that they were doing secretive things "men were meant to leave alone." Glands were a popular topic at this time both in the real research world (contemporary real life scientist Dr. Serge Voronoff claimed that by surgically grafting the glands of animals into other animals, including humans, he could correct some birth defects and help prolong life) and in the layman's world. Film examples of this are *Murders in the Rue Morgue, The Monster Maker,* and *The Ape Man*. In the 1950s—the Atomic Age—films were more focused on radiation-induced mutations where many of the monsters were caused by atomic radiation, as seen in *The Amazing Colossal Man and Them!*. Interestingly, most of these 1950s radiation-induced monsters were accidents that happened when the scientist was trying to do something else. (This is where such proverbs as "the path to hell is paved with good intentions" apply.)

As viewers became more sophisticated in their general understanding of contemporary science, the screenwriters and producers of science fiction films also kept pace and incorporated contemporary themes into their films. Science fiction films made during the 1960s and 1970s focused on such relatively new concepts as immunology (*The Brain that Wouldn't Die*), cryobiology (*Frozen Alive*), and biochemistry (*Island of Terror*). In the 1980s and 1990s, the main biological driver of science fiction films was DNA, the genetic material of life. Perhaps this period can be considered the "DNA age of science fiction cinema," and film examples include *Carnosaur*, *Jurassic Park*, and *Species*. Since then the idiom of DNA has permeated the common vernacular, so much so that it is now used to describe just about anything immutable, as in TV commercials like "SONY TV is now in our DNA." All in all, it appears that the biology in science fiction films has more or less kept up with current scientific knowledge.

Many of the standard plots of earlier science fiction films have become today's science. Glands are now understood as hormones, and atomic radiation, although something to be taken seriously, has not been the mutation-generating machine it was once thought to be. Most of the scientific silliness in these early films is just that. Even so, such silly biology can also serve as a teaching tool when we take the time to understand the implausibility of the on-screen scenarios.

Lab sets in science fiction films

While most of the lab sets of the early sci-fi films varied little (bubbling liquids in glass flasks and test tubes, Bunsen burners, and microscopes), it wasn't really until the 1970s that filmmakers aimed for more accuracy. Suddenly, science fiction film lab sets included the type of equipment working scientists actually used: centrifuge tubes, flasks, beakers, pipettes and, most significantly, syringes—all made out of plastic. Glassware could break (and was expensive to replace) and could become a liability should someone get cut or hurt, so when plastic-ware became generally available in real labs it was quickly adopted by the film industry.

Since then, lab sets have become more impressive, and many of the bench items appear to be at least functional and necessary for the work at hand. So, over the decades the lab sets of science fiction films have also become more sophisticated with more functional hardware, including computer-oriented pieces of equipment.

Influence of science fiction films on teaching biology

A question I often ask students is when was the first time they saw a human brain or a human skeleton? Just about all answer, in a cartoon, on TV, or in a movie. No one ever says the classroom. This is a simple illustration that popular media and culture are major driving forces of human knowledge and learning, all outside the traditional format of school education. Though educators might disagree, it is tempting to say that school learning supplements what is learned by life's experiences, and watching science fiction films is an important element of life's experiences. And biology lessons, accurate or not, that were picked up from viewing science fiction cinema do impact people's overall general science knowledge.

I often refer to today's generation as "the Jurassic Generation," meaning those who learn their "science" of biology from film or other elements of popular culture and not necessarily the classroom. After seeing the 1993 film, *Jurassic Park*, many thought—incorrectly—that the reanimation of dinosaurs from DNA isolated from amber-preserved insects that had bitten and fed on dinosaurs could be done. This shows that strong scientific ideas, rational or not, that appear in popular cinema take root in public consciousness quickly. Such "teachable moments" offer opportunities and are useful in highlighting the inaccuracies of science in these films by demonstrating what is wrong with the science and what it would take to actually make the science work.

Since its inception, science fiction has had a dramatic effect on how science is perceived by the general population. The world of science fiction makes everything simplistic, such as the seemingly endless array of genetic mutants that appear to be made overnight or the ability to transplant brains into any species. What are missing are the countless hours of scientists toiling away in their laboratories, where failures peppered with a few successes are more the norm. In this respect, the life of a cinema scientist has been glamorized into something more romantic than the day-to-day toil, drive, and passion required to do good and interesting science.

In summary, the world of science fiction attempts to make the impossible possible. Throughout the entire history of science fiction cinema, the biology in these films—possible or impossible—has influenced the understanding of science held by non-scientists and, conversely, contemporary popular science has influenced the biology in science fiction films. As advances in contemporary biology permeate society more and more, sophisticated themes will continue to find their way into science fiction cinema to keep pace with the general public's level of understanding. And as this happens, society's overall science IQ will also increase.

ENDNOTES

Part I

Chapter 1

1. Secretary's Advisory Committee on Genetics, Health and Society. Policy Issues Associated with Undertaking a Large U.S. Population Cohort Project on Genes, Environment, and Disease, Public Comment Draft, May 2006.
2. The Washington University v. Catalona, 437 F.Supp.2d 985, 993 (E.D. Mo. 2006).
3. Annas, George. "Privacy Rules for DNA Databanks: Protecting Coded 'Future Diaries.'" *JAMA* 270, no.19 (1993): 2346-50.
4. Supra 2, 437 F.Supp.2d 985 (2006)
5. Ibid at 1000.

Chapter 2

1. This position paper is intended to accompany the Council for Responsible Genetics' Position Paper on the Human Genome Initiative, which describes the Initiative in greater detail, evaluates its goals and methods, and its implications for expanding the number and range of predictive genetic tests.
2. A survey of corporate views about AIDS was published in the January 1988 issue of *Fortune* magazine. It revealed that 39 percent of the Chief Executive Officers surveyed would not hire individuals who were HIV-positive, while 38 percent were not sure whether they would hire such individuals. Reported in Mark A. Rothstein, *Medical Screening and the Employee Health Cost Crises,* (BNA Books 1989), 86. Job discrimination against recovered cancer patients is documented in Feldman, "Wellness and Work," in *Psychosocial Stress and Cancer,* ed. C. Cooper (1984), 173-200.
3. A brief history of employment discrimination on the basis of genetic traits is presented in Ruth Hubbard and Mary Sue Henifin, "Genetic Screening of Prospective Parents and of Workers: Some Scientific and Social Issues" in *Biomedical Ethics Review,* eds. James Humber and Robert T. Almeder (Humana Press, 1981), 99-111.

4. Individuals who have one sickle cell gene (a condition called sickle cell trait) are free of symptoms and do not know that they have the gene unless they have been tested for it. However, those who have two sickle cell genes have sickle cell anemia and may experience severe symptoms. Approximately one in five hundred African American babies is born with sickle cell anemia, and about one in ten carries the sickle cell gene. Although no scientific evidence exists to show that African Americans with sickle cell trait experience increased morbidity or mortality, their identification through screening programs in the 1970s led to job and insurance discrimination against them.

   For example, Charles Reinhart, Director of the DuPont Laboratory for Toxicology, reported in 1978 that DuPont gave pre-employment blood tests to African Americans to screen for sickle cell trait. He stated that at the Chambers Works plant, in Deepwater, New Jersey, individuals with sickle cell trait who had hemoglobin levels of less than 14 grams per 100 milliliters of blood (normal levels are usually given as 13 to 16 g per 100 mL) were restricted from work that involved handling nitro and amino compounds. Reinhart, Charles. "Chemical Hypersusceptibility." *Journal of Occupational Medicine* 20, no.5 (1978): 319-22.
5. Specifically, the Rehabilitation Act of 1973, as amended, protects individuals with disabilities who are otherwise qualified, from employment discrimination at the hands of the federal government, federal contractors and businesses receiving federal funds. 29 U.S.C. Secs. 701-795 (Supp. 1989).
6. The Americans with Disabilities Act, (Senate Bill No. 933, House Bill No. 2273) was passed by the Senate on September 7, 1989, but failed to make it through the House. The Act would bar discrimination on the basis of disability in employment, public services and public accommodations. The Act was expected to pass in Congress in 1990. One of the provisions that was targeted for deletion, but escaped amendment during 1989, would prohibit employers from using pre-employment physicals or inquiring about an applicant's disability status.
7. For example, in School Board of Nassau Co. v. Arline. 480 U.S. 273 (1987), the Supreme Court declined to decide whether the Rehabilitation Act would protect a person from employment discrimination who has no symptoms but whose medical tests indicate that a disease might develop in the future. 480 U.S. 282, n.7. In Arline, an elementary school teacher was

fired after a relapse of tuberculosis. The Court ruled that the Rehabilitation Act covers those who are able to work but "are regarded as impaired and who, as a result, are substantially limited in a major life activity."

8. In the early 1900s, women had trouble getting insurance due to misconceptions about increased female mortality due to childbirth hazards. See Note, Challenges to Sex-Based Mortality Tables in Insurance and Pensions, 6 *Women's Rights Law Reporter,* 59 (1979-1980) and Heen, Sex Discrimination in Pensions and Retirement Annuity Plans After Arizona Governing Committee v. Norris. 8 *Women's Rights Law Reporter,* 155 (1985):161. African Americans also experienced insurance discrimination. See M. James, *The Metropolitan Life: A Study in Business Growth* (1944), 338; G. Myrdal, *An American Dilemma: The Negro Problem and Modern Democracy* (1944), 316-317, 955, 1262-63, and M.S. Stuart, *An Economic Detour: A History of Insurance in the Lives of American Negroes* (1940).
9. The utility of such laws is explored by Neil A. Holtzman in *Proceed with Caution: Predicting Genetic Risks in the Recombinant DNA Era,* (Johns Hopkins University Press, 1989), 199-200.
10. The discussion of this survey has been cut since Paul Billings describes it on page 7 in this issue *of GeneWatch* (1988); 6:6. —ED.
11. According to a recent news report, the FBI is laying the groundwork for a computer information network that would contain genetic information on all violent offenders who have been incarcerated. The network would permit prosecutors to search DNA databanks to match evidence from rapes or murders against a list of DNA taken from convicted offenders. At least four states, California, Colorado, Nevada and Virginia, have drafted laws that would require that blood be taken from prisoners convicted of violent crimes so that their genetic profiles can be entered into such a DNA databank. Rorie Sherman, "On the Horizon: A DNA Data Bank," *National Law Journal,* Dec. 18, 1989, p.25. Arkansas, Georgia, Louisiana, Maryland, Arizona, Florida, Michigan and Massachusetts also enacted or introduced legislation during 1989 concerning DNA identification systems. See Table 1, "DNA Identification Bills Before State Legislatures," *GeneWatch,* vol. 6, no. 1. See also the testimony of Professor Philip Bereano before the Subcommittee on Civil and Constitutional Rights, US Senate Judiciary Committee, March 1989. Copies of Professor Bereano's testimony are available from the CRG office.

Chapter 4

1. Americans With Disabilities Act of 1990. Public Law 101-336. §§ 12101-12213. 108th Congress, 2nd session (July 26, 1990).
2. See, for example, U.S. Congress Office of Technology Assessment, "Genetic Monitoring and Screening in the Workplace," OTA-BA-455 (Washington, D.C.: U.S. Government Printing Office, October 1990.)
3. See, for example, Alper, Joseph, Paul Billings, Carol Barash, Jonathan Beckwith, Lisa Geller, and Marvin Natowicz. "Individual, Family, and Societal Dimensions of Genetic Discrimination: A Case Study Analysis." *Science and Engineering Ethics* 2, no. 1 (1996): 71-88.
4. The "Genetic Privacy Act" was announced in March 1995. It was a component of the U.S. Human Genome Project's Ethical, Legal, and Social Issues (ELSI) working group. George Annas, Leonard Glantz, and Patricia Roche (Boston University School of Public Health) authored the proposal with funding from the DOE ELSI program.
5. See the National Conference of State Legislators' (NCSL) Genetic Technologies Project database for more information on genetic discrimination state laws (http://www.ncsl.org/programs/health/genetics.htm).
6. EEOC Compliance Manual § 902, Order 915.002, 902-45 (1995).
7. For a more thorough discussion of this issue see Gruber, Jeremy "Genetic Discrimination and the Americans with Disabilities Act: An Unlikely Fit," http://www.workrights.org/issue_genetic/gd_ada.html.
8. *U.S. Senate Committee on Health, Education Labor, and Pension Hearing on Genetic Discrimination in the Workplace*, 106th Cong. (July 20th, 2000).
9. Exec. Order No. 13145, 65 Fed. Reg. 6877 (Feb. 10, 2000).
10. National Partnership for Women & Families on behalf of the Coalition for Genetic Fairness. *Faces of Genetic Discrimination: How Genetic Discrimination Affects Real People.* July, 2004. http://www.geneticalliance.org/ksc_assets/documents/facesofgeneticdiscrimination.pdf
11. House Energy and Commerce Committee, House Ways and Means Committee and House Education and Labor Committee.
12. The hold privilege is allowed by Rule VII of the Senate Standing Rules. The practice is generally used to form consensus on questionable legislation. http://rules.senate.gov/senaterules/rule07.php
13. http://www.whitehouse.gov/omb/legislative/sap/109-1/s306sap-s.pdf
14. The departments of Treasury, Labor, and Health and Human Services are charged with enforcing various aspects of Title I of GINA, and the

Equal Employment Opportunity Commission is charged with enforcing Title II.

Chapter 8

1. http://www.newscientist.com/blogs/shortsharpscience/2010/06/personal-genome-customers-sent-1.html
2. Bradley Malin and Latanya Sweeney, "Determining the Identifiability of DNA Database Entries." *J. Amer. Med Informatics Assoc*. Proceedings of the AMIA Symposium (2000), 423.
3. 23andMe Privacy Statement (accessed on 7/12/10 at https://www.23andme.com/about/privacy/)
4. See, for example, United States, Federal Trade Commission, At-home Genetic Tests: A Healthy Dose of Skepticism may be the Best Prescription (2006), online: Federal Trade Commission (http://www.ftc.gov/bcp/edu/pubs/consumer/health/hea02.shtm)

Chapter 10

1. Trinidad, Susan Brown. "Genomic Research and wide data sharing: Views of prospective participants." *Genetic Med*. 12 (2010): 486-495.
2. Greely, Henry T. "The Uneasy Ethical and Legal Underpinnings of Large-Scale Genomics Biobanks." *Annual Review of Genomics and Human Genetics* 8 (2007): 343-364.
3. Greely, Henry T. "Human Genomics Research: New Challenges for Research Ethics." *Perspectives in Biology and Medicine* 44, no. 2 (2001): 221-229.
4. Greely, Henry T. "From Nuremberg to the Human Genome: The Rights of Human Research Participants." In *Medicine After the Holocaust*. Sheldon Rubenfeld ed. (New York : Palgrave Macmillan, 2010), 85-200.
5. Greely, Henry T. "To the Barricades!" *American Journal of Bioethics* (2010); 10 (9):1-2.

Part II

Chapter 14

1. Pickle, Linda Williams. *Atlas of United States Mortality*. Hyattsville, Md.: Centers for Disease Control and Prevention, National Center for Health Statistics, 1997.
2. Goldwater LJ, Rosso AJ, Kleinfeld M. Bladder tumors in a coal tar dye plant. *Arch Environ Health* (1965); 11(6):814–817 (Dec).

Chapter 16

1. Pollack, Andrew. "F.D.A. Faults Companies on Unapproved Genetic Tests." *The N.Y. Times* (New York), June 11, 2010. http://www.nytimes.com/2010/06/12/health/12genome.html?%20scp=1&sq=fda%20genetic%20test&st=cse (accessed June 6, 2010).
2. "FDA/CDRH Public Meeting: Oversight of Laboratory Developed Tests (LDTs), Date July 19-20, 2010." U.S. Food and Drug Administration Home Page. http://www.fda.gov/MedicalDevices/NewsEvents/WorkshopsConferences/ucm212830.htm (accessed March 14, 2013).
3. "Homepage." 23 and Me. https://www.23andme.com/health/

Chapter 17

1. Weiss, Rick. "Death Points to Risks in Research; One Woman's Experience in Gene Therapy Trial Highlights Weaknesses in the Patient Safety Net." *The Washington Post*, August 6, 2007. http://www.washingtonpost.com/wp-dyn/content/article/2007/08/05/AR2007080501636_pf.html
2. Crofts, Christine, and Sheldon Krimsky. "Emergence of a Scientific and Commercial Research and Development Infrastructure for Human Gene Therapy." *Human Gene Therapy* (2005); 16:169-177 (Feb.).
3. Minutes of the RAC meeting at which Stuart Newman testified in 2000: http://www4.od.nih.gov/oba/rac/minutes/RACmin3-00.pdf
4. Gelsinger, Paul, "Jesse's Intent," http://www.circare.org/submit/jintent.pdf
5. Minutes of the RAC meeting at which the Targeted Genetics study was discussed: http://www4.od.nih.gov/oba/RAC/minutes/Sept_2003_Minutes.pdf
6. Seattle Post-Intelligencer Staff. "A moment with…H. Stewart Parker, biotech leader." *Seattle Post-Intelligencer*, August 20, 2005. http://seattlepi.nwsource.com/business/235895_momentwith10.html
7. Washburn, Jennifer. *University, Inc.: The Corporate Corruption of American Higher Education* (New York: Basic Books, 2005).

Chapter 18

1. Kraft, P., and D.J. Hunter. "Genetic risk prediction: Are we there yet?" *New Eng. J. Med.* 360, no. 17 (2009): 1701-1703.
2. Goldstein, D.B.. "Common genetic variation and human traits." *New Eng. J. Med.* 360, no. 17 (2009): 1696-1698.

Chapter 19

1. Groopman, Jerome and Pamela Hartzband. "Why 'Quality' Care Is Dangerous." *The Wall Street Journal* (New York), April 8, 2009, sec. Opinion. http://online.wsj.com/article/SB123914878625199185.html; *The N.Y. Times* (New York), "What's your underlying condition?," October 28, 2009, sec. Opinion. http://www.nytimes.com/2009/11/27/opini

Chapter 24

1. Collins, Francis S. *The Language of Life: DNA and the Revolution in Personalized Medicine,* (New York: Harper Collins, 2010), pp. xxiv-xxv.
2. Wade, Nicholas. "A decade later, genetic map yields few cures." *The New York Times*, June 12, 2012. http://www.nytimes.com/2010/06/13/health/research/13genome.html?pagewanted=all
3. United Health Canter for Health Reform and Modernization, "Personalized Medicine: Trends and Prospects for the New Science of Genetic Testing and Molecular Diagnostics." Working Paper pg. 3.
4. Association of Molecular Pathology et al. v United States Patent and Trade Office and Myriad Genetics Inc., 669 F Supp 2d 365. (2010).
5. Wadman, Meredith. "Fifty genome sequences reveal breast cancer's complexity." *Nature News* (April 2, 2011). http://www.nature.com/news/2011/110402/full/news.2011.203.html
6. Gurlinger, Marco, Andrew J. Rohan, and Stuart Horswell. "Intratumor heterogeneity and branched evolution revealed by multiregion sequencing." *New Eng. J. of Med.* 366, no. 10 (2012): 883-892.
7. Herper, Matthew. "Cancer's New Era Of Promise And Chaos." *Forbes*, June 5, 2011. http://www.forbes.com/sites/matthewherper/2011/06/05/cancers-new-era-of-promise-and-chaos/
8. Nerbert, Daniel W., Zhang Ge, and Elliot S. Vessell. "From human genetics and genomics to pharmacogenetics and pharmacogenomics: past lessons, future directions." *Drug Metabolism Review* 40, no. 2 (2008): 187-224.
9. Chiang, Alex, and Ryan P. Milton. "Personalized medicine in oncology: next generation." *Nature Reviews Drug Discovery* 10 (2011): 895-896.
10. Ibid.
11. Kwak, E.L. "Anaplastic lymphoma kinase inhibition in non-small-cell lung cancer." *New Eng. J. of Med.* 363 (2010): 1695-1703.

12. Geddes, Linda. "Daily aspirin cuts risk of colorectal cancer." *New Scientist.* October 28, 2011. http://www.newscientist.com/article/dn21100-daily-aspirin-cuts-risk-of-colorectal-cancer.html

Part III

Chapter 29

1. Guichon, J. "Don't let market forces govern human procreation." *Bionews,* November 22, 2010.
2. The Red Market Blog; "International Baby Market," blog entry by Scott Carrny. http://redmarkets.com/2010/08/international-baby-maker.html
3. Mitscherlich, Alexander, and Fred Mielke. "The Nuremberg Code." In *Doctors of infamy; the story of the Nazi medical crimes.* New York: Henry Schuman, 1949. xxiii-xxv.
4. The World Health Organization (WHO) is an organ of the United Nations and is the leading authority on health within the United Nations system.
5. The Council of Europe. "Convention for the protection of Human Rights and Dignity of the Human Being with regard to the Application of Biology and Medicine," January 12, 1999. ETS no. 164.
6. The Council of Europe. "Additional Protocol to the Convention on Human Rights and Biomedicine, on Transplantation of Organs and Tissues of Human Origin," May 1, 2006. ETS no. 186.
7. The Helsinki Declaration was developed by the World Medical Association. It is not a binding legal instrument, but its principles were drawn from worldwide regional and national legislation.
8. Such is the case in the Draft Guiding Principles on Human Organ Transplantation of 1991.
9. Ingrid Schneider, (2007).
10. European Parliament (EP), Directive 2004/23/EC, "Setting standards of quality and safety for the donation, procurement, testing, processing, preservation, storage and distribution of human tissues and cells," March 31, 2004, http://eur-lex.europa.eu/LexUriServ/LexUriServ.do?uri=OJ:L:2004:102:0048:0058:en:PDF

Chapter 31

1. See Pande, Amrita. "Commercial Surrogacy in India: Manufacturing a Perfect Mother-Worker," *Signs* (2010); 35(4): 969-992. Lisa C. Ikemoto, Reproductive Tourism: Equality Concerns in the Global Market for Fertility Services, XXVII J. Law and Inequality (2009); 27: 277-309.

2. ART self-reporting data in the United States: Victoria Clay Wright, et. al. Division of Reproductive Health, CDC, "Assisted Reproductive Technology Surveillance, United States, 2003," in Surveillance Summaries, May 22, 2006; for Canada and the United States: Edward G. Hughes, et. al. "Cross-Border Fertility Services in North America: A Survey of Canadian and American Providers," http://fertstert.org/article/S0015-0282(09)04144-2; and for Europe: F. Shenfield, et. al "Cross Border Reproductive Care in Six European Countries,"Human Reproduction, Vol. 25, no. 6, 1361-1368, 2010.
3. Legal status of ARTs: Liberal regulations in Spain, Belgium, Cyprus, Czech Republic, India, Jordan, Israel, South Africa; no regulations in US, Ukraine; strict regulations in Germany, France, the UK, Italy, Queensland and Victoria in Australia, Iceland, Sweden, Austria, Norway, Turkey; pending regulation in Thailand.
4. See Magdalina Gugucheva, Surrogacy in America. Cambridge, MA: Council for Responsible Genetics, 2010.
5. Re allegations of price-fixing: Marimer Matos, Human Egg Donor Files Antitrust Class Action, Courthouse News Service, April 14, 2011.
6. See especially the longstanding commitment of feminist organizations to protecting women's health, well-being, and reproductive rights of FINRRAGE in Australia and Bangladesh, Sama in India, Isha L'Isha in Israel and Boston, USA based Our Bodies Ourselves.
7. Author is grateful to Kathy Sloan for her assistance in providing important data for this article.

Chapter 32

1. See http://www.eggdonorneeded.com.
2. See, for example: http://www.etopiamedia.net/empnn/pdfs/norsigian1.pdf.
3. http://www.ourbodiesourselves.org/book/companion.asp?id=31&compID=97&page=4).
4. www.nature.com/nature/journal/v442/n7103/full/442607a.html.
5. http://www.nature.com/nature/journal/v460/n7259/full/4601057a.html.
6. http://www.biopoliticaltimes.org/article.php?id=4684.
7. http://www.nytimes.com/2009/06/26/nyregion/26stemcell.html.
8. http://english.ohmynews.com/articleview/article_view.asp?article_class=4&no=385258&rel_no=1.
9. http://www.dailyprincetonian.com/2008/11/21/22198/.
10. http://www.geneticsandsociety.org/article.php?id=4770.

11. IANS, (August 25th 2008), 'Surrogacy a $445 mn Business in India', *The Economic Times.*

Chapter 34

1. Robertson, John A. "Preconception Gender Selection," *Am. J. of Bioethics* 1, no. 1 (Winter 2001).
2. Raymond, F. Lucy. "Molecular Prenatal Diagnosis: the Impact of Modern Technologies." *Prenatal Diagnosis* 30, no. 7 (July 2010): 674.
3. Chachkin, Carolyn J. "What Potent Blood: Non-Invasive Prenatal Genetic Diagnosis and the Transformation of Modern Prenatal Care," *Am. J. L. & Medicine* 9, no. 9 (2007).
4. Lo, Y. M. Dennis. "Noninvasive Prenatal Diagnosis in 2020," *Prenatal Diagnosis* 30, no. 702 (2007). (hereinafter NIPD 2020).
5. Leach, Mark W. "Abortions on Disabled Babies: The Prenatal Testing Sham," *LifeNews.com*, May 25, 2011, http://www.lifenews.com/2011/05/25/abortions-on-disabled-babies-the-prenatal-testing-sham/
6. According to some scholars, legitimate reasons for sex selection exists, including adherence to certain religious beliefs and cultural traditions, Robertson, supra note 1, at 3, but these are often not medically motivated rationales.
7. Mukherjee, T., et al. "Unexpected Gender Bias Found in IVF Cycles for Sex Selection," *Fertility & Sterility (Supplement 1) S134* no. 88 (2007). (A 2007 study found that only 4 of 30 reviewed in vitro fertilizations with pre-implantation genetic diagnosis procedures were conducted to avoid sex-linked diseases. The rest were performed for elective sex selection.)
8. Dahl, Edgar, et al. "Preconception Sex Selection Demand and Preferences in the United States," *Fertility and Sterility* 85 (Feb. 2006): pp. 468, 473 (concluding, among other things, that the demand and preference among the US general population for certain fetal sex is low, and noting "[t]he results of our study are consistent with findings from prior social research.").
9. Saletan, William. "Fetal Subtraction: Sex Selection in the United States," *Slate.com*, April 3, 2008, www.slate.com/id/2188114.
10. Dahl, supra note 19, at 473 (concluding, among other things, that the demand and preference among the US general population for certain fetal sex is low, and noting "[t]he results of our study are consistent with findings from prior social research.").

11. Id.
12. Id.
13. Malinowski, Michael J. "Choosing the Genetic Makeup of Children: Our Eugenics Past, Present and Future?" *Conn. L. Rev.* 36 (2003-2004): pp. 125, 133.
14. Id.
15. Van den Heuvel, Ananda, et al. "Will the Introduction of Non-invasive Prenatal Diagnostic Testing Erode Informed Choices? An Experimental Study of Health Care Professionals," *Patient Education and Counseling* 78 (2010): pp. 24, 28.
16. Zambon, Kat. "Case Studies Illustrate the Dilemmas of Genetic Testing," *American Association for the Advancement of Science* (June 28, 2011), www.aaas.org.
17. Marteau, Theresa M. and Elizabeth Dormandy. "Facilitating Informed Choice in Prenatal Testing: How Well Are We Doing?" *Am. J. of Med. Genetics* 106 (2001): pp. 185, 189.
18. NIPD 2020, supra note 3.
19. Raymond, supra note 1, at 677. As a consequence of fetal DNA accounting for only 5 to 10 percent of the DNA in the maternal serum, current technology is best equipped at isolating fetal DNA by distinguishing paternally derived genes in fetal DNA that are not found in the maternal DNA. Id.
20. Hill, Melissa, et al. "Incremental Cost of Non-Invasive Prenatal Diagnosis Versus Invasive Prenatal Diagnosis of Fetal Sex in England," *Prenatal Diagnosis* 31 no. 3 (March 2011): p. 267.
21. Wright, Caroline F. "The Use of Cell-Free Fetal Nucleic Acids in Maternal Blood for Non-Invasive Prenatal Diagnosis," *Human Reproduction Update* 15, (2009): pp. 139, 140.,
22. Id.
23. Lo, Y. M. Dennis, et al. "Digital PCR for the Molecular Detection of Fetal Chromosomal Aneuploidy," *Proc. Nat'l Acad. Sci.* 104 (2007). pp. 13116 (Chachkin, supra note 2, at 11).
24. CVS and amniocentesis are the adopted standard of care methods for prenatal genetic diagnostics and both procedures are held to 98 to 99 percent accuracy standards. Chachkin, supra note 2, at 35.
25. See Chachkin, supra note 2, at 37.

26. Been, Peter A. and Audrey R. Chapman. "Ethical Challenges in Providing Noninvasive Prenatal Diagnosis," *Current Opinions in Obstetrics and Gynecology* 22 (2010): pp. 128, 129; see also The Genetics and Public Policy Center John Hopkins University, Reproductive Genetic Testing: A Regulatory Patchwork, ("In the United States, there is no uniform or comprehensive system for the regulation of assisted reproductive technologies, including reproductive genetic testing. The federal government does not have direct jurisdiction over the practice of medicine. Moreover, it has banned all federal funding for research involving the creation or destruction of embryos. Consequently, the regulatory framework for reproductive genetic testing in the United States is characterized by a patchwork of federal and state regulation."), available at http://www.dnapolicy.org/policy.international.php?action=detail&laws_id=63 (January 2004).
27. United States Government Accountability Office, Direct-to-Consumer Genetic Testing: Misleading Test Results Are Further Complicated by Deceptive Marketing and Other Questionable Practices (Thursday, July 22, 2010)
28. Malinowski, Michael J. "Coming into Being: Law, Ethics, and the Practice of Prenatal Genetic Screening," *Hastings L.J.* 45 (1993-1994): pp. 1435, 1494.
29. Robertson, John A. "Preconception Gender Selection," *A.J. Bioethics* 1 no. 1 (winter 2001): pp. 1, 4.
30. See Parens, supra note 44, at S15 ("According to Asch, most abortions reflect a decision not to bring any fetus to term at this time; selective abortions involve a decision not to bring this particular fetus to term because of its traits. Pro-choice individuals within and outside the disability community agree that it is morally defensible for a woman to decide [against an unwanted pregnancy] . . . The question is whether that decision is morally different from a decision to abort an otherwise-wanted fetus.")
31. Mitchell CB. The Church and the New Genetics in Genetic Ethics (Kilner JF, Pentz RD, Young FE, eds., WM B Eerdmans Publish'g Co.) (1997).
32. Chachkin, supra note 2, at 40.
33. Chachkin, supra note 2, at 40 (reasoning that the likely consequence of clinical NIPD is that the cost of bringing a child to term with predicted disabled traits may very well become disfavored, when compared to covering NIPD tests.)

34. Annas, George J. "Ethical aspects of non-invasive prenatal diagnosis: medical, market, or regulatory model?" *Early Human Development Supplement,* 47 (1996); pp. S5, S11.
35. Parens, Erik and Adrienne Asch. "The Disability Rights Critique of Prenatal Genetic Testing: Reflections and Recommendations," *The Hastings Center Report* (Special Supplement) 29 (Sep.-Oct., 1999): pp S1, S6.
36. Id. at S13.
37. Id. at S1 (emphasis added).
38. Id. at S13.
39. Id. at 40 (quoting Harriet McBryde Johnson, Unspeakable Conversations, *The New York Times Magazine* Feb. 16, 2003 Sunday, Late Edition).
40. Retsinas, Joan. "The Impact of Prenatal Technology Upon Attitudes Toward Disabled Infants," *Res. Social Health Care* 9 (1991): pp. 75, 89-90.
41. See also Wertz, Dorothy C. and Bartha Maria Knoppers, "Serious Genetic Disorders: Can or Should They be Defined," *Am. J. of Medical Genetics* 108 (2002): p. 29. (hereinafter "Serious Genetic Disorders).
42. Including individuals certified by the American Board of Medical Genetics, the American Board of Genetic Counseling, the European Society of Human Genetics, the Canadian College of Medical Genetics, and the Ibero-American Society of Human Genetics (an organization of professionals from Spanish-and Portuguese-speaking nations). The majority of the respondents were members of the Canadian College of Medical Genetics, with less than 40 percent of the respondents from Spanish-and Portuguese-speaking nations. Dorothy C. Wertz & Bartha Maria Knoppers, Serious Genetic Disorders: Can or Should They be Defined, 108 *Am. J. of Medical Genetics* 29 (2002) (hereinafter Serious Genetic Disorders).
43. "Serious Genetic Disorders", supra note 89: pp. 31-33.
44. Id.
45. Leach, Mark W. "Abortions on Disabled Babies: The Prenatal Testing Sham," *LifeNews.com,* May 25, 2011, http://www.lifenews.com/2011/05/25/abortions-on-disabled-babies-the-prenatal-testing-sham/
46. Id.
47. Robertson, John A. "Genetic Selection of Offspring Characteristics," *B.U.L. Rev.* 76 (1996): pp. 421, 446. (hereinafter "Genetic Selection of Offspring).
48. Holtzman, supra note 831, at 398.

49. Genetic Selection of Offspring Characteristics, supra note 64, at 480.
50. Id. at 479.
51. "Eugenics." Oxford American Dictionaries. (Oxford University Press, 1999).

Chapter 37

1. US Congress Office of Technology 1988 Assessment.
2. Shaw, M. W., in *American Journal of Human Genetics.* 1984, 36:1-9.

Part IV

Chapter 38

1. Emile Zola, *La Fortune des Rougons,* 1871.
2. O. Onay. "The true ramifications of genetic criminality research for free will in the criminal justice system," *Genetics, Society, and Policy* 2 (2006): pp. 80-91.
3. Hamer, D. "Rethinking behavior genetics," *Science* 298 (2008): pp. 71-72.
4. Appelbaum, P. "Behavioral genetics and the punishment of crime," *Psychiatric Services* 56 (2005): pp. 25-27.
5. Caspi et al. "Role of the genotype in the cycle of violence in maltreated children," *Science* 297 (2002): pp. 851-854.
6. Kim-Cohen et al. "MAOA, maltreatment, and gene-environment interaction predicting children's mental health: new evidence and a meta-analysis," *Molecular Psychiatry* 11 (2006): pp. 903-913.
7. Lewontin, R. *It Ain't Necessarily So: The Dream of the Human Genome and Other Illusions,* 2nd ed. (New York: The New York Review of Books, 2002).
8. Balaban, E. "Cognitive developmental biology: History, process and fortune's wheel," *Cognition* 101 (2006: pp. 298-332).
9. Redding, R.E. "The brain disordered defendant: Neuroscience and legal insanity in the twenty-first century" (Villanova School of Law working paper 61, Villanova, PA, 2006)
10. Abbot, A. "Into the Mind of a Killer," *Nature* 444 (2006): pp. 296-298.
11. Mandavilli, A. "Actions speak louder than images," *Nature* 444 (2006): pp. 664-665.
12. Raine A, Yang Y. "Neural foundations to moral reasoning and antisocial behavior," *Social, Cognitive, and Affective Neuroscience* 1 (2006): pp. 203-213.
13. Hauser, M. "The liver and the moral organ," *Social, Cognitive, and Affective Neuroscience* 1 (2006): pp. 214-220.
14. Rose, N. "The biology of culpability: Pathological identity and crime control in a biological culture," *Theoretical Criminology* 4 (2000): pp. 5-34.

15. Hacing, I. "Degeneracy, criminal behavior, and looping," in Wasserman D. Wachbroit R (eds), *Genetics and Criminal Behavior* (Cambridge University Press, NY, 2001): pp. 141-167.
16. Duster, T. "Behavorial genetics and explanations of the link between crime, violence and race," in E. Parens, A.R. Chapman, and N. Press, *Wrestling With Behavior Genetics: Science, Ethics, and Public Conversation* (Johns Hopkins University Press, Baltimore, 2006): pp. 150-175.
17. ibid, pp. 166-171.

Chapter 39

1. Zola, E. *La Bête Humaine* (Penguin Classics, 1977): pp. 368.
2. Jacobs, P. A., M. Brunton, M. M. Melville, R. P. Brittain, W. F. McClemont. *Nature* (Dec 25, 1965): pp. 208, 1351.
3. Beckwith, J. *Making Genes, Making Waves: A Social Activist in Science* (Harvard University Press, Cambridge, 2002): pp. 254.
4. Antkiewicz-Michaluk, L., M. Grabowska, L. Baran, J. Michaluk. *Arch Immunol Ther Exp* (Warsz, 1975): pp. 23, 763.
5. Brunner, H. G., M. Nelen. X. 0. Breakefield, H. H. Ropers, B. A. van Oost. *Science* (Oct 22, 1993): pp. 262, 578.
6. Brunner, H. G., et al. *Am J Hum Genet* 52, 1032 (Jun, 1993).
7. Wasserman, D. *J. Law Med Ethics* 32, 24 (Spring, 2004).
8. Cases, O., et al. *Science* 268, 1763 (Jun 23, 1995).
9. Heath, M. J., R. Hen. *Curr Biol* 5, 997 (Sep 1, 1995).
10. Caspi, A., et al. *Science* 297, 851 (Aug 2, 2002).
11. Foley, D. L., et al. *Arch Gen Psychiatry* 61, 738 (Jul, 2004).
12. Nilsson, K. W., et al. *Biol Psychiatry* 59, 121 (Jan 15, 2006).
13. Widom, C. S., L. M. Brzustowicz. *Biol Psychiatry* 60, 684 (Oct 1, 2006).
14. Haberstick, B. C., et al. *Am J Med Genet B Neuropsychiatry Genet* 135, 59 (May 5, 2005).
15. Young, S. E., et al. *Am J Psychiatry* 163, 1019 (Jun, 2006).
16. Huizinga, D., et al. *Biol Psychiatry* 60, 677 (Oct 1, 2006).
17. Huang, Y. Y., et al. *Neuropsychopharmacology* 29, 1498 (Aug, 2004).
18. Manuck, S. B., J. D. Flory, R. E. Ferrell. J. J. Mann, M. F. Muldoon. *Psychiatry Res* 95, 9 (Jul 24, 2000).
19. Beitchman, J. H., H. M. Mik, S. Ehtesham, L. Douglas, J. L. Kennedy. *Mol Psychiatry* 9, 546 (Jun, 2004).

20. Symposium, paper presented at the Symposium on Molecular Mechanisms Influencing Aggressive Behaviours, Novartis Foundation, London 2004.
21. Newman, T. K., et al. *Biol Psychiatry* 57, 167 (Jan 15, 2005).
22. Kim-Cohen, J., et al. *Mol Psychiatry*, 903 (Oct, 2006).
23. Symposium, paper presented at the Symposium on Genetics of Criminal and Antisocial Behaviour, Ciba Foundation, London, 1996–1995.
24. Jaffee, S. R., et al. *Dev Psychopathol* 17, 67 (Winter, 2005).
25. Maughan, B., G. McCarthy. *Br Med Bull* 53, 156 (Jan, 1997)
26. Kaufman, J., et al. *Proc Natl Acad Sci USA* 101,17316 (Dec 7, 2004).
27. Goleman, D. "Teen Age Risk-Taking: Rise in Deaths Prompts New Research Effort," *The New York Times,* November 24, 1987.
28. Goleman, D. "Why Do People Crave the Experience?," *The New York Times,* August 2, 1988.
29. Opinion, "Why Doctors Treat Alcoholism as a Disease," *The New York Times,* November 27, 1987.
30. Goleman, D. "Scientists Pinpoint Brain Irregularities In Drug Addicts," *The New York Times,* June 26, 1990.
31. Leary, W. E. "Brain Chemical Said to Play Role in Cigarette Addiction," *The New York Times,* February 22, 1996.
32. Angier, N. "Gene Tie to Male Violence Is Studied," *The New York Times,* October 22, 1993.
33. Morell, V. *Science* 260, 1722 (Jun 18, 1993).
34. Cowley, G., C. Hall, in *Newsweek*. (1993) pp. 57.
35. Personal communication with Xandra Breakefield.
36. Shute, N., in *U.S. News & World Report.* (2002), vol. 133. pp. 45.
37. Angier, N. "Disputed Meeting to Ask if Crime Has Genetic Roots," *The New York Times,* September 19, 1995, pp. C1.
38. (Ga., 1998).
39. (Tenn. Crim. App., 2006).
40. Radford, T. "Scientists identify gene link to violence," *The Guardian,* August 2, 2002, pp. 3.
41. DiMare, P., in *Popular Mechanics.* (2002).
42. Swan, N., in Health Minutes A. NewsRadio, Ed. (ABC NewsRadio, 2002), vol. 2006.
43. Balasubramanian, D. "How genetic makeup influences behavior," *The Hindu,* April 5, 2006 2002.

44. Hall, C. T. *The San Francisco Chronicle,* A2 (August 2, 2002).
45. Barry, E. "Study Links Past Abuse, Gene to Violent Acts," The *Boston Globe,* August 2, 2002, pp. A2.
46. Beckwith, J., in *Wrestling with Behavioral Genetics: Science. Ethics, and Public Conversation.* E. Parens, Chapman, AR, and Press, N, Ed. (Johns Hopkins University Press, Baltimore, 2005) pp. 74-99.

Chapter 40

1. Cohen, Patricia. "Genetic Basis of Crime: A New Look," *The New York Times* (June 19) 1: Arts Section, 2011.
2. Shea, Christopher. "The Nature-Nurture Debate, Redux," *Chronicle of Higher Education, The Chronicle Review* (January 9, 2009).
3. Duster, Troy. "Comparative Perspectives and Competing Explanations: Taking on the Newly Configured Reductionist Challenge to Sociology," *American Sociological Review*, 71 (February: 1-15, 2006)
4. Bittner, Egon. "The Police on Skid Row: A 'study of peace-keeping.'" *American Sociological Review* 32:699-715, 1967.
5. Cicourel, Aaron V. *The Social Organization of Juvenile Justice*. New York: Wiley, 1967.
6. Sudnow, David. "Normal Crimes: Sociological Features of the Penal Code in a Public Defender's Office." *Social Problems* 12:255-76, 1965.
7. Goffman, Erving. "The Moral Career of the Mental Patient." *Psychiatry* 22(2):123-42, 1959.
8. Skolnick, Jerome H. "Corruption and the Blue Code of Silence." Police Practice and Research 3:7-19, 2002; Skolnick, Jerome H. and James J. Fyfe. 1993. *Above the Law: Police and the Excessive Use of Force*. New York: Free Press.
9. Jackall, Robert. *Street Stories: The World of Private Detectives.* Cambridge, MA: Harvard University Press, 2005.
10. Fullwiley, Duana. "The Biologistical Construction of Race: Admixture Technology and the New Genetic Medicine," *Social Studies of Science* 38(5): 695-735, 2008; Fullwiley, Duana, "The Molecularization of Race: Institutionalizing Human Difference in Pharmacogenetics Practice," *Science as Culture*, 16 (1): 1-30, 2007.
11. Blow, Charles. "Smoke and Mirrors." *The New York Times* (Oct. 22, 2010) at: http://www.nytimes.com/2010/10/23/opinion/23blow.html also see graph at:

http://www.nytimes.com/imagepages/2010/10/23/opinion/23blow_chart.html?ref=opinion); Levine, Harry and Deborah P. Small 2008, Marijuana Arrest Crusade: Racial Bias and Police Policy in New York City, 1997-2007. New York Civil Liberties Union, final version, 2010a at http://www.nyclu.org/files/Marijuana-Arrest-Crusade_Final.pdf; Levine et al. 2010b, "Arresting Blacks for Marijuana in California," Drug Policy Alliance, Los Angeles, Ca, Oct 2010.

Chapter 42

1. Crow, T. J. "The emperors of the schizophrenia polygene have no clothes," *Psychological Medicine* 38 (2008): pp. 1681-1685.
2. Merikangas, K. R., & Risch, N. "Will the genomics revolution revolutionize psychiatry?" *American Journal of Psychiatry* 160 (2003): pp. 625-635.
3. Joseph, J. "The Equal Environment Assumption of the Classical Twin Method: A Critical Analysis," *Journal of Mind and Behavior* 19 (1998) pp. 325-358; Joseph, J. "Twin studies in psychiatry and psychology: Science or pseudoscience?" *Psychiatric Quarterly* 73 (2002) pp. 71-82; Joseph, J., *The Gene Illusion: Genetic Research in Psychiatry and Psychology Under the Microscope.* New York: Algora Publishing, 2004); Joseph, J. *Genetic Research in Psychiatry and Psychology: A Critical Overview* in K. Hood, C. Tucker Halpern, G. Greenberg, & R. Lerner, *Handbook of Developmental Science, Behavior, and Genetics.* (Malden, MA: Wiley-Blackwell, 2010): pp. 557-625.
4. Plomin, R., DeFries, J. C., McClearn, G. E., & McGuffin, P. *Behavorial Genetics,* 5th ed. New York: Worth Publishers: 2008.
5. Joseph, 2004, 2006, 2010. Kamin, L.J. *The Science and Politics of I.Q.* Potomac, MD: Larence Erlbaum Associates, 2006.
6. Turkheimer, E. "Three Laws of Behavior Genetics and What They Mean," *Current Directions in Psychological Science* 9 (2000): pp. 160-164.
7. Turkheimer, E. "Commentary: Variation and Causation in the Environment and Genome." *International Journal of Epidemiology* 40 (2011): pp. 598-601.
8. Gershon, E. S., Alliey-Rodriguez, N., & Liu, C. "After GWASL Searching for Genetic Risk for Schnizophrenia and Bipolar Disorder,"

American Journal of Psychiatry 168 (2011): pp. 253-256; Hwaorth, C.M.A., and Plomin, R. "Quantitative genetics in the era of molecular genetics: Learning abilities and disability as an example," *Journal of the American Academy of Child and Adolescent Psychiatry* 49 (2010): pp. 783-793; Manolio, T.A., Collins, F.S., Cox, N.J., Goldstein, D.b., Hindorff, L.A., et al. "Finding the missing heritability of complex diseases," *Nature* 461 (2009): pp. 747-753; Plomin, R. "Commentary: Why are children in the same family so different? Non-shared environment three decades later," *International Journal of epidemiology* 40 (2011): pp. 582-592; For a critical appraisal of "missing hertiability" see Latham, J., & Wilson, A. "The great DNA data deficit: Are genes for disease a mirage?" *The Bioscience Research Project* (Dec 18, 2010), http://www.bioscienceresource.org/commentaries/article.php?id=46.

9. Joseph, 2010; Latham & Wilson, 2010.
10. Sullivan, P. F. "Don't Give Up On GWAS," *Molecular Psychiatry* (Aug 9, 2011).
11. 5th edition; Plomin, DeFries, McClearn, & McGuffin, 2008.
12. DeFries, J. C., & Plomin, R. "Behavioral Genetics," *Annual review of Psychology* 29 (1978): pp. 473-515.
13. Loehlin, J. C., Willerman, L., & Horn, J. M. "Human Behavior Genetics," *Annual review of Psychology* 39 (1988): pp. 101-133.
14. Plomin, R., DeFries, J. C., & McClearn, G. E. *Behavioral Genetics: A Primer,* 2nd ed. New York: W.H. Freeman and Company, 1990.
15. Plomin, R. "The role of inheritance in behavior," *Science* 248 (1990): pp. 183-188.
16. Plomin, R., McClearn, G. E., Smith, D. L., Vignetti, S., Chorney, M. J., Chorney, K., Venditti, C. P., Kasarda, S., Thompson, L. A., Detterman, D. K., Daniels, J., Owen, M., & McGuffin, P. "DNA Markers Associated with High Verus Low IQ: The IQ Quantitative Trait Loci (QTL) Project," *Behavior Genetics* 24 (1994): pp. 107-118.
17. Deary, I. J., Penke, L., & Johnson, W. "The neuroscience of human intelligence differences," *Nature* 11 (2010): pp. 201-211.
18. Plomin, R., Owen, M. J., & McGuffin, P. "The genetic basis of complex behaviors," *Science* 264 (1994): pp. 1733-1739.

19. Plomin, R., DeFries, J. C., McClearn, G. E., & Rutter, M. *Behavioral Genetics,* 3rd ed. New York: W.H. Freeman and Company, 1997.
20. Rutter, M., & Plomin, R. "Opportunities for Psychiatry from Genetic Findings," *British Journal of Psychiatry* (1997): pp. 171, 209-219.
21. Plomin, R., & Rutter, M. "Child development, molecular genetics, and what to do with genes once they are found," *Child Development* 69 (1998): pp. 1223-1242.
22. Plomin, R., Corley, R., Caspi, A., Fulker, D. W., & DeFries, J. C. "Adoption results for self-reported personality: Evidence from non-additive genetic effects?" *Journal of Personality and Social Psychology* 75 (1998): pp. 211-218.
23. Plomin, R., & Crabbe, J. "DNA," *Psychological Bulletin* 126: pp. 806-828.
24. Plomin, R. "Behavioral genetics in the twenty-first century," *International Journal of Behavioral Development* 24 (2000): pp. 30-34.
25. McGuffin, P., Riley, B., & Plomin, R. "Toward behavioral genomics," *Science* 291 (2001): pp. 1232-1249.
26. Plomin, R., DeFries, J. C., McClearn, G. E., & McGuffin, P. *Behavioral Genetics,* 4th ed. New York: Worth Publishers, 2001.
27. Plomin, R., & McGuffin, P. "Psychopathology in the postgenomic era," *Annual Review of Psychology* 54 (2003): pp. 205-228.
28. Plomin, R. "General Cognitive Ability" in R. Plomin, J. DeFries, I. Craig, & P. McGuffin. *Behavioral Genetics in the Postgenomic Era.* Washington D.C., American Psychological Association Press, 2003.
29. McGuffin, P., & Plomin, R. "A decade of the Social, Genetic, and Developmental Psychiatry Centre at teh Institute of Psychiatry," *British Journal of Psychiatry* (2004): pp. 185, 280-282.
30. Plomin, R. "Genetics and Developmental Psychology," *Merrill-Palmer Quarterly* 50 (2004): pp. 341-352.
31. Plomin, R., & Spinath, F. M. "Intelligence: Genetics, Genes, and Genomics," *Journal of Personality and Social Psychology* 86 (2004): pp. 112-129.
32. Plomin, R. "Finding Genes in Child Psychology and Psychiatry: When are we going to be there?" *Journal of Child Psychology* 46 (2005): pp. 1030-1038. (In this quote, Plomin was referring to his "What To Do with Genes Once They Are Found" 1998 publication co-authored with Rutter.)
33. Plomin, 2005.
34. Haworth & Plomin, 2010.
35. Plomin, 2011.

36. Plomin, Corley, Caspi, Fulker, & DeFries, 1998.
37. Horgan, J. "Eugenics Revisited," *Scientific American* 268 no. 6 (1993): pp. 122-131.
38. Turkheimer, 2011.

Chapter 44

1. Ripke, S., Sanders, A. R., Kendler, K. S., Levinson, D. F., Sklar, P., Holmans, P. A., Lin, D. Y. et al. "Genome-wide association study identifies five new schizophrenia loci," *Nat Genet*. 43, 969-76; Sklar, P., Ripke, S., Scott, L.J., Andreassen, O.A., Cichon, S., Carrddock, N., Edenberg, H.J., Nurnberger, J.I. et al. "Large-scale genome-wide association analysis of bipolar disorder identifies a new susceptibility locus near ODZ4," *Nat Genet.* 43 (2011): pp. 977-983.
2. http://www.healthcanal.com/genetics-birth-defects/22589-Researchers-most-powerful-genetic-studies-psychosis-date.html
3. Sidanius, J., Pratto, F. *Social Dominance: an intergroup theory of social hierarchy and oppression*. (Cambridge University Press, Cambridge, UK, 1999).
4. Thornhill, Randy, and Craig Palmer. *A natural history of rape biological bases of sexual coercion*. Cambridge, Mass.: MIT Press, 2000.
5. Müller, G. B. "Evo-devo: extending the evolutionary synthesis," *Nat Rev Genet.* 8 (2007): pp. 949-949.
6. Gilbert, S. F., Epel, D. *Ecological Developmental Biology: Integrating epigenetics, medicine, and evolution*. Sinauer, Sunderland, Mass., USA, 2009.
7. West-Eberhard, M. J. *Developmental Plasticity and Evolution*. Oxford University Press, New York, 2003.
8. Newman, S. A., Bhat, R. "Dynamical patterning modules: a 'pattern language' for development and evolution of multicellular form," *Int. J. Dev. Biol.* 53 (2009): 693-705.
9. Baldwin, J. M. "A new factor in evolution," *The American Naturalist* 30 (1896): pp. 441-451, 536-553.
10. Smee, F. John, Kevin N. Laland, and Marcus W. Feldman. *Niche construction: the neglected process in evolution*. Princeton: Princeton University Press, 2003.
11. Palmer, A. R. "Symmetry breaking and the evolution of development," *Science* 306 (2004): pp. 828-833.
12. Trut, L., Oskina, I., Kharlamova, A. "Animal evolution during domestication: the domesticated fox as a model," *Bioessays* 31 (2009): pp. 349-360.

13. Weaver, I. C., Cervoni, N., Champagne, F. A., D'Alessio, A. C., Sharma, S., Seckl, J. R., Dymov, S., Szyf, M., Meaney, M. J. "Epigenetic programming by maternal behavior," *Nat Neurosci* 7 (2004): pp. 847-854.
14. Kashimada, K., Koopman, P. "Sry: the master switch in mammalian sex determination," *Development* 137 (2011): 3921-3920.
15. Kuroiwa, A., Handa, S., Nishiyama, C., Chiba, E., Yamada, F., Abe, S., Matsuda, Y., 2011. Additional copies of CBX2 in the genomes of males of mammals lacking SRY, the Amami spiny rat (Tokudaia osimensis) and the Tokunoshima spiny rat (Tokudaia tokunoshimensis). Chromosome Res. 19, 635-44.
16. Graves, J. A. M. "Weird animal genomes and the evolution of vertebrate sex and sex chromosomes," *Ann rev genetics* 42 (2008): pp. 565-586.

Chapter 45

1. Fisher, R.A. "Limits to intensive production in animals," *Journal of Heredity* 4 (1951): 217-218.
2. Dreary, I.J. *Intelligence: a short introduction.* (Oxford, Oxford University Press, 2001).
3. Zimmer, C. "Searching for intelligence in our genes," *Science American* (October, 2002).
4. Téglás, E., Vul, E., Girotto, V., Gonzalez, M., Tenenbaum, J.B., and Bonatti, L.L. "Pure Reasoning in 12-Month-Old Infants as Probabilistic Inference," *Science* 332 (2011): pp. 1054-1059.
5. Vygotsky, L.S. "The genesis of higher mental functions," in Richardson, K. & Sheldon, S. *Cognitive Development to Adolescence.* Hove: Erlbaum.
6. Joseph, J. "Genetic research in psychology and psychiatry: a critical overview," in K.E.Hood, C.T Halpern, G. Greenberg & R.M. Lerner (Eds.). *Handbook of developmental science, behavior and genetics.* New York: Wiley-Blackwell, 2010: pp. 557-625.
7. Evans, D. M., & Martin, N. G. "The validity of twin studies," *GeneScreen 1* (2000): pp. 77-79.
8. Christe, P., Moller, A.P., Saino. N. & De Lope, F. "Genetic and environmental components of phenotypic variation in immune response and body size of a colonial bird, *Delichon urbica* (the house martin)," *Heredity* 86 (2000): pp. 75-83.

9. Golan, D. & Rosset, S. "Accurate estimation of heritability in genomewide studies using random effects model." *Bioinformatics* 27 (2011): pp. i317-i323.
10. Turkheimer, E. "Commentary: variation and causation in the environment and genome," *International Journal of Epidemiology* 40 (2011): pp. 598-691.

Chapter 46

1. "Intelligence and Genetic Determinism," *GeneWatch* 19 (2006): pp. 9-12.
2. Gould, S.J. *The Mismeasure of Man.* W.W. Norton & Company, New York, 1981.
3. Cooper, R.S. "Race and IQ: Molecular Genetics as Deus ex Machina," *American Psychologist* 60 (2005): pp. 42-44.
4. Hernstein, R.J. & Murray, C. *The Bell Curve: The Reshaping of American Life by Difference in Intelligence* (Free Press, New York, 1994).
5. Fisher, C.S. et al. (eds.) *Inequality By Design: Cracking the Bell Curve Myth,* (Princeton University Press, Princeton, NJ, 1996); Ceci, S.J. "How much does schooling influence general intelligence and its cognitive components? A reassessment of the evidence," *Developmental Psychology* 27 (1991): pp 703-722.
6. Evans, G.W. & Schamberg, M.A. Childhood poverty, chronic stress, and adult working memory. Proceedings of the National Academy of Sciences, Early Edition 10.1073/pnas.0811910106: 1 - 5 (2009).
7. Zax, J.S. & Rees, D.I. "IQ Academic Performance, Environment, and Earnings," *The Review of Economics and Statistics* 84 (2002): pp. 600-616.
8. Nisbett, R. *Intelligence and How to Get It: Why Schools and Cultures Count* (W.W. Norton & Company, New York, 2009).
9. Dickens, W.T. & Flynn, J.R. "Black Americans Reduce the Racial IQ Gap," *Psychological Science* 17 (2006): pp. 913-920.

Chapter 47

1. Fowler, JH, Baker LA, Dawes CT. "Genetic Variation in Political Participation," *American Political Science Review* 102 (2008): 233-248. doi:10.1017/S0003055408080209
2. Miller, G, Zhu G, Wright MJ, et al. "The Heritability and Genetic Correlates of Mobile Phone Use: A Twin Study of Consumer Behavior," *Twin Research and Human Genetics* 15 no. 1(2012).

3. Sapra, S, Beavin LE, Zak PJ. "A Combination of Dopamine Genes Predicts Success by Professional Wall Street Traders," *PLoS ONE* 7 no.1(2012); 7(1):e30844. doi:10.1371/journal.pone.0030844
4. Garcia, JR, MacKillop J, Aller EL, et al. "Associations between dopamine D4 receptor gene variation with both infidelity and sexual promiscuity," *PLoS ONE* 5 no. 11 (2010).
5. Bernet, W, Vnencak-Jones CL, Farahany N, et al. "Bad nature, bad nurture, and testimony regarding MAOA and SLAC6A4 genotyping at murder trials," *Journal of Forensic Sciences* 52 no.6 (2007) pp. 1362-1371.
6. Fowler, JH, Dawes CT. "Two Genes Predict Voter Turnout," *The Journal of Politics* 70 no.3 (2008): pp. 579-594. doi:10.1017/S0022381608080638
7. Park, J-H, Gail MH, Weinberg CR, et al. "Distribution of allele frequencies and effect sizes and their interrelationships for common genetic susceptibility variants," *Proceedings of the National Academy of Sciences* 108 no.44 (2011): pp. 18026-31. doi:10.1073/pnas.1114759108
8. Zwarts, L, Magwire MM, Carbone MA, et al. "Complex genetic architecture of *Drosophila* aggressive behavior," *Proceedings of the National Academy of Sciences* 108 no.41 (2011): pp. 295-302. doi:10.1073/pnas.1113877108
9. Mackay, T. "The genetic architecture of complex behaviors: lessons from *Drosophila*," *Genetica* 136 no. 2 (2009): pp. 295-302.
10. Edwards, A, Zwarts L, Yamamoto A, et al. "Mutations in many genes affect aggressive behavior in *Drosophila melanogaster.*" *BMC Biol.* 7 no.1 (2009): pp. 29.

Part V

Chapter 49

1. Bieber, Frederick R. "Science and Technology of Forensic DNA Profiling: Current Use and Future Directions, in DNA and the Criminal Justice System: The Technology of Justice" (David Lazer ed., 2004); see also Ben Mitchell, "Police Warning to Criminals over DNA Breakthrough," *The Scotsman,* Nov. 19th, 2004.
2. Williams, Robin, "Making Do with Partial Matches: DNA Intelligence and Criminal Investigations in the United Kingdom," Presentation for DNA Fingerprinting and Civil Liberties: Workshop #2, American Society for Law, Medicine & Ethics, 17-18 September 2004.
3. FBI, CODIS Bulletin, "Interim Plan for the Release of Information in the Event of a Partial Match at NDIS," July 20, 2006.

4. Farahany, Nita A. and William Bernet. "Behavioral Genetics in Criminal Cases: Past, Present, and Future," *Genomics* 2, *Soc'y & Pol'y* 72 (2006).
5. DNAPrint Genomics Is Encouraging Law Enforcement Agencies To Include DNAWitness TM In Their NIJ Grant Proposals, http://www.dna-print.com/welcome/press/press_recent/200 4/august_16/.
6. DNAPrint Announces The Release Of Retinome[tm] For The Forensic Market: Eye Color Prediction From Crime Scene DNA, http://www.dna-print.com/welcome/press/press_recent/200 4/august_17/.
7. Law Enforcement Exemptions to the HIPAA Regulations: Testimony Before the Subcomm. on Privacy and Confidentiality of the National Comm. on Vital Statistics (Feb. 18, 2004) (statement of Chris Calabrese, Counsel to the American Civil Liberties Union's Technology and Liberty Program).
8. Affymetrix, http://www.affymetrix.com/index.affx.
9. Bioengineers develop smallest DNA sequencer, http://bioeng. berkeley. edu/content/view/307/157/.
10. Singer, Emily. "The Personal Genome Project: What Would Happen if Genetic and Medical Records Were Freely Available to Anyone Who Wanted Them?" *Tech. Rev.,* Jan. 20, 2006, http://www.technologyreview. com/Biotech/16169/.
11. Thompson, Carolyn. "Police DNA Collection Sparks Questions," *Associated Press,* March 17, 2007, http://www.usatoday.com/news/nation/2007-03-17-dna-collection_ N.htm?csp=34.
12. Brief for American Civil Liberties Union of Washington as Amici Curiae Supporting Defendant, State v. Athan, 158 P.3d 27 (Wash. 2007) (No. 75312-1); see also Richard Willing, "Police Dupe Suspects into Giving up DNA," *U.S.A. Today,* Sept. 11, 2003, A03.
13. For a detailed analysis of the concerns associated with the collection of so-called "abandoned" DNA, see Elizabeth E. Joh, Reclaiming 'Abandoned' DNA: The Fourth Amendment and Genetic Privacy, 100 Nw. U. L. Rev. 857 (2006).
14. See Rothstein & Talbot
15. GeneWatch UK, The Police National DNA Database: An Update (Human Genetics Parliamentary Briefing No. 6, July 2006), available at http://www.genewatch.org/uploads/f03c6d66a9b35453573 8483c1c3d49e4/ MPSBrief_1.pdf.

16. Police DNA 'Sweeps' Extremely Unproductive, a report by the Police Professionalism Initiative, Department of Criminal Justice, University of Nebraska at Omaha, Sept. 2004.
17. Belluck, Pam. "Slow DNA Trail Leads to Suspect in Cape Cod Case," *The New York Times,* Apr. 16, 2005; see also Eileen McNamara, "Not Making His Case," *Boston Globe,* Apr. 17, 2005.
18. Harmon, Rockne. Assistant Dist. Att'y of Alameda County, Post-Conviction Review: A Prosecutor's Viewpoint, Remarks at the DNA Fingerprinting and Civil Liberties Workshop hosted by the American Society for Law, Medicine & Ethics (Sept. 2005).
19. California Commission on the Fair Administration of Justice, Emergency Report and Recommendations Regarding DNA Testing Backlogs, Feb. 20, 2007.
20. Puit, Glenn. Man Files Lawsuit in False Imprisonment, Las Vegas Rev.-J., July 6, 2002.
21. Thompson, W. C., F. Taroni, and C.G.G. Aitken. "How the probability of a false positive affects the value of DNA evidence," *J. of Forensic Sci.* (Jan. 2003).
22. Spencer, Buffy. "Four Men Charged with DNA Tampering," *The Republican,* Mar. 17, 2007.
23. Genetic Information Nondiscriminaton Act of 2007: Hearing on H.R. 493 Before the Subcomm. on Health of the H. Comm. on Energy and Commerce, 110th Cong. (2007) (statement of the Honorable Francis S. Collins, Director, National Human Genome Research Institute).
24. Table I prepared with Joanne Kang, ACLU Washington Legislative Office.
25. In re Welfare of C.T.L., 722 N.W.2d 484 (Minn. Ct. App., 2006) (declaring the statute unconstitutional, "because Minn.Stat. § 299C.105, subd. 1(a)(1) and (3) (2005), direct law enforcement personnel to conduct searches without first obtaining a search warrant based on a neutral and detached magistrate's determination that there is a fair probability that the search will produce contraband or evidence of a crime, and because the privacy interest of a person who has been charged with a criminal offense, but who has not been convicted, is not outweighed by the state's interest in taking a biological specimen from the person for the purpose of DNA analysis, the portions of Minn. Stat. § 299C.105, subd. 1(a)(1)

and (3), that direct law enforcement personnel to take a biological specimen from a person who has been charged but not convicted violate the Fourth Amendment to the United States Constitution and Article I, Section 10 of the Minnesota Constitution").

26. Tanner, Robert. "More state back taking DNA from arrestees," *Deseret News,* June 30, 2006 (finding that state Senator Ron Ramsey's amendment to remove database provisions "won wide support, but was delayed for a year" to add six additional DNA analysts to state lab to address backlog issues).

Chapter 50

1. Roberts, D. *Killing the Black Body: Race, Reproduction, and the Meaning of Liberty* (New York: Pantheon, 1997). However, a recent study indicates that expressed attitudes stemming from Tuskegee's aftermath do not translate into sharp differences in behavior, by race, at least insofar as willingness to participate in clinical trials is concerned. D. Wendler, et al., "Are Racial and Ethnic Minorities Less Willing to Participate in Health Research?" PLOS Medicine 3, no. 2 (2006): 1-10, available at <http://medicine.plosjournals. org/perlserv/?request=get-document&doi=10.1371/journal. pmed.0030019> (last visited February 16, 2006).
2. Cannon, L. "One Bad Cop," *The New York Times Magazine,* October 1, 2000, 2.
3. Glover, S. and M. Lait. "Lack of Funds Stalls Rampart Probe: The LAPD Seeks Private Donations so that an Independent Panel Can Begin Investigating the Department's Handling of The Scandal," *Los Angeles Times,* November 6, 2003, B1.
4. See Mark Harrison, "Dallas Police Frame and Deport Hispanics," *The Razor Wire* 6, no. 1, The November Coalition <http://www.november.org/razorwire/rzold/27/page03.html> (last visited February 16, 2006).
5. Gold, S. "35 Are Pardoned in Texas Drug Case," *Los Angeles Times,* August 23, 2003, A11.
6. Parenti, C. "Police Crime," at <http://zmag.org/ZMag/articles/ mar96parenti.htm> (last visited February 16, 2006).
7. Ibid., at 2.
8. Personal communication with Dan Krane, September 11, 2005. As noted in the text, Krane is one of the nation's leading experts on DNA forensic

technology. Reference notes 9 and 10 explain his role. He was an expert witness at this trial, and sent me extensive notes in the e-mail noted in the "note" that follows in the full references.

9. Paoletti, D. R., T. E. Doom, M. L. Raymer, and D. E. Krane. "Assessing the Implications for Close Relatives in the Event of Similar but Non-Matching DNA Profiles," *Jurimetrics* 3, no. 2 (2006): pp. 161-175.
10. Ibid.
11. Krane, D. personal communication with the author, Sept. 11, 2005.
12. The Federal DNA Act and most state DNA collection statutes require that the state expunge (from the DNA databank) the profiles of convicted persons whose convictions are reversed. However, in a glaring gap in logic, these statutes do not address what to do with profiles from persons who are not even suspects. The police often retain these DNA profiles in their own, private "suspect databases." For example, Chicago, Miami, and London, Ohio, all keep private police suspect databases. A. B. Chapin, "Arresting DNA: Privacy Expectations of Free Citizens versus Post-Convicted Persons and the Unconstitutionality of DNA Dragnets," Minnesota Law Review 89, no. 6 (2005): 1842-1874.
13. This case was the subject of a full hour documentary by the television news program, *48 Hours,* which aired November 26, 2005.
14. The next segment is drawn from my paper, "Comparative Perspectives and Competing Explanations: Taking on the Newly Configured Reductionist Challenge to Sociology," *American Sociological Review* (2006): 1-15.
15. See J. H. Skolnick, "Corruption and the Blue Code of Silence," Police Practice and Research 3, no. 1 (2002): 7-19; J. H. Skolnick, and J. J. Fyfe, Above the Law: Police and the Excessive Use of Force (New York: Free Press, 1993).
16. Jackall, R. R. *Street Stories: The World of Private Detectives* (Cambridge, MA: Harvard University Press, 2005).
17. See Chapin, supra note 12, at 1847.
18. Ibid., at 1843, discussing Griffin v. Wisconsin, 107 S. Ct. 3164 (1987).
19. Ibid., at 1854.
20. Sanger, D. "In Address, Bush Says He Ordered Domestic Spying," *The New York Times,* December 18, 2005, 1.
21. Wambaugh, Joseph. *The Blooding,* (New York: Morrow, 1989).

22. Hanson, M. "DNA Dragnet," *American Bar Association Journal* 90 (2004): 38-43, 42
23. Boeschenstein, N. "The Charlottesville Dragnet, Part I: Just the Facts, Ma'am," *Archipelago,* 1-8, available at <http:/www.archipelago. org/vol8-2/ boeschenstein.htm> (last visited March 13,2006).
24. Ibid.
25. Joh, E. E. "Reclaiming 'Abandoned' DNA: The Fourth Amendment and Genetic Privacy," *Northwestern University Law Journal* 100, no. 2 (2006): 857-884.
26. Miller-El v. Cockrell, 123 S Ct. 1029 (2003).
27. Shriver, M. D., et al. "Ethnic-Affiliation Estimation by Use of Population-Specific DNA Markers," *American Journal of Human Genetics* 60 (1997): 957-964; A. L. Lowe, et al., "Inferring Ethnic Origin by Means of an STR Profile," *Forensic Science International* 119 (2001): 17-22.
28. The website "ancestrybydna.com" is one of several where one can apply for a kit, and then send in a DNA sample. The company then does an analysis and sends back a report with estimates of the proportion of one's ancestry that is purportedly from one of several large continental groupings.
29. Touchette, N. "Genome Test Nets Suspected Serial Killer," *Genome News Network,* June 13, 2003.
30. Tang, H., et al. "Genetic Structure, Self-Identified Race/Ethnicity, and Confounding in Case-Control Association Studies," *American Journal of Human Genetics* 76 (2005): 268-275.
31. Zhang, J. "New Study Links Race and DNA Material," Friday, February 4, 2005, available at <http://daily.stanford.edu/tempo?page=content&id=15971&repository=0001_article> (last visited March 13, 2006.)
32. Gavel, D. "Fight Crime through Science," *Harvard Gazette,* November 30, 2000.
33. "DNA Fingerprint Act of 2005" was signed into law on January 6, 2006 as Title X of the Violence Against Women Act and Department of Justice Reauthorization Act, Pub. Law 109-162, 119 Stat. 2960.
34. SEC. 1004. Authorization to Conduct DNA Sample Collection from Persons Arrested or Detained Under Federal Authority. (a) In General-Section 3 of the DNA Analysis Backlog Elimination Act of 2000 (42 U.S.C. 14135a) is amended-(1) in subsection (a) (A) in paragraph (1), by striking 'The Director' and inserting the following: (A) The Attorney General may, as

prescribed by the Attorney General in regulation, collect DNA samples from individuals who are arrested or from non-United States persons who are detained under the authority of the United States.

35. Kimmelman, J. "Risking Ethical Insolvency: A Survey of Trends in Criminal DNA Databanking," *Journal of Law, Medicine & Ethics* 28 (2000): 209-221.
36. Simoncelli, T. "Dangerous Excursions: The Case Against Expanding Forensic DNA Databases to Innocent Persons," *Journal of Law, Medicine & Ethics* 34 (2006): 390-397.
37. Pub. Law 109-162, 119 Stat. 2960, codified at 42 U.S.C. §14132(a)(1) (2005). The amendment retains the prohibition on inclusion of samples voluntarily submitted solely for the purpose of elimination. Id.
38. Wacquant, L. "Deadly Symbiosis: When Ghetto and Prison Meet and Mesh" *Punishment and Society* 3, no. 1 (2000): 95-134. Reprinted in D. Garland, ed., Mass Imprisonment: Social Causes and Consequences (London: Sage, 2001): 82-120.
39. Gross, S. R., et al. "Exonerations in the United States, 1989 through 2003," Unpublished manuscript, University of Michigan Law School (2004) available at <http://www.law.umich.edu/newsandinfo/exonerations-in-us.pdf> (last visited March14, 2006).

Chapter 51

1. In general, as the number of alleles in a DNA profile decreases, the probability that a randomly chosen person will, by coincidence, happen to match that profile increases. Because the alleles vary greatly in their rarity, however, it is possible for a profile containing a few rare alleles to be rarer overall that a profile containing a larger number of more common alleles. Consequently, when discussing the likelihood of a coincidental match it is more helpful to focus on the estimated frequency of the profile than the number of loci or alleles encompassed in the profile.

1. Koehler, J. J. "Error and exaggeration in the presentation of DNA evidence," *Jurimetrics,* 34: 21-39, 1993.
2. Thompson, W. C. "Forensic DNA Evidence," In B. Black & P. Lee (Eds.), *Expert Evidence: A Practitioner's Guide to Law, Science and the Manual,* St. Paul, Minn.: West Group, 1997 pp. 195-266.

3. Aronson, Jay D. "Genetic Witness: Science, Law and Controversy in the Making of DNA Profiling," *Rutgers University Press* (2007).
4. Ibid.
5. National Research Council. "The Evaluation of Foreign DNA Evidence," *National Academy Press* (1996): p. 2.
6. Lynch, Michael, Simon Cole, Ruth McNally & Kathleen Jordan. "Truth Machine: The Contentious History of DNA Fingerprinting," *University of Chicago Press* (2008).
7. Ibid.
8. Cole, Simon A. "How much justice can technology afford? The impact of DNA technology on equal criminal justice." *Science and Public Policy,* 34(2) 95-107, March 2007; Simon A. Cole, "Double Helix Jeopardy," *IEEE Spectrum,* 44-49, August 2007
9. Levine, Harry G., Jon Gettman, Craig Reinarman & Deborah P. Small. "Drug arrests and DNA: Building Jim Crow's Database." Paper produced for the Council forResponsible Genetics (CRG) and its national conference, Forensic DNA Databases and Race: Issues, Abuses and Actions held June 19 20, 2008, at New York University. Available at www.gene-watch.org.
10. Butler, John M. Forensic DNA Typing: Biology, Technology and Genetics of STR Markers (2nd Ed.). Elsevier/Academic Press, 2005.24
11. Willing, Richard. "Suspects get snared by a relative's DNA," *USA Today,* June 8, 2005, at 1A; David R. Paoletti, Travis E. Doom, Michael L. Raymer & Dan Krane, "Assessing the implications for close relatives in the event of similar but no matching DNA profiles," *Jurimetrics Journal* 46: 161-175 (2006).
12. Willing 2005.
13. Thompson, W. C. "Beyond bad apples: Analyzing the role of forensic science in wrongful convictions." *Southwestern Law Review* 37:101-124 (forthcoming).
14. Thompson, W. C., F. Taroni & C.G.G. Atiken. "How the probability of a false positive affects the value of DNA evidence." *Journal of Forensic Sciences,* 48(1): 47-54 (2003).
15. Thompson, W. C. "Tarnish on the 'gold standard': Understanding recent problems in forensic DNA testing." The Champion, 30(1): 10-16 (January 2006).

16. Ekblom, Paul. "Can we make crime prevention adaptive by learning from other ecological struggles?" *Studies on Crime and Crime Prevention* 8 (1998): pp. 27-51.
17. Ekblom, Paul. "How to police the future: Scanning for scientific and technological innovations which generate potential threats and opportunities in crime, policing and crime reduction," In. M. Smith and N. Tilley (Eds.) Crime Science: New Approaches to Preventing and Detecting Crime. Cullompton: Willan, 2005.

Chapter 52

1. The same technology, produced by DNAPrint Genomics, is also packaged as Ancestry by DNA for recreational genealogical ancestry testing. It is also used in biomedical research settings for purposes of admixture mapping for disease traits and to prevent confounding in 'mixed' populations in case-control studies for complex disease traits. See http://www.dnaprint.com/welcome/productsandservices/index2.php (Accessed March 28, 2008).

References

1. Roberts, P. *Sunday Advocate,* 1 June 2003, p. 1A.
2. Daubert v. Merrell Dow Pharmaceuticals, 509 US 579 (1993).
3. Notes to Fed. R. Evid. 702, http://www.law.cornell.edu/rules/fre/ACRule702.htm.
4. Imwinkleried, E., in "DNA and the Criminal Justice System: The Technology of Justice," D. Lazer, Ed. (Cambridge: MIT Press, 2004), chap. 5, p. 98-99.
5. Imwinkleried, E., in "DNA and the Criminal Justice System: The Technology of Justice," D. Lazer, Ed. (Cambridge: MIT Press, 2004), chap. 5, p. 99.
6. National Research Council. *The Evaluation of Forensic DNA Evidence, Committee on DNA Forensic Science: An Update* (Washington, D.C.: National Academy Press, 1996)
7. Frudakis, T. "Molecular Photofitting: Predicting Ancestry and Phenotype Using DNA," (Burlington, MA: Academic Press, 2007), p. 429.

Chapter 53

1. "Cuomo Seeks Genetic Data of Offenders," *The New York Times* 1992.
2. http://www.fbi.gov/about-us/lab/codis/ (accessed August 20, 2011).
3. Please see CRG's Guide to Forensic DNA Databases at http://www.councilforresponsiblegenetics.org/dnadata
4. Resolution No AGN/67/RES/8.
5. http://www.interpol.int/Public/Forensic/dna/default.asp (accessed August 21, 2011).

Chapter 56

1. Anderson v. Commonwealth of Virginia, 274 Va. 469 (2007); In re Welfare of C.T.L., 722 N.W.2d 484, (Minn.App. 2006).
2. In the Matter of Bojorquez, (Pima Cnty. Juvenile Ct. No. 168544-04). An Arizona appellate court considered the issue in a separate set of cases consolidated under the name, Mario W. v. Kaipio, No. 1CA-SA 11-0016.
3. From the French for "in the bench," referring to the full bench of judges.
4. United States v. Mitchell, --- F.3d ----, 2011 WL 3086952 (3rd Cir. Jul. 25, 2011).
5. On September 19, 2011, shortly before this article was published, the Ninth Circuit dismissed the Pool appeal, because Mr. Pool had plead guilty to a lesser charge and the case therefore no longer presented the question of whether the government could force a person merely accused of a crime to provide a DNA sample. The court vacated all the prior opinions in this case, as is standard when a case becomes moot on appeal.
6. Haskell v. Brown, 677 F.Supp.2d 1187 (N.D. Cal. 2009).
7. People v. Buza 197 Cal.App.4th 1424 (Cal.App. 2011).
8. For a more detailed legal analysis of the issue, see Risher, Warrantless Collection of DNA From People Merely Accused of a Crime Raises Not Only Privacy Concerns But Also Questions About Efficacy, 88 Criminal Law Reporter 320 (Bureau of Ntl. Affairs 2010).

Chapter 60

1. The video can be viewed at http://www.lvrj.com/news/dna-related-error-led-to-wrongful-conviction-in-2001-case-125160484.html

2. See Willliam C. Thompson, "Tarnish on the 'Gold Standard': Recent Problems in Forensic DNA Testing," *The Champion,* January/February 2006.
3. See William C. Thompson, et al., "Evaluating Forensic Evidence: Essential Elements of a Competent Defense Review," *The Champion,* April 2003. (Part 2 of this article appeared in the May 2003 issue of *The Champion.*)
4. Levy, Harlan. "Caught Up in DNA's Growing Web," *The New York Times* (oped), March 17, 2006.
5. McCartney, Carole. "The DNA Expansion Programme and Criminal Investigation," *British Journal of Criminology,* October 25, 2005 [Published by Oxford University Press on behalf of the Centre for Crime and Justice Studies (ISTD)].
6. Liptak, Adam. "Justices Rule Lab Analysts Must Testify on Results," *The New York Times,* June 26, 2009.
7. See William C. Thompson, et al., "Evaluating Forensic Evidence: Essential Elements of a Competent Defense Review," *The Champion,* April 2003. (Part 2 of this article appeared in the May 2003 issue of The Champion.)
8. Melendez-Diaz v. Massachusetts, 557 U.S. _____ (2009)

Part VI

Chapter 65

1. Hefner, Philip. "Technology and Human Becoming," *Fortress* (2003).
2. Pepperell, Robert. "The Posthuman Condition," *Intellect* (1995).
3. Monsanto Corporation. "Frankenstein Food? Take Another Look," (online), available at http://www.searchmonsanto.com/monsanto-uk/frankensteinfoods. html, 2pp, [accessed 23 November 1999].
4. Kaku, Michio. *Visions How Science will Revolutionize the Twenty-First Century and Beyond,* Oxford University Press, 1998: pp. 14-15.
5. Warwick, Kevin. *I, Cyborg,* (Century, Australia, 2002), 275.
6. "The Extropian Principles: A Transhumanist Declaration," version 3.0, Max More Online, accessed March 19, 1999, http://www.maxmore.com/extprn3.htm.
7. "The Transhumanist FAQ," TransHumanist Online, accessed August 20, 2000, http://www.transhumanist.org

Chapter 69

1. See Jill Scott, ed., Artists-in-Labs: Processes of Inquiry (Springer Vienna Architecture) 2006
2. See Symbiotica, http://www.symbiotica.uwa.edu.au/
3. See Linda Candy, "Guide to Practice-Based Research:A Guide," www.creativityandcognition.com/.../PBR-Guide-1.1-2006.pdf
4. See Genspace.org
5. www.biofaction.com/synth-ethic/
6. Ibid.
7. Cooking As Alchemy:Homaro Cantu + Ben Roche TED:IdeasWorthSpreadinghttp://www.ted.com/search?q=homaro+cantu
8. fringejoyride.com/2012/06/14/farmification/; DLDwomen 2012: Farmification – YouTube
9. Regine Debatty. Claire L. Evans. Pablo Garcia. Andrea Grover, Thumb. (Pittsburg, New Art/Science Affinities with Studio for Creative Inquiry and Miller Gallery at Carnegie Mellon University, 2011

INDEX

C

E

G

H

I

M

O

P

Q

R

S

T

U

V

W

X

Z

To subscribe to print and electronic versions of *GeneWatch*, please send a subscription request with your full name, mailing address and email address, along with a check* or credit card (MasterCard or Visa) information, to Council for Responsible Genetics, 5 Upland Road, Suite 3, Cambridge, MA 02140.

Orders are also welcome online at www.councilforresponsiblegenetics.org or by phone at (617) 868-0870. Additional donations are welcome. Thank you!

*Annual print subscription rates are as follows:

Individual: $35
Nonprofit: $50
Library: $70
Corporation: $100